THERMODYNAMICS
A Short Undergraduate Introduction

M.J. Rice
Virginia Polytechnic and State University VirginiaTech Invent the Future®

Edited by C.L. Dancey
Virginia Polytechnic and State University VirginiaTech Invent the Future®

KLAR ENGINEERING

A KLAR ENGINEERING PUBLICATION

Published by KLAR ENGINEERING, Blacksburg, Virginia.
Typeset in glorious LaTeX.

Printed in the Odenton, Maryland, USA. GASCH PRINTING

10 9 8 7 6 5 4 3 2 1

Between the sun and the moon,
dreams and fantasy,
folly and wisdom,
the savage and divine,
stands strength, tenderness, grace and beauty ...

A lady born to honor,
with auburn hair,
purposeful eyes,
dressed in white from head to toe.

The fruit of those moments we shared together,
between the bells, drums and shouts,
within the thick green walls,
the reader beholds.

Now in another land,
she's ready to fight and live,
dressed in black,
with a heart of silver,
a mind of gold.

These pages are dedicated to you.
Thank you for our time together E.C.R.

Yours sincerely, Dr. Matthew Rice

PREFACE

This book is intended for the undergraduate student in his, or her, second year of undergraduate instruction. It is assumed that the student, perhaps studying mechanical engineering, is most interested in mastering the techniques and skills required for doing engineering problems, and less inclined to demand mathematical proofs and theorems: Hence, this book presents an undergraduate introduction to engineering thermodynamics.

Given the large number textbooks on the thermodynamics "scene", it is right and proper to ask the question: "Why another text on undergraduate thermodynamics?" To answer this question, we must have a look at the current state of instruction in engineering subjects in general and thermodynamics in particular. As one who's studied engineering in the late 1990's and early 2000's, and taught the subject for over a decade now, one is struck by the radical change in instruction technique and outcomes over the last decade. The current approach to teaching thermodynamics is often a "cookbook" approach where the students are instructed as to which equations are relevant to a particular device or application, followed by the relevant mathematical procedure for obtaining the solution. Provided the "cookbook" is relatively complete and the students have ready access to course material (written notes, textbook, cheat-sheet, etc.) this approach can lead to a sort of competence whereby the student can solve a given number of highly specific problems. In-line with this methodology is the notion that students should have, or will demand, material be available "at their finger-tips". The logical evolution of this approach are lectures delivered using pre-recorded media (e.g. slides and Power-Point) in conjunction with pre-written notes. Exams and other methods of assessment are necessarily "open-book" , etc. The benefits of this approach are twofold: It allows for the covering of vast amounts of material often of a highly specialized nature over a relatively short period of time. It also allows for the minimization of effort required on the instructor's part per-unit of course content. One concrete outcome of this approach is the compression of an undergraduate course in thermodynamics, one of the most challenging topics conceptually for the undergraduate, from a two semester into a single semester course in many research intensive institutions.

While the benefits of the current approach are formidable, I believe there are significant costs:

1. Students are often left bereft and "at sea" so to speak regarding fundamental concepts.
2. The resulting lack of general knowledge of thermodynamic principles results in a dearth of effectiveness in solving novel or new problems.
3. The lack of general and effective tools (i.e. concepts) in analyzing thermodynamic problems tends to stunt any spontaneous and natural curiosity on the student's part and reduces the learning process to an exercise in "pattern solving".
4. The use of pre-recorded media and the resulting large volume of information delivered to the student tends to hinder the development of independent analytical skills and problem solving ability on the student's part: Pre-written course notes and slides, given the shear volume of material, become a necessary crutch to the student.
5. A lack of engagement (e.g. regular attendance, completion of reading assignments, etc.) and mutual respect between the students *and* instructor. Students, like their professors, appreciate and respond best when they perceive that value is created within the classroom on a daily basis.

Thus, we arrive at the current state of affairs with respect to thermodynamics text-books. With the increasing volume of specialized material covered in a relatively short period of time, there has been a marked increase in specificity and length of textbooks. A glance at the most widely used texts in engineering thermodynamics has them pegged at nearly one thousand pages. With the adoption of teaching solution procedures to specific problems (the "cook-book" approach) we have an inordinate amount of space devoted to examples, most of which, would be unnecessary if the student possessed a thorough and general (principled) understanding of the subject. In addition, with the increase in specificity comes an increase in tedium, whereby long tracts of text are devoted to explaining the applicability of some relationship where a more general approach would require less duplication and exhaustive explanation. Referring to the thermodynamics texts of years ago one notices the contrast between the modern expansive treatise and their more compact and general elders. Given these observations, it would appear to be no accident that low student readership and participation in reading assignments is such a problem today.

It is in the context of this change in teaching technique and associated texts that this textbook will strike out on an alternate path:

- To address the lack of compactness in modern thermodynamic textbooks and to counter-act a lack of student engagement with reading assignments, the current text will keep the main-body of material to a minimum ($\approx$ 250pgs).
- To counter-act the tendency for modern students to study by scanning textbook examples (the result of a "cook-book" approach), the number of examples are purposely maintained at a modest amount: It is left up to the instructor to present additional examples of his or her choosing in lecture, and for the student to gain knowledge and skills in "working problems" via coursework instead of scanning pre-worked solutions.
- In the place of providing every conceivable example of a specific application or permutation of the governing equations, general relationships are to be applied to course-work problems which will be used to illustrate specific applications.

The previous guidelines lead to the following approach to coursework and exams (the reader will forgive me if he or she can conceive of others):

- Homework problems are to be divided into various grades both in difficulty as well as the provision of guidance. Specifically, since the use of concepts is taught, re-enforced and retained by the application of general principles to specific problems, the student will need help or guidelines to set-up and solve problems (at least initially). Therefore, coursework problems are rated in terms of the amount guidance given on a scale of 1 → 3 (3 giving the least guidance). In addition, problems are rated in terms of difficulty on a scale of 1 → 3 (3 giving the most difficulty).

- Coursework problems will not be provided in the main text so as to discourage students from seeking out solutions through various external sources (i.e. the Internet). Instead, instructors will have access to supplementary problems which can be assigned individually and without pre-set assignment numbers (again, to discourage the seeking out of solutions externally). EES solutions to the problems included are provided to facilitate modification of problem statements, etc.

- Given the approach of using general principles to attack specific problems, sample exams and quizzes are provided on a supplementary basis. These exams will test the student on her knowledge of fundamental concepts and are written assuming a closed notes/book format. Obscure facts or information are provided within the problem statements as needed.

Finally, to ensure compactness, certain topics will be either emphasized or de-emphasized compared to the treatment given in typical texts. Specifically, the link between open and close-system conservation laws (for mass, energy, entropy and momentum) will be made explicit via use of the Reynolds Transport Theorem wherever necessary. The treatment of reacting systems will rely on the concept of the reaction mechanism and reaction coordinate (in place of the rather haphazard and inelegant approach currently taken in modern texts). Given the emphasis on general principles, the analysis of exotic cycles (e.g. Sterling, Erricson, Dual cycles, etc.) are relegated to individual course-work problems. The same goes for the analytical equivalent of "curve-fitting", the Polytropic process.

With respect to the structure of the text, topics are arranged as follows:

- Chapter 1:Review of Conservation of Mechanical Energy with reference to the concepts of Heat, Internal Energy and the Second-Law of Thermodynamics. Introduction to basic concepts.

- Chapter 2: Properties of pure substances

- Chapter 3: Energy inter-actions for closed and open-systems

- Chapter 4: Energy-balance for closed and open-systems

- Chapter 5: The Second-Law of Thermodynamics

- Chapter 6: Derivation of the property entropy, closed and open-system entropy-balance, property relationships, isentropic efficiency & Availability

- Chapter 7: Power cycles

- Chapter 8: Refrigeration cycles

- Chapter 9: Ideal gas mixtures and Psychrometrics
- Chapter 10: Reacting systems

For those instructors teaching undergraduate Thermodynamics over a two semester period, the first semester's content would include materials up through power cycles (Chapter 7). A single semester course may treat Psychrometrics as optional.

DR. M. RICE

Lexington, Virginia
April, 2013

TO THE STUDENT

We are here to teach you *and you will learn!*

—Regimental Executive Officer,

As an undergraduate who has enrolled in Engineering Thermodynamics I have no doubt that you are now experiencing acute symptoms of what one can describe only as "Thermodynamics Angst". After all, you have heard the stories: 'Don't you know, half the class fails that course". Or "When my room-mate hit entropy, he decided to change major". And possibly "I felt like I was flying blind, I couldn't visualize or see anything the instructor was talking about". And of course "Two-thirds of the class failed the first exam and half failed the second. You don't know what you're getting into!" As adults who are spending time and good money to acquire a highly marketable skill, I feel I owe you the truth. Yes, Thermodynamics is difficult. Yes, it does deal with quantities which cannot be directly measured or visualized but are, nonetheless real. And yes, you must have the fortitude and maturity to accept temporary failure to grasp the material. It has been said that programming builds patience, attention to detail and character: The same can be said for Thermodynamics.

Having said this, one must also note the tremendous potential benefits of mastering the subject. Thermodynamics will open up a world of devices and phenomena utilizing the inter-action between work, heat and energy: Jet and rocket engines, turbochargers and turbines, internal combustion engines, refrigerators, etc. are all subject to the laws of Thermodynamics. In short, there is no subject to be learned in your undergraduate career which has more general relevance to everyday modern life than Thermodynamics.

To get you started off on the right track I want to provide some words of advice:

- Do not cut scholastic or intellectual corners. You will be hearing terms such as "conservation of mass", "conservation of energy" and the "entropy-balance". These terms refer to principles which depend on concepts such as "states", "control volumes", "open/closed systems", "processes", etc. These principles and concepts are the conceptual tools which will allow you to independently solve a broad range of problems from the simplest to the most complex and are to be taken seriously. Do not ignore them!

- If you value your time and peace of mind, jettison any pre-existing tendency to learn a subject by simple pattern-following. From experience, the hallmark of a student in trouble is one who starts every assignment by looking through the book for a *similar* example problem to the one he or she is attempting to solve. This approach, while minimizing short-term effort which would have been used to learn the basic concepts, eventually results in ever increasing expenditure of time searching for relevant pre-worked examples. Being reduced to a state of utter helpless grasping for pre-worked examples from the text during an exam, or worse, a departmental meeting at a future employer is not in your best interest.
- Many of the principles you will learn, once properly understood, are reasonable and indeed make sense. As a result, resist the temptation to blindly accept calculated solutions. Since Thermodynamics deals with observables (pressure, temperature, volume, etc.) as well as non-observables (energy, heat, entropy, etc.), students sometimes throw their arms up in ignorance when it comes time to interpret the *meaning* of a calculated solution. An incorrect solution is one thing, a non-physical (i.e. non-nonsensical) solution can be dangerous and should not pass without comment on your part.
- Be goal oriented in your work: I have seen many students solve a particular problem by *calculating everything in sight*! Your first step in working a problem should be to write a general expression for the desired quantity and then progressively break the expression down into component terms. It is the discovery and calculation of these contributing terms which should be driving your solution procedure. Resist the temptation to calculate all conceivable quantities in the hope you will stumble onto your desired result. Instead, think ahead, work through the problem consciously on scratch paper, reverse or change course if a particular solution procedure proves fruitless. When, and only when, you have the solution at hand, neatly record your work.
- Finally, expect to work hard mastering one of the most important, difficult and potentially satisfying disciplines in engineering. You will be exercising a new set of intellectual muscles and just as in any sport, the key is practice, practice and more practice.

Good luck ladies and gentlemen!

Dr. Matthew J. Rice

CONTENTS

CHAPTER 1

MOTIVATION AND BASIC CONCEPTS

I am going to begin by making a shocking statement:

> "The laws of mechanics, conservation of mechanical energy, you hitherto learned are incomplete!"

1.1 MOTIVATION: WHY WHAT YOU ALREADY KNOW IS INCOMPLETE!

We begin by writing conservation of mechanical energy developed in college or secondary-school physics

$$Work = \Delta E_{mechanical} = \Delta KE + \Delta PE \tag{1.1}$$

where work refers to a mechanical transfer of energy to a system and the symbol Δ refers to the change in some property (in this case total mechanical energy $E_{mechanical} = PE + KE$). You shall now see that this expression actually fails to represent some of the simplest of phenomena.

Experiment 1.1 (Moving block on a stationary table with friction) *Assume a block with initial velocity ${}_{Block}V_i \neq 0$ moves along a table-top surface. The table and block both have equal initial temperatures (${}_{Block}T_i = {}_{Table}T_i = T_i$). After some time the block decelerates due to the effect of friction between the block and table surface until the motion ceases (i.e. ${}_{Block}V_f = 0$). Application of conservation of mechanical energy (1.1) to the $Block + Table$ system as shown in Figure 1.1, (noting that the gravitational force is perpendicular to the block motion, hence external work is*

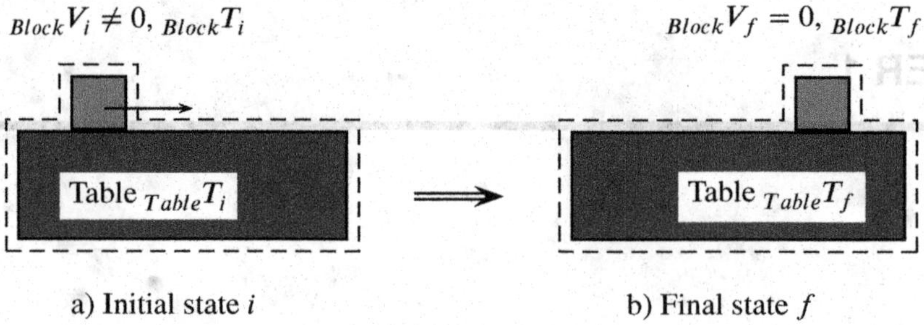

Figure 1.1 Table and block system (system signified by – – –)

zero), yields

$$\cancelto{0}{Work_{i\to f}} = \Delta({}_{Block+Table}E_{i\to f}) = \Delta({}_{Block+Table}KE_{i\to f}) + \cancelto{0}{\Delta({}_{Block+Table}PE_{i\to f})} =$$

$$\Delta_{Block}KE_{i\to f} + \cancelto{0}{\Delta_{Table}KE_{i\to f}} = \Delta_{Block}KE_{i\to f} = \cancelto{0}{\frac{{}_{Block}m_{Block}V_f^2}{2}} - \frac{{}_{Block}m_{block}V_i^2}{2}$$

$$= -\frac{{}_{Block}m_{block}V_i^2}{2} \tag{1.2}$$

or

$$\frac{{}_{Block}m_{Block}V_i^2}{2} = 0 \tag{1.3}$$

Since $_{Block}m \neq 0$ *then* $_{Block}V_i = 0$.

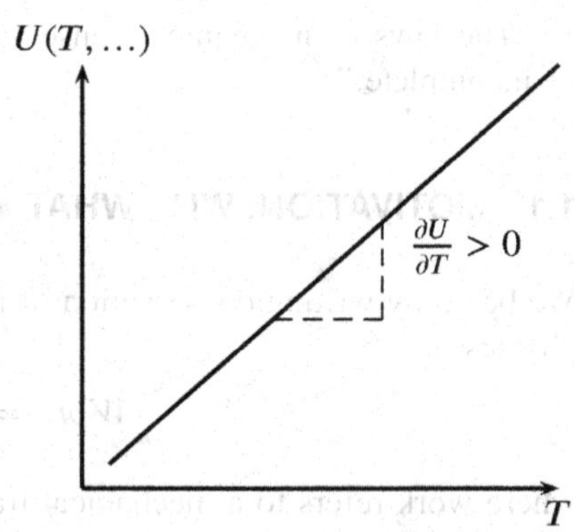

Figure 1.2 Functional dependence of internal energy U and temperature T

But $_{Block}V_i \neq 0$ so we have the curious result $0 = {}_{Block}V_i \neq 0$. What has gone wrong? Where did the kinetic energy of the block go? Well, after conducting this experiment you would find that, given sufficiently sensitive measuring equipment, *the block plus table system gets warmer*! Conclusion: We define a new form of energy termed "internal energy" U which is an increasing function of temperature T as shown in Figure 1.2: $dU/dT > 0$ or in plain English, the hotter the system, the greater the internal energy U. Thus, from this simple experiment we arrive at a new (and improved) version of conservation of energy

$$Work = \Delta E_{total} = \overbrace{\Delta U}^{\text{Change in "Internal Energy"}} + \overbrace{\Delta KE + \Delta PE}^{\text{Change in "Mechanical Energy"}} \tag{1.4}$$

Now, after having arrived at a new and more complete version of conservation of energy you might have thought our troubles are over. Unfortunately that is not the case.

Experiment 1.2 (Hot and cold block in thermal contact) *As shown in Figure 1.3 we have two blocks "A" and "B" of some material, one initially hot, the other cold (e.g. ${}_AT_i > {}_BT_i$) in physical (and thermal) contact. Note that over time, we expect the block temperatures to equalize to some temperature between ${}_AT_i$ and ${}_BT_i$ (i.e. ${}_BT_i < {}_BT_f$). Now, let us apply energy conservation (1.4) derived in the previous example to the* initially *cold block "B" indicated by the dashed line in Figure 1.3.*

$$\cancelto{0}{{}_BWork} = \Delta_B E_{total} = \Delta_B U + \cancelto{0}{\Delta_B KE} + \cancelto{0}{\Delta_B PE} \tag{1.5}$$

or

$$\Delta_B U = 0 \tag{1.6}$$

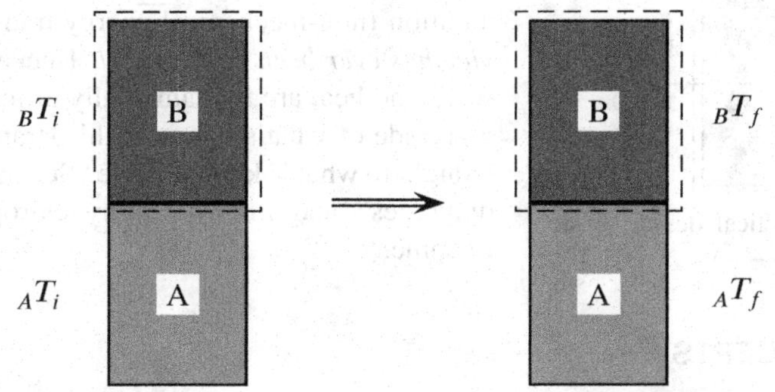

a) Initial state i (${}_BT_i < {}_AT_i$)   b) Final state f (${}_BT_f = {}_AT_f$)

Figure 1.3 Hot bar A and cold bar B in physical contact

But, as indicated from the previous example, $dU/dT > 0$, thus if $\Delta_B U = 0$ then this implies $\Delta_B T = 0$ as well and there is no change in the cold block temperature in spite of the fact that it is in physical (and thermal) contact with the hot block. This contradicts physical intuition, as it should: We conclude that the previous conservation of energy equation (1.4) includes only mechanical forms of energy transfer (i.e. Work). We therefore define a new form of non-mechanical energy transfer induced by a temperature difference between objects called "Heat" (or heat transfer) Q. Thus, our second revision of the energy conservation equation is

$$\overbrace{Q}^{\text{Non-mech. E-transfer due to "Heat"}} + \overbrace{W}^{\text{Mech. E-transfer due to "Work"}} = \Delta E_{total} = \overbrace{\Delta U}^{\text{Change in "Internal Energy"}} + \overbrace{\Delta KE + \Delta PE}^{\text{Change in "Mech. Energy"}} \tag{1.7}$$

Note, the revised energy-balance (1.7) would seem, in a mathematical sense, to imply an equivalence between heat Q and work W. Is this true? Given a temperature (and a corresponding internal energy) change of some stationary object, absent any additional information, would one be able to (separately) determine the heat or work transfer?

Experiment 1.3 (Non-equivalency of heat Q and work W) *Assume the existence of a device as shown in Figure 1.4 whose internal mechanism is such that there are no changes in potential, kinetic or internal and thus total energy ($\Delta E_{total} = \Delta U + \Delta PE + \Delta KE$), nor any other property of the system (e.g. Temperature T, pressure P, etc.) over some period of time. The device receives an energy transfer due to heat Q from some heat source while producing a certain amount of work W. Application of our revised energy conservation equation (1.7) yields*

$$Q + W = \Delta E_{total} = \cancelto{0}{\Delta U} + \cancelto{0}{\Delta KE} + \cancelto{0}{\Delta PE} = 0 \Rightarrow Q = -W_{on\ device} \quad (1.8)$$

or

$$Q = W_{by\ device} \quad . \quad (1.9)$$

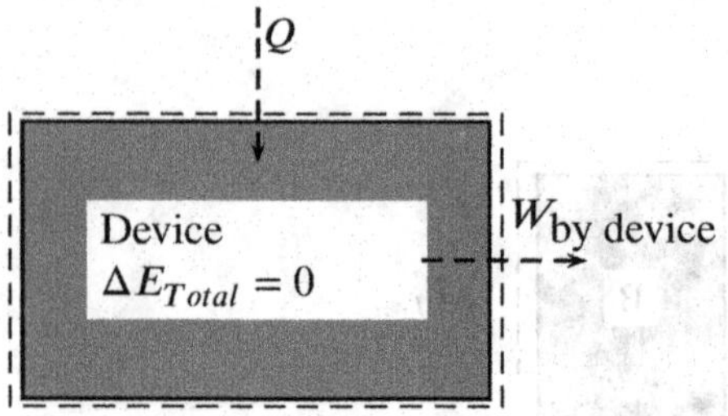

Figure 1.4 Hypothetical device (system boundary shown via – – –)

Thus, the work produced by the device is equal to the heat addition (non-mechanical energy transfer): *No such device has ever been constructed!* Hence, we conclude that work and heat are fundamentally different quantities. As a prelude of what's to come, this strange result will lead straight to what is known as the "Second-Law of Thermodynamics" and a new property, entropy, in the chapters to come.

1.2 BASIC CONCEPTS

We will now introduce basic concepts and terms which will be central to the material from this point forward. It is best to endeavor to commit each and everyone of these terms and concepts to memory. Lucky for you, most of them a quite intuitive.

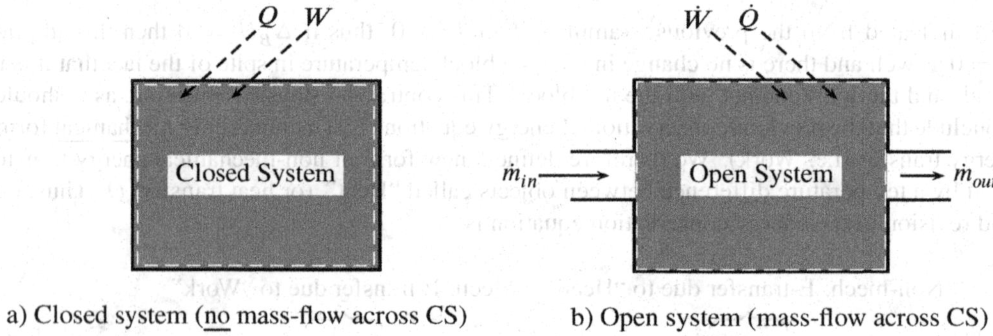

Figure 1.5 Closed vs. open systems illustrated (control surface CS indicated by – – –)

Definition 1.1 ***System:*** *A macroscopic thermodynamic "system" is any region in space (the control volume) bounded by either a real or imaginary surface (the system boundary or control surface). The system may be stationary or moving, or change shape and volume. As shown in Figure 1.5 a), systems which consist of only the same fixed quantity of mass are term "closed systems". In contrast, as shown in Figure 1.5 b), systems exhibiting mass-flow across the system boundary are referred to as "open systems".*

Table 1.1 Extensive properties and intensive counter-parts

Extensive Property	Intensive Property
Volume V	$v = V/m$=Vol./Mass
Total Energy E	$e = E/m$=Total Energy/Mass
Kinetic Energy KE	$ke = KE/m$=Kinetic Energy/Mass

It is important to understand that the terms "system" and "control volume" are synonymous as are "system boundary" and "control surface". In addition, the distinction between open and closed systems as displayed in Figure 1.5 might seem trivial but, will in fact, be critical to solving problems. Learn it!

Definition 1.2 ***Thermodynamic properties extensive and intensive:*** *A "thermodynamic property" is an observable (measurable) or non-observable characteristic of a system. Examples include pressure P, temperature T, volume V, mass m, amongst others. Properties can be sub-divided into "intensive" properties which are not dependent on the size or extent of the system (e.g. pressure P) vs. "extensive" properties which are dependent on the size or extent of the system (e.g. mass m or volume V). All extensive properties (with the exception of mass m) are denoted by upper-case symbols while intensive properties are expressed in lower-case (with the exception of pressure P and temperature T).*

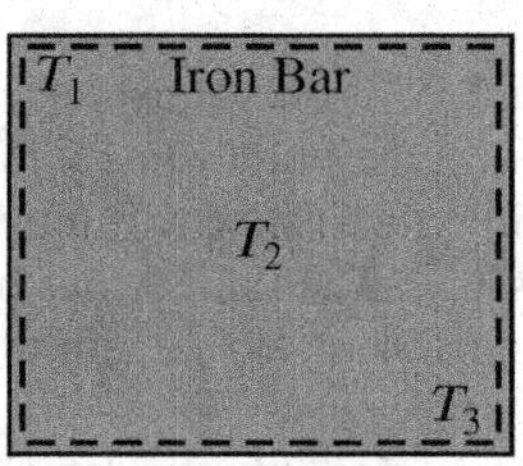

Figure 1.6 Non-equilibrium system with $T_1 \neq T_2 \neq T_3$

Note that extensive properties of a system (volume V, total energy E, etc.) can always be transformed into an *intensive* counter-part by dividing by the mass m of the system: "extensive property/mass" = "intensive property". Examples are given in Table 1.1. In addition, the intensive counter-part of an extensive property is preceded by the term "specific" or followed by "per unit-mass". Hence, $v = V/m$ is referred to as "specific volume" while $u = U/m$ would be referred to as "specific internal energy" or "internal energy per-unit mass". For example, a system consisting of an iron bar which is cold at one corner and hot on the other cannot be assigned a (single) property temperature T since it is not in "thermal equilibrium". This is not however to say that an *average* temperature $T_{average}$ cannot be assigned. Note, that over time, the bar will achieve a uniform temperature due to energy (heat) transfer within the bar, thus achieving a state of "thermal equilibrium". Intensive thermodynamic properties, such as temperature T and pressure P, can only be assigned to a system which is in an "equilibrium state".

Definition 1.3 ***Temperature:*** *Two bodies placed in thermal contact will come to mutual thermal equilibrium when they share the same temperature.*

Although temperature is an abstract concept (of which we have day-to-day experience), it is measured with devices call thermometers using the various scales $°K$ (Kelvin), $°R$ (Rankine), $°C$ (Centigrade) and $°F$ (Fahrenheit): The former two are *absolute* ($0°K$ and $0°R$ correspond to the cold of deep space), while the latter, $°C$ and $°F$ are not. Note that $°K$ and $°C$ are used in the metric system while $°R$ and $°F$ are used in the Imperial system. Conversions between the various scales are given by (1.10)

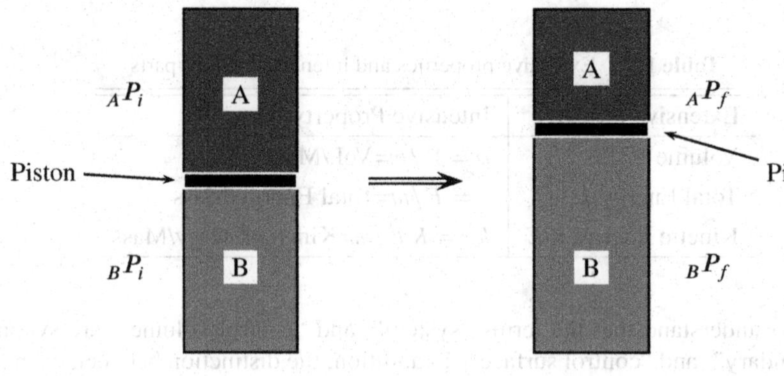

a) Initial state i (${}_BP_i > {}_AP_i$) b) Final state f (${}_BP_i = {}_AP_i$)

Figure 1.7 Two gases A and B in non-equilibrium a) and equilibrium b)

$$T(^\circ F) = 32 + \frac{9}{5}T(^\circ C)$$
$$T(^\circ K) = 273 + T(^\circ C) \tag{1.10}$$
$$T(^\circ R) = 459.67 + T(^\circ F)$$

Definition 1.4 ***Pressure:*** *When two systems are in thermo-mechanical equilibrium they are said to have the same pressure.*

The above definition may appear to be rather cryptic so let's refer to Figure 1.7 which illustrates a container of gas separated by a movable weightless and frictionless piston. In the initial state i, the pressure of the separated gases A and B are not equal. However, over time the piston will move until the system is in equilibrium: The two gases will have the same pressure. The dimensions of pressure are $force/area$ with Pascals $Pa = N/m^2$ used in the metric and $Psi = lbf/in^2$ in the Imperial unit system.

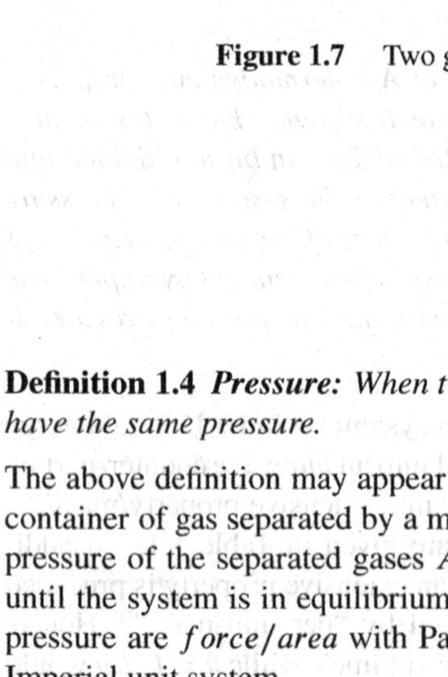

Figure 1.8 Relationship between absolute, gauge and reference pressure.

Like temperature, pressure P is experienced on a day-to-day basis and is measured by various mechanical devices (i.e. *pressure gauges*). Pressure can be expressed as "gauge" or "absolute"

$$P_{absolute} = P_{gauge} + P_{ref} \tag{1.11}$$

where P_{ref} is a reference pressure equal to the pressure outside the measurement device (gauge), usually but not always $1atm$. This relationship is illustrated in Figure 1.8. In addition, pressure is often expressed as a height h (often termed the "Hydraulic head") with units $mmHg$, $inHg$ or mmH_2O, inH_2O, etc. and is calculated via the conversion formula

$$h = \frac{P}{\rho g} \tag{1.12}$$

where ρ is the density of the measurement fluid and g is the gravitational acceleration. If ℓ, m and t represent dimensions of length, mass and time, $\rho \sim m/\ell^3$, $h \sim \ell$, $g \sim \ell/t^2$, $F \sim \frac{m\ell}{t^2}$ (i.e. $F = ma$)

and the right-hand-side of (1.12) has dimensions $\frac{F/\ell^2}{(m/\ell^3)(\ell/t^2)} = \frac{\frac{\cancel{m}\cancel{\ell}}{\cancel{t^2}}/\cancel{\ell^2}}{(\cancel{m}/\ell^3)(\cancel{\ell}/\cancel{t^2})} = \frac{\ell^{\cancel{3}}}{\cancel{\ell^2}} = \ell.$

EXAMPLE 1.1 Calculation of pressure

Problem Statement:
A small tank containing a gas under pressure is contained within a larger tank which itself contains a gas under pressure. A pressure gauge attached to the smaller tank reads $140kPa$ while the gas inside this tank is known to have an (absolute) pressure of $267kPa$. In addition, the large tank is connected to a manometer containing Mercury. The large tank and connected manometer are placed in a room at an (absolute) pressure of $1bar$. Determine the pressure within the large tank and the manometer height.

Given:
$P_A = 267kPa$,
$P_{A_{gauge}} = 140kPa$,
$P_{room} = 1bar$

Find:
P_B,
h

Assumptions:
$\rho_{Hg} = 13.5(10^3)kg/m^3$ Table B.8

Schematic:

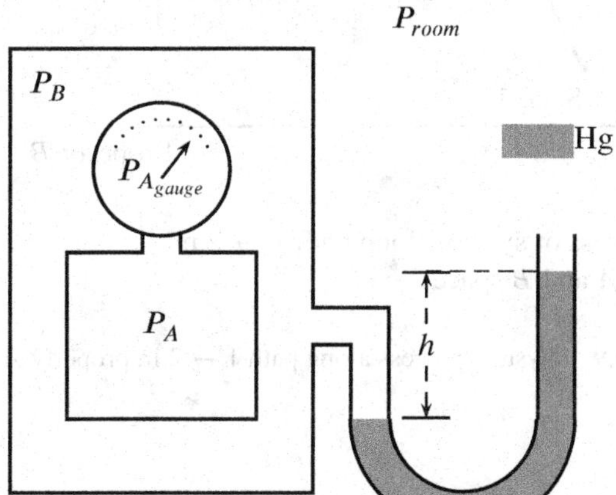

Analysis:
Applying the absolute vs. gauge pressure definition (1.11) to tank A gives

$$P_A = P_{A_{gauge}} + \overset{P_B}{\cancel{P_{A_{ref}}}} \Rightarrow P_B = P_A - P_{A_{gauge}} = 267kPa - 140kPa = 127kPa \overset{P_B}{\Longleftarrow} \quad (1.13)$$

Next, applying the absolute vs. gauge pressure definition (1.11) to tank B while noting that, from the above figure, the manometer is measuring gauge pressure, we have via (1.12)

$$P_B = \overset{\rho gh}{\cancel{P_{B_{gauge}}}} + \overset{P_{room}}{\cancel{P_{B_{ref}}}} \Rightarrow \rho gh = P_B - P_{room} \quad (1.14)$$

And thus

$$h = \frac{P_B - P_{room}}{\rho g} = \frac{127kPa - 1\cancel{bar}\frac{100kPa}{1\cancel{bar}}}{13.5(10^3)\frac{kg}{m^3}9.8\frac{m}{s^2}} = 0.000205\frac{\frac{\cancel{kN}}{\cancel{m^2}}}{\frac{kg}{\cancel{m^2}-s^2}}\frac{1000N}{1\cancel{kN}}$$

$$= 0.205\frac{N}{kg/s^2} = 0.205\frac{\cancel{kg}-m/\cancel{s^2}}{\cancel{kg}/\cancel{s^2}} = 0.205mHg \overset{h}{\Longleftarrow} \quad (1.15)$$

$$\boxed{P_B = 127kPa,\ h = 0.205mHg}$$

□

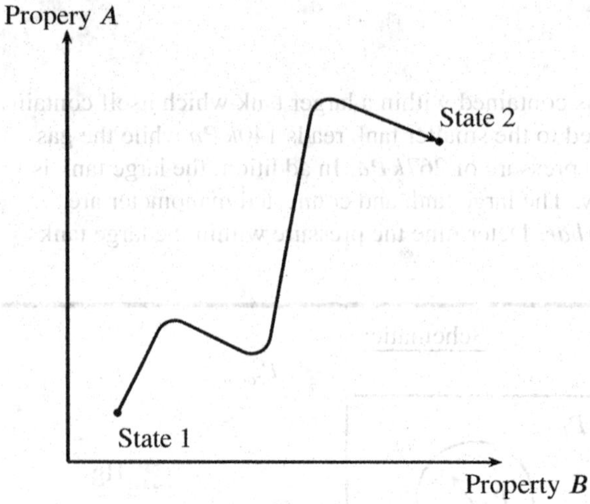

a) Process of system along path 1 → 2 in Prop. *A* and *B* space

Term	Process type
Isobaric	$P = Constant$
Isothermal	$T = constant$
Isochoric	$V = Constant$

b) Process types

Figure 1.9 System process along path 1 → 2 in property *A* and *B* space, and process definitions.

Definition 1.5 ***Process:*** *The path (i.e. succession of equilibrium states) through (over) which the system passes is called the "process". Processes proceed from some initial to a final state. The process is the path taken by the system between two end-states.*

Figure 1.9 illustrates the movement of the system from end-state "1" to "2" along the path 1 → 2. Note, the axes indicate two properties "A" and "B" (e.g. T and P, P and V, etc.). For processes whereby one property is kept fixed Table 1.9 b) gives the terminology for various constant-property processes. For example, noting that the term "iso" means same or constant and "therm" refers to heat or temperature, the term "isothermal" implies a constant temperature process.

EXAMPLE 1.2 Thermodynamic process

Problem Statement:
A system composed of a gas contained in a piston-cylinder device is initially at a temperature of $273°K$ and a volume $1m^3$. The gas is heated slowly to a temperature of $373°K$ but the piston is kept fixed. Next, the piston is allowed to move slowly upward and the gas cools back to a temperature of $273°K$. During this cooling process $TV^{0.4} = constant$. Finally, the gas is compressed via a slow downward motion of the piston, while at the same time, energy (heat) is removed from the gas till the volume returns to $1m^3$. The effect is to maintain a constant gas temperature. Sketch the processes on a $T - V$ diagram and determine the maximum volume during the combined process $1 \longrightarrow 2 \longrightarrow 3 \longrightarrow 1$.

Schematic:

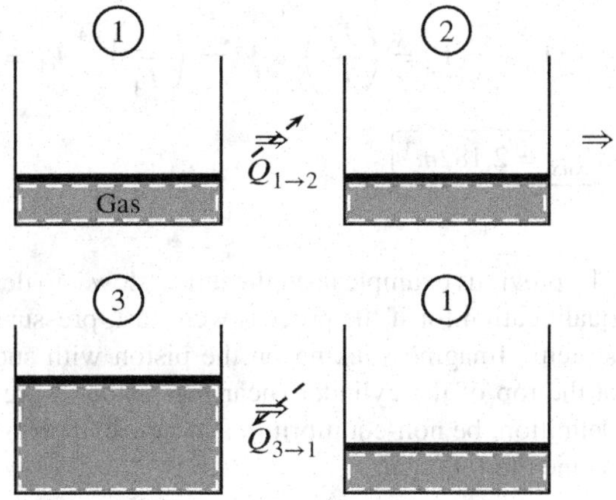

Given:
Process 1 → 2 (isochoric, $V = constant$ heating),
Process 2 → 3 (expansion, $TV^{0.4} = constant$),
Process 3 → 1 (isothermal $T = constant$ compression),
$T_1 = 273°K$,
$T_2 = 373°K$,
$T_3 = 273°K$,
$V_1 = 1m^3$

Find:
Process(es) $T - V$ diagram,
V_{max}

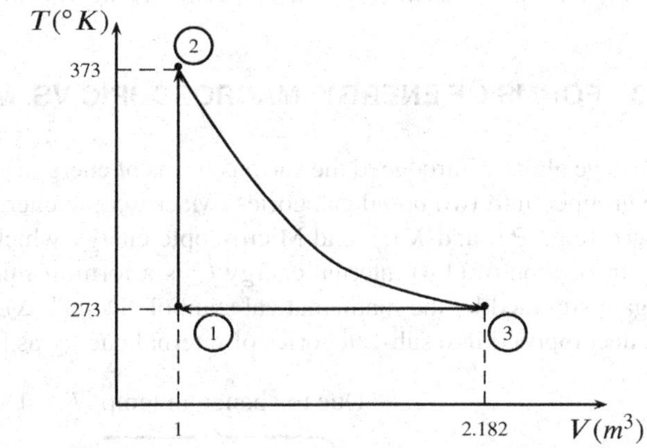

Assumptions:
None

Analysis:
Process 1 → 2:
Constant volume (isochoric) process (piston position is fixed). Thus, process 1 → 2 is a vertical line from states $T_1 = 273°K$ and $V_1 = 1m^3$ to $T_2 = 373°K$ and $V_2 = V_1 = 1m^3$ on the $T - V$ plot. Energy transfer (heat) to the gas is signified by $Q_{1\to2}$.

Process 2 → 3:
Expansion where $TV^{0.4} = constant = c$. Note, $T(V) = \frac{c}{V^{0.4}}$, thus $\frac{dT(V)}{dV} = \frac{-0.4c}{V^{1.4}} < 0$ and $\frac{d^2T(V)}{dV^2} = \frac{-0.4(-1.4)c}{V^{2.4}} = \frac{0.56c}{V^{2.4}} > 0$, hence gas temperature as a function of volume $T(V)$ is downward sloping and convex on the $T - V$ plot.

Process 3 → 1:
Isothermal compression (T is constant). Thus, process 3 → 1 is a horizontal line from $T_3 = T_1 = 273°K$ and V_3 to $T_1 = 273°K$ and $V_1 = 1m^3$ on the $T - V$ plot.

Note, from the $T - V$ plot, state 3 corresponds to the maximum volume state: $V_{max} = V_3$. From the process equation for process 2 → 3 we have $TV^{0.4}$ = constant = c. Thus

$$TV^{0.4} = c \Rightarrow T_2V_2^{0.4} = T_3V_3^{0.4} \Rightarrow V_3^{0.4} = \frac{T_2}{T_3}V_2^{0.4} \tag{1.16}$$

or

$$V_{max} = V_3 = \left(\frac{T_2}{T_3}\right)^{\frac{1}{0.4}} \overset{V_1}{\cancel{V_2}} = \left(\frac{T_2}{T_3}\right)^{\frac{1}{0.4}} V_1 = \left(\frac{373°\cancel{K}}{273°\cancel{K}}\right)^{\frac{1}{0.4}} (1m^3) = 2.182m^3 \xLeftarrow{V_{max}} \tag{1.17}$$

$$\boxed{V_{max} = 2.182m^3}$$

□

The previous example used the term "slow" to describe the processes. This is a subtle but important qualification for if the process were fast, pressure would likely not be spatially uniform within the system: Imagine yanking on the piston with such force and rapidity that the pressure of the gas at the top of the cylinder (near the piston) were less than at the bottom. Such a state would, by definition, be non-equilibrium and hence, a pressure could not be assigned to the system. Thus we define the following:

Definition 1.6 ***Quasi-equilibrium process:*** *A quasi-equilibrium process from* $1 \rightarrow 2$ *is a sufficiently slow process such that all properties of the system are defined at any given time.*

1.3 FORMS OF ENERGY: MACROSCOPIC VS. MICROSCOPIC

We have already introduced the various forms of energy U, PE, KE, etc. These forms of energy can be grouped into two broad categories: Macroscopic energy which depends on system position and speed (e.g. PE and KE), and Microscopic energy which includes all remaining forms of energy. By inspection of (1.4) internal energy U is a form of microscopic energy. Furthermore, a change (again signified by the mathematical symbol "Delta" Δ) in internal energy over some process can be decomposed into sub-categories of internal energy as illustrated in (1.18):

$$\Delta U(T,\ldots) = \overbrace{\Delta\text{“Sensible Energy”}}^{\text{Due to change in temp. } T} + \overbrace{\Delta\text{“Latent Energy”}}^{\text{Due to change in phase}} + \overbrace{\Delta\text{“Chemical Energy”}}^{\text{Due to chemical reactions}} + \overbrace{\Delta\text{“Nuclear Energy”}}^{\text{Due to atomic reactions}} \tag{1.18}$$

A change in <u>sensible</u> internal energy results in a temperature change of the system (think cooling or heating of a solid or a gas). A change in material phase (i.e. melting, solidification, vaporizing or condensing) produces a change in <u>latent</u> internal energy. Chemical reactions (combustion) result in a change in <u>chemical</u> (internal) energy of the system. And finally, nuclear reactions (fission or fusion) result in a change in <u>nuclear</u> (internal) energy of the system. In the course of your study, the first three terms, sensible, latent and chemical internal energy, will be investigated in detail, while the latter would be reserved for a course in Nuclear Physics.

CHAPTER 2

PROPERTIES OF PURE SUBSTANCES

Whether our thermodynamic system is composed of a container of liquid water, or a gas-turbine with inlet flows of air and fuel, and outflow of exhaust, all thermodynamic problems refer to systems composed of a chemical substance. Given this fact, let us jump right in and investigate the physical properties of a chemically pure material.

2.1 MATERIAL PHASES OF SINGLE COMPONENT SUBSTANCE

Definition 2.1 ***Material Phase:*** *A material phase represents a single, homogenous component or sub-component of a chemical substance. If the chemical substance is pure (i.e. single chemical species), the phase will also be chemically pure.*

Examples of pure chemical phases include ice-cubes in a glass of pure water or liquid propane coexisting with propane vapor in a storage tank. Examples of multi-chemical phases include liquid gasoline (a chemical mixture of Octane and other hydrocarbons) coexisting with gasoline vapors in a fuel tank. Notice that all the preceding examples involve multiple phases coexisting together (i.e. liquid-vapor, solid-liquid, etc.). Thus, we present the simple intuitive definitions of vapor (or gas), liquid and solid phases:

Definition 2.2 ***Solid Phase:*** *A material composed of a solid phase, when placed within a container, will not conform to the container interior space.*

Definition 2.3 ***Liquid Phase:*** *A material composed of a liquid phase, when placed within a container, will conform to the container interior space and form a free-surface.*

Definition 2.4 ***Vapor Phase:*** *A material composed of a vapor phase (gas), when placed within a container, will conform to the container interior space but will not form a free-surface.*

The physical mechanism responsible for these distinct behaviors is the strength of the chemical bonds of the material molecules: Rigid inter-atomic/molecular bonds are associated with solids, while very weak inter-atomic/molecular bonds are present in gases. Table 2.1 illustrates the behavior of the different material phases of H_2O.

Table 2.1 H_2O material phase

Solid	Liquid	Vapor

Experiment 2.1 (Slow isobaric vaporization of a solid in a piston-cylinder device) *Let us place a quantity of solid material within a piston-cylinder device which completely encompasses the space within. The system, composed of a solid material is at state "1" with system temperature, pressure and volume T_1, P and V_1. Note that T_1 is below the melting temperature of the material at the given pressure P.*

Process 1 → 2: *We transfer a certain amount of heat $Q_{1\to2}$ to the material which has the effect of raising the temperature T until melting just begins to occur (no liquid phase is present, but the addition of more heat would result in a solid-to-liquid phase transformation). Thus, state "2" is a "saturated solid" and the temperature at state "2" is the melting temperature at the given pressure: $T_2 = T_{melt}|_P$. Note, that if the material were H_2O and the system pressure 1atm, $T_2 = T_{melt}|_{P=1atm} = 0°C$. Using whatever means necessary, we measure the volume and find that the solid has slightly expanded: $V_2 > V_1$.*

Process 2 → 3: *Next, to the saturated solid, we slowly add additional heat $Q_{2\to3}$ until some, but not all, of the solid is melted: This is state "3". We measure the temperature and find that it remains unchanged: $T_3 = T_2 = T_{melt}|_P$ while the volume may have increased or decreased depending on the material (solid H_2O ice decreases in volume upon melting, most other substances expand). Let us assume the material expands upon melting, thus $V_3 > V_2$. The experimental process thus far is shown overleaf:*

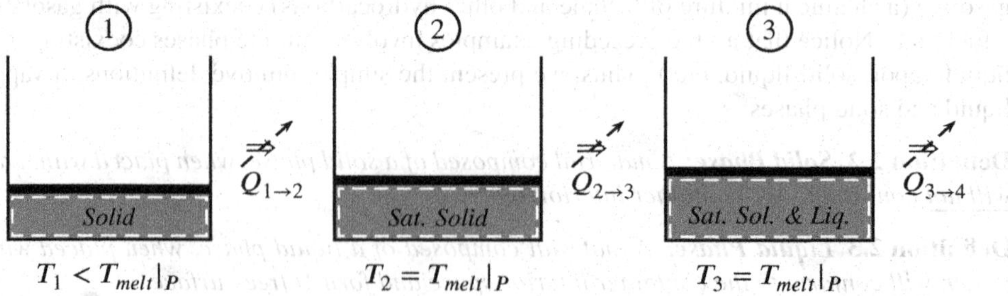

Process $3 \rightarrow 4$**:** *To the mixture of solid and liquid at state "3" we now slowly add just enough additional heat* $Q_{3\rightarrow 4}$ *until all the remaining solid has been liquified and the material is a compressed liquid. Measuring the temperature again we find that it has not changed but volume has further increased:* $T_4 = T_3 = T_2 = T_{melt}|_P$ *and* $V_4 > V_3$*. Thus, the isobaric phase transformation process* $2 \rightarrow 3 \rightarrow 4$ *is also isothermal.*

Process $4 \rightarrow 5$**:** *Next, we slowly add heat* $Q_{4\rightarrow 5}$ *until the liquid is just about to boil (vaporize) yielding a saturated liquid at state "5". Measuring the temperature and volume of the liquid we find that they have both increased:* $T_5 = T_{sat}|_P > T_4$ *and* $V_5 > V_4$ *where* $T_{sat}|_P$ *is the vaporization or boiling temperature at the given pressure. If the system material were chemically pure* H_2O*, the boiling temperature at* $1atm$ *would be* $100°C$ *or* $T_{sat}|_{P=1atm} = 100°C$*.*

Process $5 \rightarrow 6$**:** *As in the case of the saturated solid, we slowly add additional heat* $Q_{5\rightarrow 6}$ *until some of the liquid vaporizes: The system is now a mixture of saturated vapor and saturated liquid. Measuring temperature and volume we find that the material has remained at the same temperature, but the volume has increased:* $T_6 = T_5 = T_{sat}|_P$ *and* $V_6 > V_5$*.*

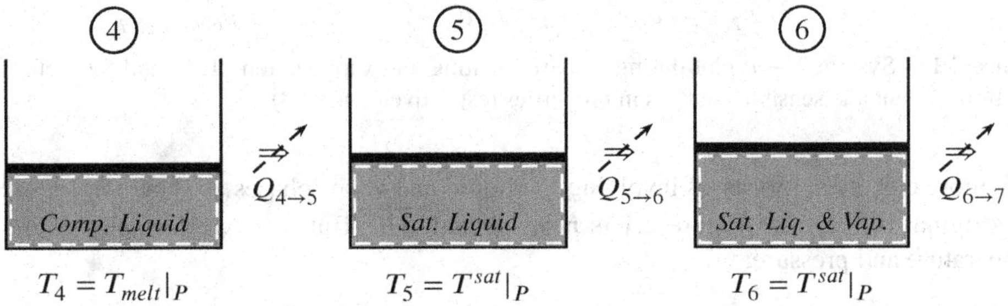

Process $6 \rightarrow 7$**:** *Adding more heat* $Q_{6\rightarrow 7}$ *to this mixture of saturated liquid and saturated vapor until all the liquid is vaporized, we again find that the temperature has remained constant, but volume has increased:* $T_7 = T_6 = T_{sat}|_P$ *and* $V_7 > V_6$*. Thus, the isobaric vaporization process is also isothermal:* $T_7 = T_6 = T_5 = T_{sat}|_P$ *and the material is now a saturated vapor.*

Process $7 \rightarrow 8$**:** *Finally, we add more heat* $Q_{7\rightarrow 8}$ *to the vapor resulting in an increase in temperature and volume:* $T_8 > T_7$ *and* $V_8 > V_7$*. The final state of the material is a super-heated vapor.*

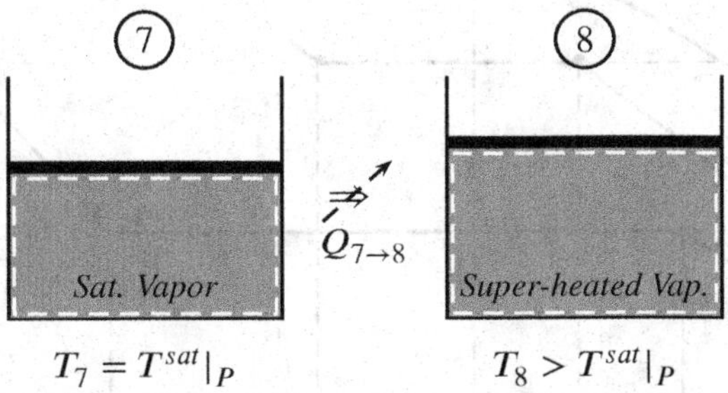

Plotting temperature vs. the mixture specific volume υ *(system volume* V *divided by system mass* m*) for the overall experiment yields Figure 2.1:*

Let us make a few comments about the previous experiment: First Figure 2.1 includes references to process exhibiting either "sensible" or "latent" changes in properties (the first resulting in a change in temperature, the latter a change in phase) as introduced in §1.3. Processes such as melting and boiling involving latent changes in properties and, if isobaric, are *also isothermal*. Next, let's

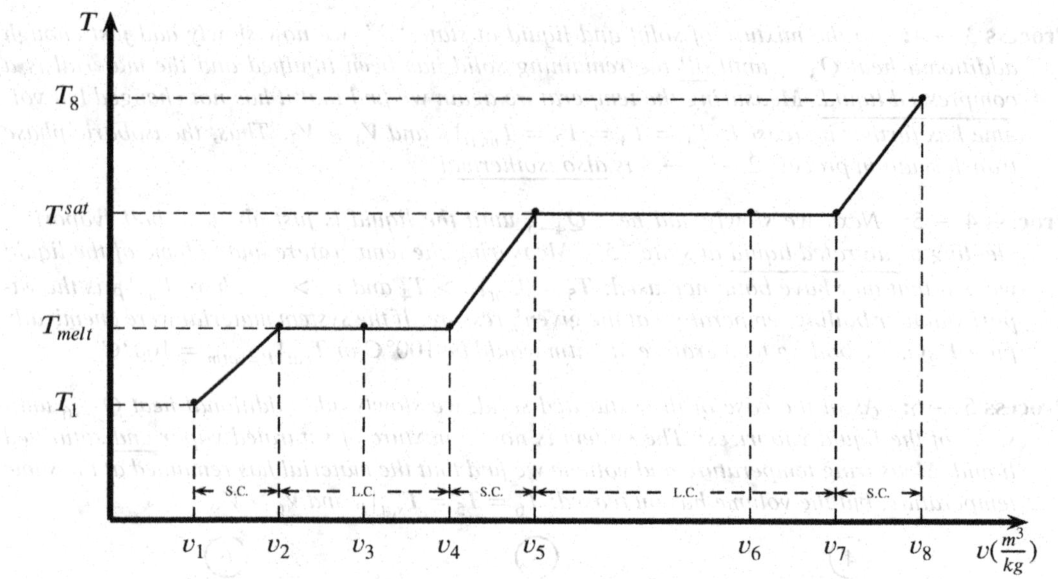

Figure 2.1 System $T - v$ plot during isobaric melting and vaporization. L.C. and S.C refer to processes exhibiting latent and sensible changes in properties respectively (see 1.3).

look more closely at processes involving the liquid and vapor phases as shown in Figure 2.2 where our original isobar from Figure 2.1 is now labeled $\overline{BB}$. But first, we formally define saturation temperature and pressure:

Definition 2.5 ***Saturation Temperature*** T^{sat}: *The "saturation temperature"* T^{sat} *of a substance is the temperature at a given pressure P at which vapor and liquid coexist:* $T^{sat}\big|_P = T_{liquid\leftrightarrow vapor}\big|_P$

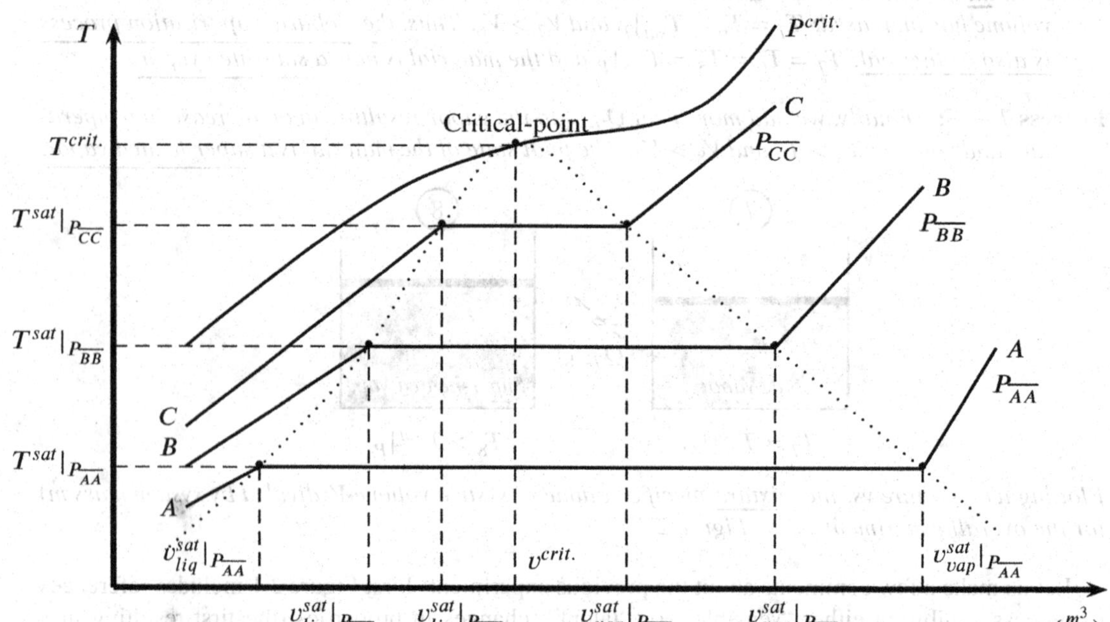

Figure 2.2 System $T - v$ plot with various isobars. Critical-point and saturation dome included (dotted line).

Definition 2.6 ***Saturation Pressure*** P^{sat}: *The "saturation pressure" P^{sat} of a substance is the pressure at a given temperature T at which vapor and liquid coexist:* $P^{sat}\Big|_T = P_{liquid \leftrightarrow vapor}\Big|_T$

Table B.2 and B.3 gives properties for saturated H_2O liquid and vapor along with saturation temperature $T^{sat}(P)$ (or saturation pressure $P^{Sat}(T)$) as a function of pressure P (or temperature T). For example, from Table B.2 $T^{sat}_{H_2O}\Big|_{P=1atm} = 100°C$ and $T^{sat}_{H_2O}\Big|_{P=0.5atm} \approx 80°C$. Conversely, from Table B.3 $P^{sat}_{H_2O}\Big|_{T=100°C} = 1atm$ and $P^{sat}_{H_2O}\Big|_{T=80°C} \approx 0.5atm$.

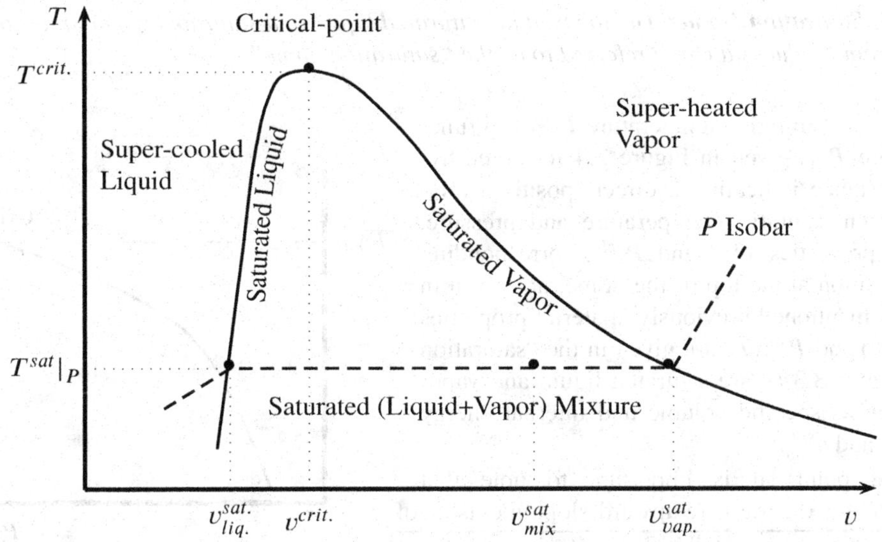

a) $T - v$ plot

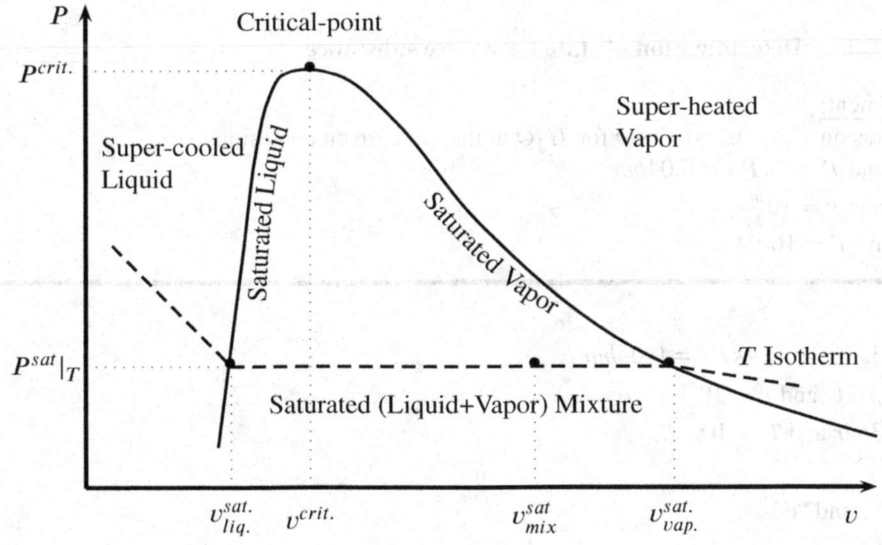

b) $P - v$ plot

Figure 2.3 System $T - v$ and $P - v$ plots with saturation dome and the various phases present.

Returning to the previous experiment, let us change the original system pressure $P_{\overline{BB}}$ and repeat the heat addition process. For example, reducing the pressure to $P_{\overline{AA}}$ yields the $T-v$ curve (isobar) given by $\overline{AA}$. If instead we now increase the pressure to $P_{\overline{CC}}$ and re-run the experiment we have the isobar $\overline{CC}$. Running a number of experiments at various pressures allows us to connect all the locations on the $T-v$ plot which are saturated liquid or saturated vapor. These connected locations on the $T-v$ diagram define the "saturation dome" illustrated in Figure 2.3 a). Likewise, if instead, the heat addition were carried out isothermally in the experiment, the connected locations of all saturated states on the $P-v$ diagram gives the saturation dome shown in Figure 2.3 b).

Definition 2.7 ***Saturation Dome:*** *The locus of all saturated liquid and saturated vapor states on a property diagram produces a curve referred to as the "saturation dome".*

A sketch of the saturation temperature T^{sat} as a function of pressure P is given in Figure 2.4 for some hypothetical substance indicating a direct (positive) relationship between saturation temperature and pressure. Critical-point properties T^{crit} and P^{crit}, corresponding to the state location at the top of the dome, are given in Table B.1. As mentioned previously, material properties such as $T^{sat}(P)$ and $P^{sat}(T)$ are given in the "saturation Tables" B.2 and B.3 as are saturated liquid and vapor properties such as specific volume and internal energy v^{sat}_{liq}, v^{sat}_{vap}, u^{sat}_{liq} and u^{sat}_{vap}.

As a final point, it is important to note that isobars in the $T-v$ diagram are upward-sloping outside of the dome (e.g. as a gas is heated isobarically, it expands), while isotherms in the $P-v$ diagram are downward-sloping.

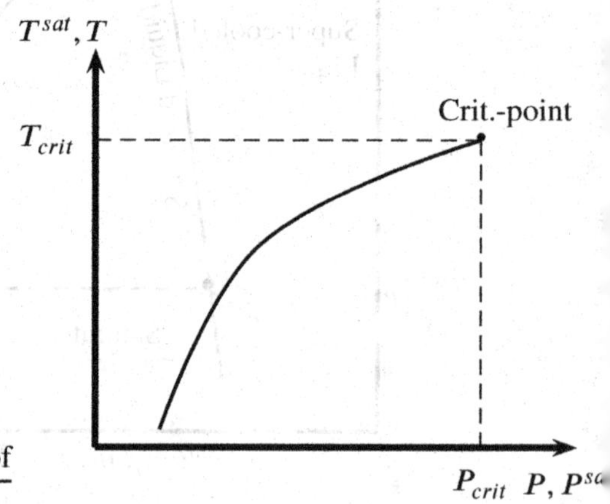

Figure 2.4 $T^{sat}(P)$ and $P^{sat}(T)$.

EXAMPLE 2.1 Determination of state for a pure substance

Problem Statement:

Locate the states on a $P-v$ and $T-v$ for H_2O at the three given conditions:

a) $T = 35°C$ and $P = 4kPa = 0.04bar$

b) $T = 35°C$ and $v = 10\frac{m^3}{kg}$

c) $P = 3bar$ and $T = 100°C$.

Given:

a) $H_2O, T = 35°C, P = 4kPa = 0.04bar$

b) $H_2O, T = 35°C$ and $v = 10\frac{m^3}{kg}$

c) $H_2O, P = 3bar$ and $T = 100°C$

Find:

States "a", "b" and "c"

Assumptions:

Saturation Tables (H_2O) B.3 and B.2

Schematic:

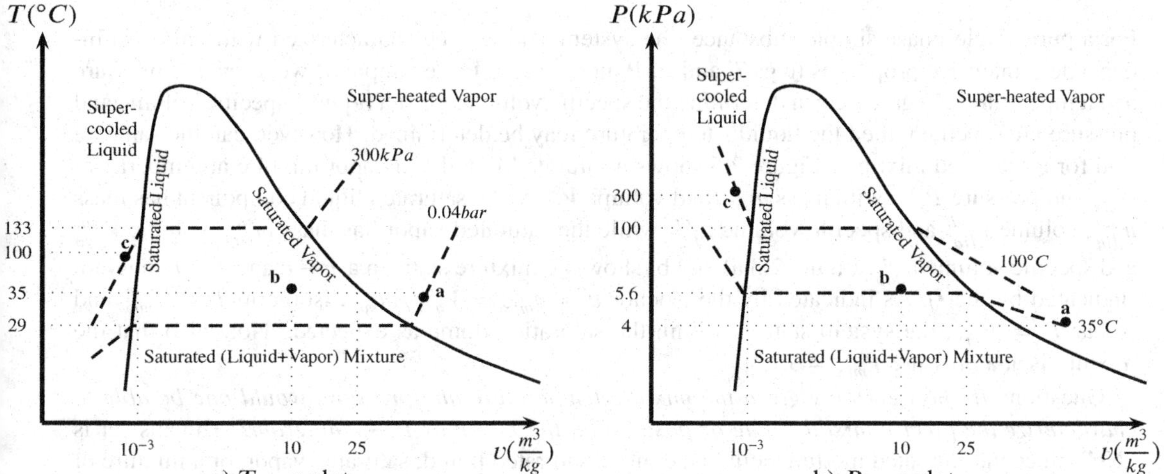

a) $T - v$ plot b) $P - v$ plot

Analysis:

a) $T^{sat}\big|_{P=0.04bar} = 29°C$ via H_2O saturation pressure Table B.3. Inter-section of $P = 0.04bar$ $= 4kPa$ isobar and $T = 35°C$ isotherm on $T - v$ plot at location "a" is in the super-heated vapor region: $T = 35°C > 29°C = T^{sat}\big|_{P=0.04bar}$ or $T > T^{sat}\big|_{P=0.04bar}$.

Note, the $P - v$ diagram can also be used: $P^{sat}\big|_{T=35°C} = 0.05628bar$ (via H_2O saturation temperature Table B.2). Inter-section of $T = 35°C$ isotherm and $P = 0.04bar$ isobar on $P - v$ plot at location "a" in the super-heated vapor region: $P^{sat}\big|_{T=35°C} = 0.05628bar > 0.04bar = P$ or $P < P^{sat}\big|_{T=35°C}$.

H_2O at $T = 35°C$ and $P = 0.04bar$ is super-heated vapor.

b) $v^{sat.}_{liq.}\Big|_{T=35°C} = 10^{-3}\frac{kg}{m^3}$, $v^{sat.}_{vapor}\Big|_{T=35°C} = 25\frac{kg}{m^3}$ (via H_2O saturation temperature Table B.2).

Inter-section of $T = 35°C$ isotherm and $v = 10\frac{m^3}{kg}$ isochor on $T - v$ diagram at location "b" in saturated mixture region ($v^{sat.}_{vapor}\Big|_{T=35°C} > v > v^{sat.}_{liq.}\Big|_{T=35°C}$).

Using the $P - v$ diagram: Inter-section of isobar $P = P^{sat}\big|_{T=35°C}$ ($T^{sat}\big|_{P^{sat}|_{T=35°C}} = 35°C$) and isochor $v = 10\frac{m^3}{kg}$ in saturated mixture region: $v^{sat.}_{vapor}\Big|_{P=P^{sat}|_{T=35°C}} > v > v^{sat.}_{liq.}\Big|_{P=P^{sat}|_{T=35°C}}$

H_2O at $T = 35°C$ and $v = 10\frac{m^3}{kg}$ is a mixture of saturated liquid and vapor.

c) $T^{sat.}\big|_{P=300kPa} = 133.6°C$ (via H_2O saturation pressure Table B.3). Inter-section of $T = 100°C$ isotherm and $P = 300kPa$ isobar on $T - v$ diagram at location "c" in super-cooled liquid region: $T = 100°C < 133.6°C = T^{sat.}\big|_{P=300kPa}$.

With respect to the $P - v$ diagram: Inter-section of the $T = 100°C$ isotherm and the isobar $P = 300kPa$ (with $T^{sat.}\big|_{P=300kPa} = 133.6°C$) in super-cooled liquid region.

H_2O, $P = 300kPa$ and $T = 100°C$ is a super-cooled liquid.

□

2.2 PROPERTIES OF SATURATED MIXTURES: QUALITY x

For a pure single-phase simple substance, the system state can be characterized with only two independent intensive properties (e.g. T and v, P and v, etc.). For example, if we know the pressure and temperature of a gas, we can determine the specific volume. Or if a liquid's specific volume and pressure are specified, then the liquid's temperature may be determined. However, can the same be said for a saturated mixture? Figure 2.5 shows a saturated liquid and vapor mixture at temperature T_{mix} and pressure P_{mix}, with mass m_{mix} and volume V_{mix}. The saturated liquid component has mass m^{sat}_{liq}, volume V^{sat}_{liq} and specific volume v^{sat}_{liq}, while the saturated vapor has mass m^{sat}_{vap}, volume V^{sat}_{vap} and specific volume v^{sat}_{vap}. Figure 2.3 a) and b) show the mixture state on a $P-v$ and $T-v$ diagram (indicated by the •). As indicated by the isochor $v = v_{mix} = V_{mix}/m_{mix}$, isotherm $T = T_{mix}$ and isobar $P = P_{mix}$, the system state is within the saturation dome as expected. Note also that the mixture is at a pressure $P_{mix} = P^{sat}|_{T_{mix}}$.

Question: *If one were only given the mixture temperature and pressure, would one be able to characterize the precise mixture state or position on the $T-v$ or $P-v$ diagram?* The answer is "No", since the saturated mixture could be either a saturated liquid, saturated vapor, or a mixture of the two. Indeed, the reason the two properties T and P are not enough to characterize the state of a saturated mixture is that within the saturated dome temperature and pressure are related according to Figure 2.4 and thus are *not* independent. Hence, we introduce a new property applicable only to saturated mixtures: Quality

Definition 2.8 ***Quality* x:** *Mixture quality* $x \equiv \frac{m^{sat}_{vap}}{m^{sat}_{mix}}$ *(used to describe saturated mixture states along with the mixture pressure* P_{mix} *or temperature* T_{mix}*).*

Specifically, the expression for quality in terms of mixture and component specific volumes is

$$x = \frac{m^{sat}_{vap}}{m^{sat}_{mix}} = \frac{v^{sat}_{mix} - v^{sat}_{liq}}{v^{sat}_{vap} - v^{sat}_{liq}} \tag{2.1}$$

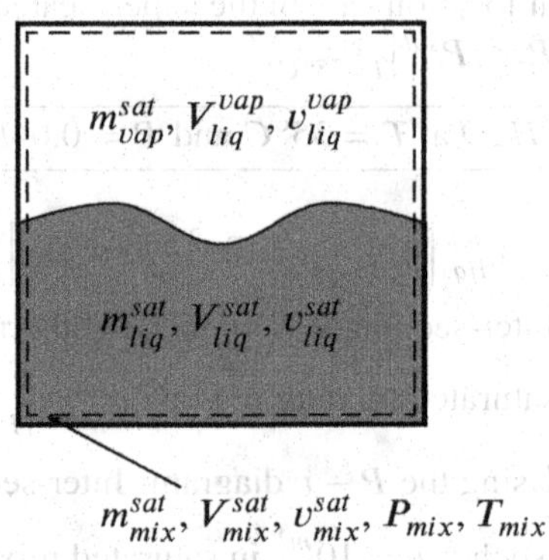

Figure 2.5 Two-phase system: CV indicated by – – ––.

Proof: Application of a mass-balance to the mixture system shown in Figure 2.5 yields

$$m^{sat}_{mix} = m^{sat}_{vap} + m^{sat}_{liq} \tag{2.2}$$

Likewise, a volume-balance on the mixture system gives

$$V^{sat}_{mix} = V^{sat}_{vap} + V^{sat}_{liq} \tag{2.3}$$

Noting the definition of specific volume $v = V/m$ we can re-write (2.3) as

$$V^{sat}_{mix} = V^{sat}_{vap} + V^{sat}_{liq} \Rightarrow v_{mix} m^{sat}_{mix} = v^{sat}_{vap} m^{sat}_{vap} + v^{sat}_{liq} m^{sat}_{liq} \Rightarrow$$

$$v^{sat}_{mix} = v^{sat}_{vap} \underbrace{\frac{m^{sat}_{vap}}{m^{sat}_{mix}}}_{x} + v^{sat}_{liq} \frac{m^{sat}_{liq}}{m^{sat}_{mix}} = v^{sat}_{vap} x + v^{sat}_{liq} \frac{m^{sat}_{liq}}{m^{sat}_{mix}} \tag{2.4}$$

Now, from the mass-balance (2.2)

$$m^{sat}_{mix} = m^{sat}_{vap} + m^{sat}_{liq} \Rightarrow m^{sat}_{liq} = m^{sat}_{mix} - m^{sat}_{vap} \Rightarrow \frac{m^{sat}_{liq}}{m^{sat}_{mix}} = \overbrace{\cancel{\frac{m^{sat}_{mix}}{m^{sat}_{mix}}}}^{1} - \overbrace{\cancel{\frac{m^{sat}_{vap}}{m^{sat}_{mix}}}}^{x} = 1 - x \Rightarrow \frac{m^{sat}_{liq}}{m^{sat}_{mix}} = 1 - x \tag{2.5}$$

Substituting the result from (2.5) into the modified volume-balance (2.4) gives

$$v^{sat}_{mix} = v^{sat}_{vap} x + v^{sat}_{liq} \overbrace{\cancel{\frac{m^{sat}_{liq}}{m^{sat}_{mix}}}}^{1-x} = v^{sat}_{vap} x + v^{sat}_{liq}(1 - x) = x(v^{sat}_{vap} - v^{sat}_{liq}) + v^{sat}_{liq} \Rightarrow x = \frac{v^{sat}_{mix} - v^{sat}_{liq}}{v^{sat}_{vap} - v^{sat}_{liq}} \tag{2.6}$$

Thus

$$\boxed{x = \frac{v^{sat}_{mix} - v^{sat}_{liq}}{v^{sat}_{vap} - v^{sat}_{liq}}} \tag{2.7}$$

■

A simple geometric interpretation for quality x can be formulated from (2.1): Let a saturated mixture at pressure P^{sat}_{mix} be represented by state location "B" indicated on the $T - v$ diagram in Figure 2.6. In addition, the saturated liquid and vapor states at P_{mix} are indicated by points "A" and "C" respectively. Constructing the segments $\overline{AB}$ and $\overline{AC}$, (2.1) can be rewritten as

$$\boxed{x = \frac{v^{sat}_{mix} - v^{sat}_{liq}}{v^{sat}_{vap} - v^{sat}_{liq}} = \frac{\overline{AB}}{\overline{AC}}} \tag{2.8}$$

Note, (2.8) implies the following limiting behavior:

$$x = \begin{cases} 0, & \overline{AB} = 0 \text{ (All saturated liquid)} \\ 1, & \overline{AB} = \overline{AC} \text{ (All saturated vapor)} \end{cases} \tag{2.9}$$

2.2.1 GENERALIZATION TO OTHER MIXTURE PROPERTIES

Referring to saturation Tables B.2 and B.3 we see that saturated liquid and vapor have an associated specific internal energy as well as specific volume: Measuring the internal energy and mass of the saturated liquid or vapor components alone one could calculate $u^{sat}_{liq} = U^{sat}_{liq}/m^{sat}_{liq}$ and $u^{sat}_{vap} = U^{sat}_{vap}/m^{sat}_{vap}$. Thus, letting $v \to u$, our previous $T - v$ and $P - v$ diagrams become $T - u$ and $P - u$ diagrams while retaining the same qualitative appearance. The same goes for the property enthalpy H defined as

$$\boxed{H = U + PV, \text{ (Enthalpy)}} \tag{2.10}$$

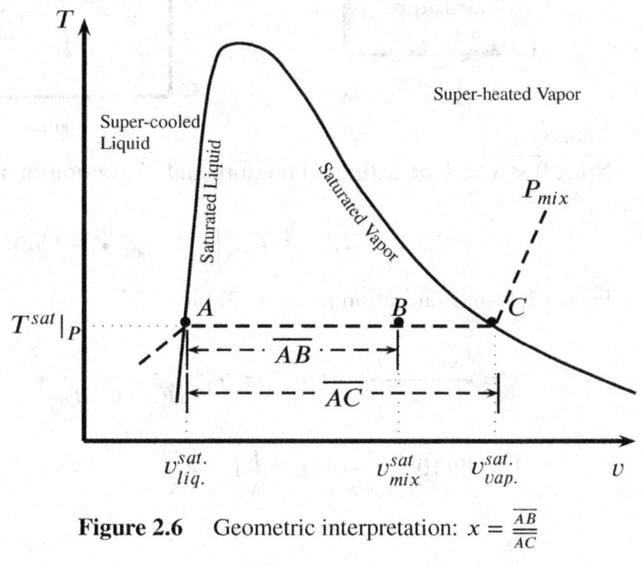

Figure 2.6 Geometric interpretation: $x = \frac{\overline{AB}}{\overline{AC}}$

where, like v and u, specific enthalpies h are given in saturation tables. With respect to our expression for quality, letting $v \longrightarrow u$ or $v \longrightarrow h$ (2.1) becomes

$$x = \frac{u^{sat}_{mix} - u^{sat}_{liq}}{u^{sat}_{vap} - u^{sat}_{liq}} = \frac{h^{sat}_{mix} - h^{sat}_{liq}}{h^{sat}_{vap} - h^{sat}_{liq}} \tag{2.11}$$

where $h^{sat}_{mix} = H^{sat}_{mix}/m^{sat}_{mix}$, $h^{sat}_{liq} = H^{sat}_{liq}/m^{sat}_{liq}$ and $h^{sat}_{vap} = H^{sat}_{vap}/m^{sat}_{vap}$.

A final word with respect to using saturation Tables B.2 and B.3: To save an arithmetic step, the quantity $h_{fg} = h^{sat}_{vap} - h^{sat}_{liq}$ the vapor-to-liquid enthalpy change is calculated beforehand and tabulated separately for your convenience.

EXAMPLE 2.2 Determination of state for a saturated substance using quality

Problem Statement:
Saturated H_2O at $100kPa$ has a quality of 0.25. Find the mixture specific volume, internal energy and temperature. Also, if the total mixture volume is $1m^3$, determine the mixture, vapor and liquid mass.

Given:
H_2O, $P_{mix} = 100kPa$, $x = 0.25$, $V^{sat}_{mix} = 1m^3$

Find:
v^{sat}_{mix}, u^{sat}_{mix}, T_{mix},
m^{sat}_{mix}, m^{sat}_{liq}, m^{sat}_{vapor}

Assumptions:
Saturation pressure Tables B.3

Schematic:

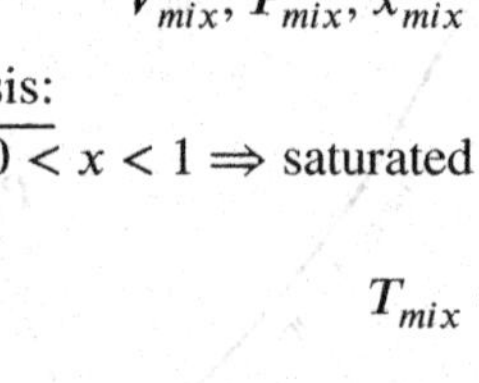

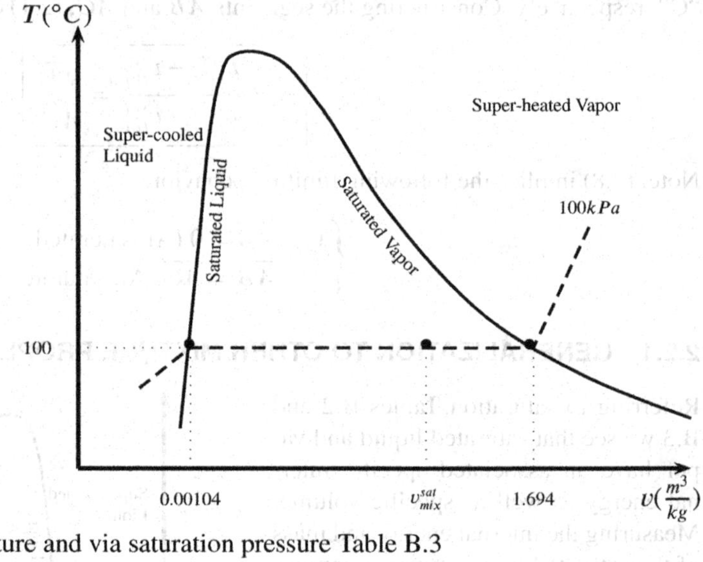

Analysis:
Note, $0 < x < 1 \Rightarrow$ saturated mixture and via saturation pressure Table B.3

$$T_{mix} = T^{sat}_{mix}\big|_{P_{mix}=100kPa} = 99.63°C \overset{T_{mix}}{\Longleftarrow} \tag{2.12}$$

From (2.1) and saturation pressure Table B.3

$$x = \frac{v^{sat}_{mix} - v^{sat}_{liq}}{v^{sat}_{vap} - v^{sat}_{liq}} \Rightarrow v^{sat}_{mix} = v^{sat}_{liq}\Big|_{P_{mix}=100kPa} + x\left(v^{sat}_{vap} - v^{sat}_{liq}\right)\Big|_{P_{mix}=100kPa} =$$

$$1.0432(10^{-3})\frac{m^3}{kg} + 0.25\left(1.694\frac{m^3}{kg} - 1.0432(10^{-3})\frac{m^3}{kg}\right) = 0.4242\frac{m^3}{kg} \overset{v^{sat}_{mix}}{\Longleftarrow} \tag{2.13}$$

From (2.11) and saturation pressure Table B.3

$$x = \frac{u^{sat}_{mix} - u^{sat}_{liq}}{u^{sat}_{vap} - u^{sat}_{liq}} \Rightarrow u^{sat}_{mix} = u^{sat}_{liq}\Big|_{P_{mix}=100kPa} + x\left(u^{sat}_{vap} - u^{sat}_{liq}\right)\Big|_{P_{mix}=100kPa} =$$

$$417.36\frac{kJ}{kg} + 0.25\left(2506.1\frac{kJ}{kg} - 417.36\frac{kJ}{kg}\right) = 939.545\frac{kJ}{kg} \xLeftarrow{u^{sat}_{mix}} \quad (2.14)$$

Next, from the definition of specific volume

$$v = \frac{V}{m} \Rightarrow m = \frac{V}{v} \Rightarrow m^{sat}_{mix} = \frac{V_{mix}}{v^{sat}_{mix}} = \frac{1\cancel{m^3}}{0.4242\frac{\cancel{m^3}}{kg}} = 2.358kg \xLeftarrow{m^{sat}_{mix}} \quad (2.15)$$

From the definition of quality (2.1)

$$x = \frac{m^{sat}_{vapor}}{m^{sat}_{mix}} \Rightarrow m^{sat}_{vapor} = xm^{sat}_{mix} = 0.25(2.358kg) = 0.589kg \xLeftarrow{m^{sat}_{vapor}} \quad (2.16)$$

And hence via a mass-balance on the mixture

$$m^{sat}_{mix} = m^{sat}_{vap} + m^{sat}_{liq} \Rightarrow m^{sat}_{liq} = m^{sat}_{mix} - m^{sat}_{vap} = 2.358kg - 0.589kg = 1.768kg \xLeftarrow{m^{sat}_{liq}} \quad (2.17)$$

$T_{mix} = 99.63°C, v^{sat}_{mix} = 0.4242\frac{m^3}{kg}, u^{sat}_{mix} = 939.545\frac{kJ}{kg}, m^{sat}_{mix} = 2.358kg, m^{sat}_{vapor} = 0.589kg,$
$m^{sat}_{liq} = 1.768kg,$

□

2.3 PROPERTIES OF SUPER-COOLED LIQUIDS

From the $T-v$ diagram (say of H_2O) shown in Figure 2.7, specific volume v of a super-cooled liquid is a *very* weak function of pressure (i.e. $\frac{\partial v}{\partial P} \approx 0$), and a weakly positive function of temperature (i.e. $\frac{\partial v}{\partial T} > 0$): As a compressed liquid is heated, volume expands slightly. Using these two facts let us compare the specific volume of a compressed liquid at temperature T and pressure P with that of a saturated liquid at temperature $T^{sat} = T$ and pressure $P^{sat}\big|_T$:

$$v^{sup.-cooled}_{liq}\Big|_{T,P} = \overbrace{v^{sat}_{liq.}\Big|_{T^{sat}=T}}^{v \text{ sat. liquid at temp. } T} + \overbrace{\cancel{\Delta v\Big|^{P}_{P^{sat}|_T}}}^{\Delta v \text{ from sat. to sup.-cooled liquid} \;\to\; \approx 0}$$

$$\Rightarrow v^{sup.-cooled}_{liq.}\Big|_{T,P} \approx v^{sat}_{liq.}\Big|_{T^{sat}=T} \quad (2.18)$$

The same also goes for specific internal energy u. Thus

$$\begin{aligned} v^{sup.-cooled}_{liq}\Big|_{T,P} &\approx v^{sat}_{liq}\Big|_{T^{sat}=T}, \text{ (Super-cooled liquid)} \\ u^{sup.-cooled}_{liq}\Big|_{T,P} &\approx u^{sat}_{liq}\Big|_{T^{sat}=T}, \text{ (Super-cooled liquid)} \end{aligned} \quad (2.19)$$

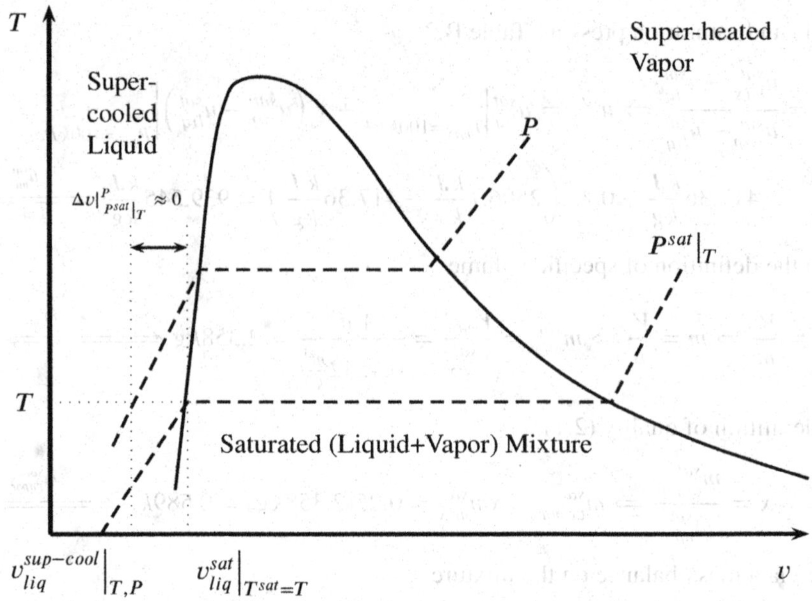

Figure 2.7 $T - v$ diagram illustrating approximate properties for super-cooled liquid.

In addition, like u and v, an approximate expression for the specific enthalpy h of a super-cooled liquid also exists.

$$h_{liq}^{sup.-cooled}\Big|_{T,P} = \overbrace{h_{liq}^{sat}\Big|_{T^{sat}=T}}^{h \text{ sat. liquid at temp. } T} + \overbrace{\Delta h|_{P^{sat}|_T}^{P}}^{\Delta h \text{ from sat. to sup.-cooled liq.}}$$

$$= h_{liq}^{sat}\Big|_{T^{sat}=T} + \Delta(u+Pv)|_{P^{sat}|_T}^{P} = h_{liq}^{sat}\Big|_{T^{sat}=T} + \Delta(u)|_{P^{sat}|_T}^{P} + \Delta(Pv)|_{P^{sat}|_T}^{P} \quad (2.20)$$

Now, from (2.19) we found $v_{liq}^{super-cooled}\Big|_{T,P} \approx v_{liq.}^{sat}\Big|_{T^{sat}=T}$ and $u_{liq}^{super-cooled}\Big|_{T,P} \approx u_{liq}^{sat}\Big|_{T^{sat}=T}$. Thus, (2.20) reduces to

$$h_{liq}^{sup.-cooled}\Big|_{T,P} = h_{liq}^{sat}\Big|_{T^{sat}=T} + \cancelto{\approx 0}{\Delta(u)|_{P^{sat}|_T}^{P}} + \Delta(Pv)|_{P^{sat}|_T}^{P}$$

$$\approx h_{liq}^{sat}\Big|_{T^{sat}=T} + v_{liq}^{sat}\Big|_{T^{sat}=T} \Delta(P)|_{P^{sat}|_T}^{P} \quad (2.21)$$

or

$$h_{liq}^{sup.-cooled}\Big|_{T,P} \approx h_{liq}^{sat}\Big|_{T^{sat}=T} + v_{liq.}^{sat}\Big|_{T^{sat}=T} (P - P^{sat}|_T), \text{ (Super-cooled liquid)} \quad (2.22)$$

EXAMPLE 2.3 Determination of state for a super-cooled liquid

Problem Statement:

One and a half liters of water (super-cooled liquid H_2O) at $100kPa$ has a temperature of $4°C$. Find the mass, internal energy and enthalpy of the water.

Given:
H_2O, $P = 100kPa$,
$T = 4°C$, $V = 1.5L$

Find:
m, U, H

Assumptions:
Saturation Tables B.2 and B.3

Schematic:

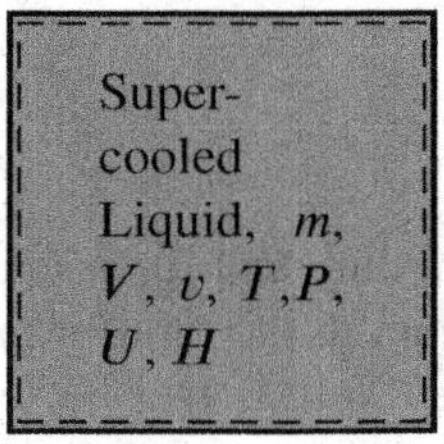

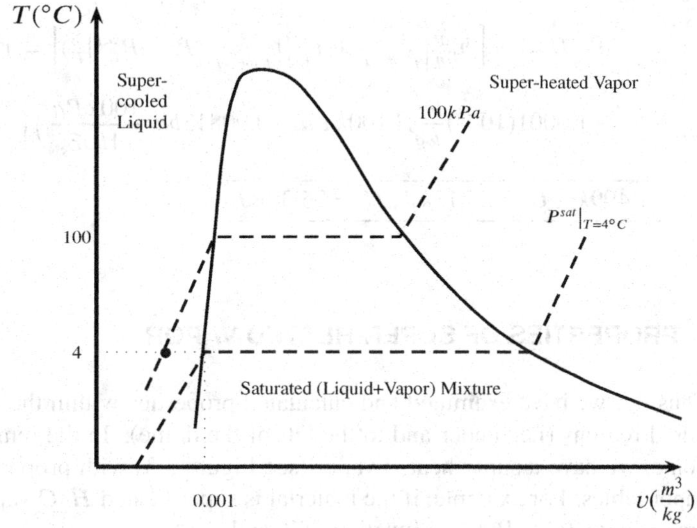

Analysis:
Note, from the $T - v$ diagram and $T = 4°C < 99.63°C = T^{sat}\big|_{P=100kPa} \Rightarrow$ super-cooled liquid. From the definition of specific volume and (2.19)

$$v = \frac{V}{m} \Rightarrow m = \frac{V}{v_{liq}^{super-cooled}\Big|_{\substack{T=4°C,\\ P=100kPa}}} \approx \frac{V}{v_{liq}^{sat}\Big|_{T^{sat}=T=4°C}} \tag{2.23}$$

where $v_{liq}^{sat}\Big|_{T^{sat}=4°C} = 1.0001(10^{-3})\frac{m^3}{kg}$ (saturation temperature Table B.2). Thus

$$m \approx \frac{V}{v_{liq}^{sat}\Big|_{T^{sat}=4°C}} = \frac{1.5L}{1.0001(10^{-3})\frac{m^3}{kg}} = 1499.85\frac{\cancel{L}-kg}{\cancel{m^3}}\frac{1\cancel{m^3}}{1000\cancel{L}} = 1.499kg \stackrel{m}{\Longleftarrow} \tag{2.24}$$

Now, from (2.19)

$$U = mu_{liq}^{super-cooled}\Big|_{\substack{T=4°C,\\ P=100kPa}} \approx mu_{liq}^{sat}\Big|_{T^{sat}=T=4°C} \tag{2.25}$$

where from Table B.2 $u_{liq}^{sat}\Big|_{T^{sat}=T=4°C} = 16.77\frac{kJ}{kg}$. Returning to (2.25)

$$U \approx mu_{liq}^{sat}\Big|_{T^{sat}=T=4°C} = 1.499\cancel{kg}\left(16.77\frac{kJ}{\cancel{kg}}\right) = 25.13kJ \stackrel{U}{\Longleftarrow} \tag{2.26}$$

Finally, from (2.22)

$$H = mh_{liq}^{super-cooled}\Big|_{\substack{T=4°C,\\ P=100kPa}} \approx m\left[h_{liq}^{sat}\Big|_{T^{sat}=T} + v_{liq.}^{sat}\Big|_{T^{sat}=T}(P - P^{sat}\big|_T)\right] \tag{2.27}$$

From Table B.2 $h_{liq}^{sat}\Big|_{T^{sat}=T=4°C} = 16.78\frac{kJ}{kg}$ and $P_{liq}^{sat}\Big|_{T^{sat}=T=4°C} = 0.00813bar$. Returning to (2.27)

$$H \approx m\left[h_{liq}^{sat}\Big|_{T^{sat}=T} + v_{liq}^{sat}\Big|_{T^{sat}=T}(P - P^{sat}|_T)\right] = 1.499\cancel{kg}\left[16.78\frac{kJ}{\cancel{kg}}\right.$$
$$\left. +1.0001(10^{-3})\frac{m^3}{\cancel{kg}}\left(100kPa - 0.00813\cancel{bar}\frac{100kPa}{1\cancel{bar}}\right)\right] = 25.30kJ \xLeftarrow{H} \quad (2.28)$$

$m = 1.499kg, U = 25.13kJ, H = 25.30kJ$

□

2.4 PROPERTIES OF SUPER-HEATED VAPOR

Thus far, we have examined and calculated properties within the saturated mixture and super-cool liquid regions (i.e. under and to the left of the dome). In the remaining region, to the right of the dome, we have a super-heated vapor (see Figure 2.3) with properties provided in the super-heated vapor tables. For example, if the material is super-heated H_2O vapor (i.e. steam), properties v, u, h are given in Table B.4 as a function of T and P.

EXAMPLE 2.4 Determination of properties for a super-heated vapor

Problem Statement:
Steam at $1bar$ and $120°C$ flows in small tube at a volume flow-rate of $0.1m^3/min$. Find the mass flow-rate in kg/sec, the specific internal energy and enthalpy of the steam.

Given:
Steam (H_2O), $P = 1bar$, $T = 120°C$, $\dot{V}_{steam} = 0.1\frac{m^3}{min}$

Find:
$\dot{m}_{steam}$, u_{steam}, h_{steam}

Assumptions:
Super-heated vapor/steam Table B.4

Schematic:

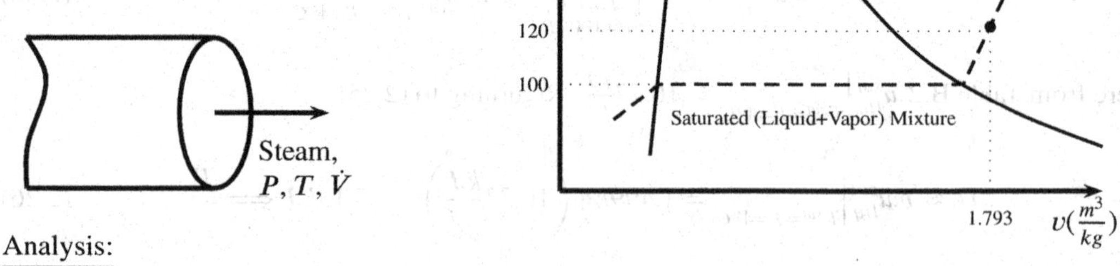

Analysis:
From the definition of specific volume (where t refers to time in the following)

$$v = \frac{V}{m} \Rightarrow m = \frac{V}{v} \Rightarrow \dot{m} = m\frac{1}{t} = \frac{V}{v}\frac{1}{t} = \frac{V\frac{1}{t}}{v} = \frac{\dot{V}}{v} \Rightarrow \dot{m}_{steam} = \frac{\dot{V}_{steam}}{v_{steam}} \quad (2.29)$$

or

$$\dot{m}_{steam} = \frac{\dot{V}_{steam}}{v_{steam}} \tag{2.30}$$

Now $T = 120°C > T^{sat}|_{P=1bar} = 99.63°C$ hence the H_2O is a super-heated vapor. Thus, via steam Table B.4 $v_{steam} = v_{vapor}^{super-heated}\Big|_{T=120°C, P=1bar} = 1.793\frac{m^3}{kg}$ and from (2.30)

$$\dot{m}_{steam} = \frac{\dot{V}_{steam}}{v_{steam}} = \frac{0.1\frac{m^3}{min}}{1.793\frac{m^3}{kg}} = \frac{0.1\frac{m^3}{min}\frac{1min}{60s}}{1.793\frac{m^3}{kg}} = 9.29(10^{-4})\frac{kg}{s} \overset{\dot{m}_{steam}}{\Longleftarrow} \tag{2.31}$$

Finally, using steam Table B.4 we have

$$u|_{P=1bar, T=120°C} = 2537\frac{kJ}{kg} \overset{u_{steam}}{\Longleftarrow} \tag{2.32}$$

$$h|_{P=1bar, T=120°C} = 2716\frac{kJ}{kg} \overset{h_{steam}}{\Longleftarrow} \tag{2.33}$$

$$\boxed{\dot{m}_{steam} = 9.29(10^{-4})\tfrac{kg}{s},\ u_{steam} = 2537\tfrac{kJ}{kg},\ h_{steam} = 2716\tfrac{kJ}{kg}}$$

□

2.5 IDEAL GAS BEHAVIOR (I.G.B.)

Up to now we have used tables to determine properties of a super-heated vapor. However, it has been found that for a super-heated vapor of *sufficiently* high temperature and low pressure, a very simple equation of state, the ideal gas law, is applicable:

$$\boxed{PV = NR_UT, \text{ (Ideal gas law (I.G.B.))}} \tag{2.34}$$

where N is the number of moles of gas and R_U is the Universal Gas Constant. The Universal Gas Constant R_U is a new quantity which, from (2.34), has (metric) units of $R_U = \frac{PV}{NT} \sim \frac{Pa-m^3}{mol-°K} = \frac{N/m^2-m^3}{mol-°K} = \frac{N-m}{mol-°K} = \frac{J}{mol-°K}$ (or $\frac{kJ}{kmol-°K}$). Hence, R_U has units of $\frac{Energy}{mol-°K}$ and, as the name implies, is a *universal constant* with a value of $8.3\frac{J}{mol-°K}$. Thus, (2.34) is valid for any substance provided Ideal Gas Behavior (IGB) is valid (which we shall discuss in the following sub-section §2.5.1). Now, there are several remarks to be made regarding (2.34): First, T is in units of absolute temperature such as $°K$ (Kelvin) in metric and $°R$ (Rankine) in the Imperial system. Secondly, (2.34) can take on a number of equivalent forms:

$$PV = NR_UT = NR_UT\frac{m}{m} = mR_UT\frac{N}{m} = mR_UT\frac{1}{MW} = m\left(\frac{R_U}{MW}\right)T = mRT \tag{2.35}$$

Thus

$$\boxed{PV = mRT, \text{ (Ideal gas law (I.G.B.))}} \tag{2.36}$$

where the gas constant R (not the universal gas constant R_U!!!) is

$$\boxed{R = \frac{R_U}{MW}, \text{ (Gas constant)}} \tag{2.37}$$

Also, in terms of density ρ and specific volume v we have

$$PV = mRT \Rightarrow P = \frac{m}{V}RT = \rho RT \Rightarrow \boxed{P = \rho RT},$$

$$PV = mRT \Rightarrow P\frac{V}{m} = RT \Rightarrow \boxed{Pv = RT} \tag{2.38}$$

Note that the gas constant R used in (2.36) and (2.38), and the universal gas constant R_U are <u>not</u> the same constants as indicated by (2.37). Indeed, from (2.37) we have in metric units $R = \frac{R_U}{MW} \sim \frac{J/(\cancel{mol}-°K)}{kg/\cancel{mol}} = \frac{J}{kg-°K}$. Thus, R_U is a universal constant while <u>*R is materially dependent*</u>. Molecular weights MW for various chemical species are given in Table B.1.

EXAMPLE 2.5 Determination of state for an ideal gas

<u>Problem Statement:</u>
Air in a room is found to have a pressure of $100kPa$ and a temperature of $27°C$. Assuming ideal gas behavior, determine the air density within the room.

<u>Given:</u>
Air, $P = 100kPa, T = 27°C$, I.G.B.

<u>Find:</u>
ρ_{air}

<u>Assumptions:</u>
$MW_{air} = 28.97\frac{kg}{kmol}$ for air equivalent (via Table B.1)

<u>Analysis:</u>
From the ideal gas law (2.38)

<u>Schematic:</u>

Room containing air

$$P = \rho RT \Rightarrow \rho_{air} = \frac{P}{R_{air}T} \tag{2.39}$$

Now from (2.37) and Table B.1

$$R_{air} = \frac{R_U}{MW_{air}} = \frac{8.3\frac{J}{\cancel{mole}-°K}}{(28.97\frac{kg}{\cancel{kmol}})\frac{1\cancel{kmol}}{1000\cancel{mol}}} = 286\frac{J}{kg-°K} \tag{2.40}$$

Thus, from (2.39) and (2.40)

$$\rho_{air} = \frac{P}{R_{air}T} = \frac{100kPa}{286\frac{J}{kg-\cancel{°K}}(273+27)\cancel{°K}} = \frac{100\frac{\cancel{kN}}{m^2}}{286\frac{N-m}{kg}(300)}\frac{1000N}{1\cancel{kN}} =$$

$$\frac{10^5}{286(300)}\frac{\frac{\cancel{N}}{m^2}}{\frac{\cancel{N}-m}{kg}} = 1.16\frac{kg}{m^3} \xLeftarrow{\rho_{air}} \tag{2.41}$$

$$\boxed{\rho = 1.16\frac{kg}{m^3}}$$

□

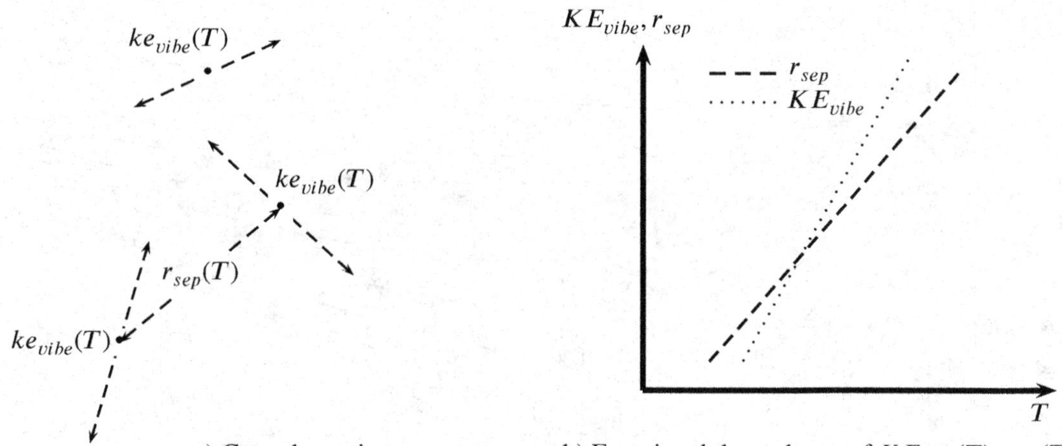

a) Gas schematic

b) Functional dependence of $KE_{vibe}(T)$,$r_{sep}(T)$.

Figure 2.8 Physical behavior of gas composed of molecules • as a function of temperature.

2.5.1 VALIDITY OF IDEAL GAS BEHAVIOR

The simplicity of the ideal gas model (I.G.B.) is self-evident from the previous example. The question now is "when can we assume I.G.B.?". Well, a super-heated vapor (a gas) is composed of molecules vibrating randomly about some average position as shown in Figure 2.8 a). This vibration corresponds to a vibrational kinetic energy KE_{vib} which is related to the gas temperature: As the gas temperature increase, molecular osculations and thus KE_{vib} increase. In addition to molecular oscillations, the average distance separating molecules r_{sep} also increases with temperature. Thus, as $T_{gas} \uparrow \Rightarrow KE_{vib} \uparrow \Rightarrow r_{sep} \uparrow$ as indicated in Figure 2.8 b). Conversely, as gas pressure increases the molecular separation distance decreases ($P \uparrow \Rightarrow r_{sep} \downarrow$) and the gas molecules are forced closer together. Ideal gas behavior (I.G.B.) is valid for sufficiently large molecular separation distance r_{sep} implying that for sufficiently high temperatures and low pressures I.G.B. (2.34) is valid.
So, to what extent does a gas under a given pressure and temperature exhibit ideal gas behavior? To answer this we introduce the "compressibility factor" Z:

$$\boxed{Z(T_{red}, P_{red}) = \frac{PV}{NR_U T} = \frac{PV}{mRT} = \frac{Pv}{RT} = \frac{P}{\rho RT}} \tag{2.42}$$

which is a function of a *normalized* or reduced temperature T_{red} and pressure P_{red}

$$\boxed{T_{red} = \frac{T}{T_{crit}}, P_{red} = \frac{P}{P_{crit}}} \tag{2.43}$$

If the compressibility factor Z is equal to 1, I.G.B. is exactly valid! Figure 2.9 shows curve-fits (based on measurements of a variety of gaseous materials) of $Z(T_{red}, P_{red})$ and indicates $Z \approx 1$ for very low reduced pressure and high reduced temperatures. Hence, if one specifies a range of $0.95 < Z < 1.05$ for *approximate* validity of the ideal gas equation then, from Figure 2.9:[1]

$$\boxed{P_{red} \lessapprox 0.2, T_{red} \gtrapprox 2 \Rightarrow Z \approx 1, \text{(I.G.B approximately valid)}} \tag{2.45}$$

[1]A more accurate, but complicated equation of state valid for liquids and gases (but not two phase mixtures) is the Van der Waals equation:

$$(P + \frac{N^2 a}{V^2})(V - Nb) = NR_U T \tag{2.44}$$

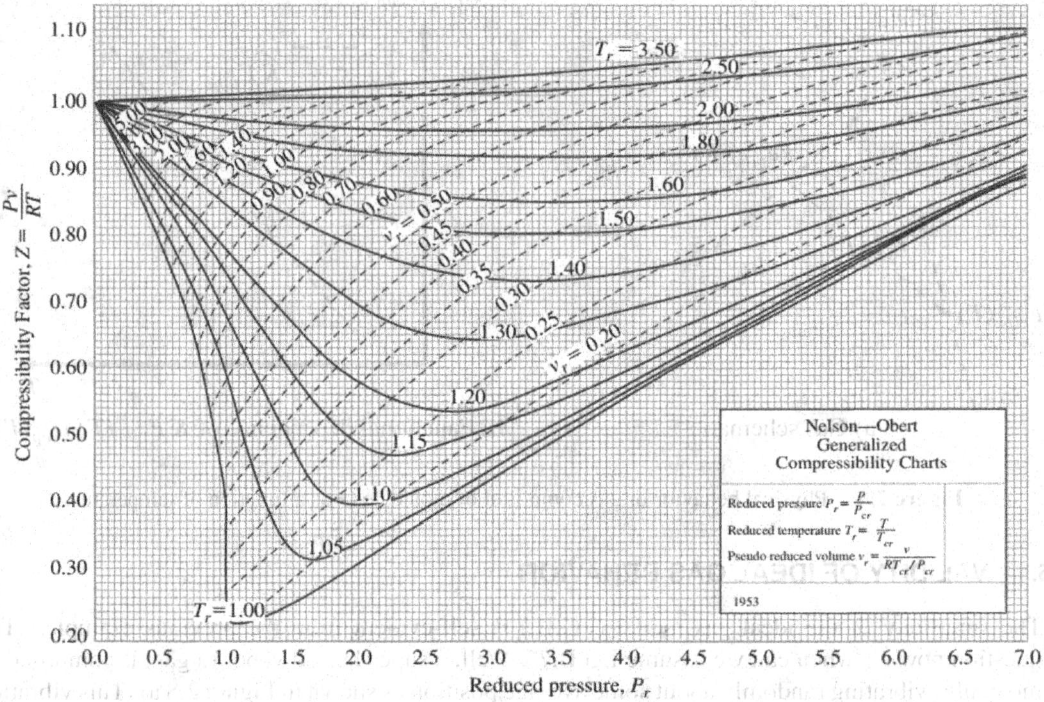

Figure 2.9 Generalized compressibility chart $Z(T_{red}, P_{red})$. Dashed lines indicated isochors of reduced (specific) volume $v_{red} = v/v_{crit}$. *Source:* https://www.chemistry.stackexchange.com

Note, that (2.45) is *sufficient, but not necessary* to ensure the validity of the Ideal Gas Law. In other words, if the range of reduced pressure and temperature corresponds to (2.45), you may safely assume I.G.B., otherwise, you must resort to the compressibility chart (Figure 2.9) to determine the value of Z (which may indeed reside in the range $0.95 < Z < 1.05$).

EXAMPLE 2.6 Validity of ideal gas behavior (I.G.B.)

Problem Statement:

500 grams of air initially with a volume of 500 Liters is compressed to half its volume while temperature is held constant at 300° K. Determine the initial and final air pressures in kPa. Sketch the process on a $T - v$ plot.

Given:

Air, $m = 500g$, $V_1 = 500L$, $V_2 = V_1/2$, $T_1 = 300° K$, isothermal process

Find:

P_1 and P_2 in kPa

Assumptions:

Data for air equivalent material from Tables B.1,
I.G.B. (subject to validation)

where a and b are physical constants. Note, in the limit at high temperatures, low pressure and thus *very high specific volume*, (2.44) reduces to the ideal gas law (2.34).

Schematic:

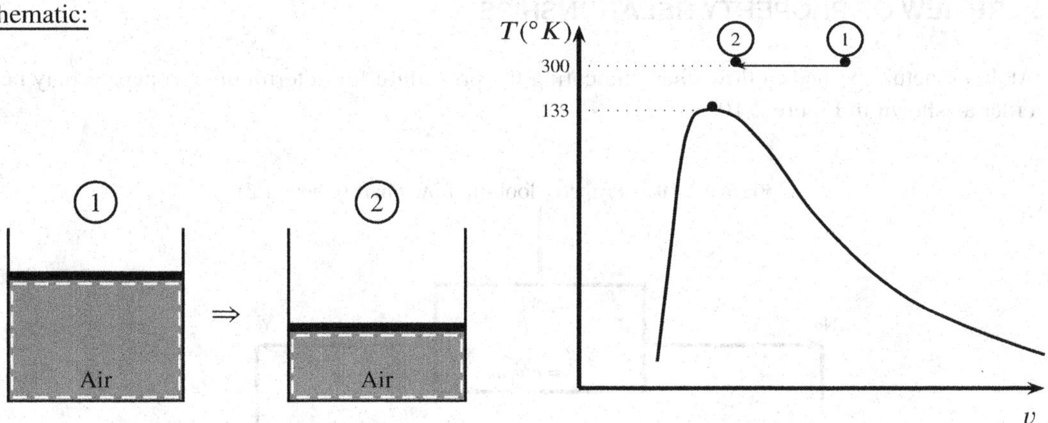

a) Piston-cylinder device b) $T - v$ plot

Analysis:
At state 1 via I.G.B. (2.36)

$$P_1 V_1 = mR_{air}T_1 \Rightarrow P_1 = \frac{mR_{air}T_1}{V_1} \tag{2.46}$$

From the expression (2.37) for the gas constant R and via Table B.1

$$R_{air} = \frac{R_U}{MW_{air}} = \frac{8.3\frac{kJ}{\cancel{kmol}-{}^\circ K}}{28.97\frac{kg}{\cancel{kmol}}} = 0.2865\frac{kJ}{kg-{}^\circ K} \tag{2.47}$$

Thus, combining (2.46) and (2.47) gives

$$P_1 = \frac{mR_{air}T_1}{V_1} = \frac{500g\left(0.2865\frac{kJ}{kg-{}^\circ\cancel{K}}\right)300^\circ\cancel{K}}{500L} = 85.950\frac{\cancel{g}-kJ}{\cancel{kg}-\cancel{L}}\frac{1000\cancel{L}}{1m^3}\frac{1\cancel{kg}}{1000\cancel{g}}$$

$$= 85.950\frac{kJ}{m^3} = 85.950\frac{kN-m}{m^3} = 85.950\frac{kN}{m^2} = 85.95kPa \xLeftarrow{P_1} \tag{2.48}$$

For process $1 \to 2$, $T_2 = T_1$ (isothermal) and via I.G.B. (2.36) we have

$$\frac{P_1V_1}{mR_{air}} = T_1 = T_2 = \frac{P_2V_2}{mR_{air}} \Rightarrow \frac{P_1V_1}{\cancel{mR_{air}}} = \frac{P_2V_2}{\cancel{mR_{air}}} \Rightarrow P_1V_1 = P_2V_2 \tag{2.49}$$

or

$$P_2 = P_1\frac{V_1}{V_2} = P_1\frac{\cancel{V_1}}{0.5\cancel{V_1}} = 2P_1 = 2(85.95kPa) = 171.9kPa \xLeftarrow{P_2} \tag{2.50}$$

Check validity of I.G.B.: Via Table B.1 $T_{red}\big|_1 = T_{red}\big|_2 = \frac{T_2}{T_{crit}} = \frac{300^\circ\cancel{K}}{133^\circ\cancel{K}} = 2.255 \gtrsim 2\ \surd$ Also, $P_{red}\big|_2 = \frac{P_2}{P_{crit}} = \frac{171.9\cancel{kPa}}{3770\cancel{kPa}} = 0.0455 \lessapprox 0.2$ and since $P_1 < P_2$ this implies $P_{red}\big|_1 < P_{red}\big|_2 \lessapprox 0.2\ \surd$.
$\Rightarrow$ I.G.B. valid over process!

$$\boxed{P_1 = 85.95kPa, P_2 = 171.9kPa}$$

□

2.6 REVIEW OF PROPERTY RELATIONSHIPS

At this juncture perhaps a flow-chart indicating the procedure for determining properties may be in order as shown in Figure 2.10.

Figure 2.10 Property look-up flow-chart (Chapter 2)

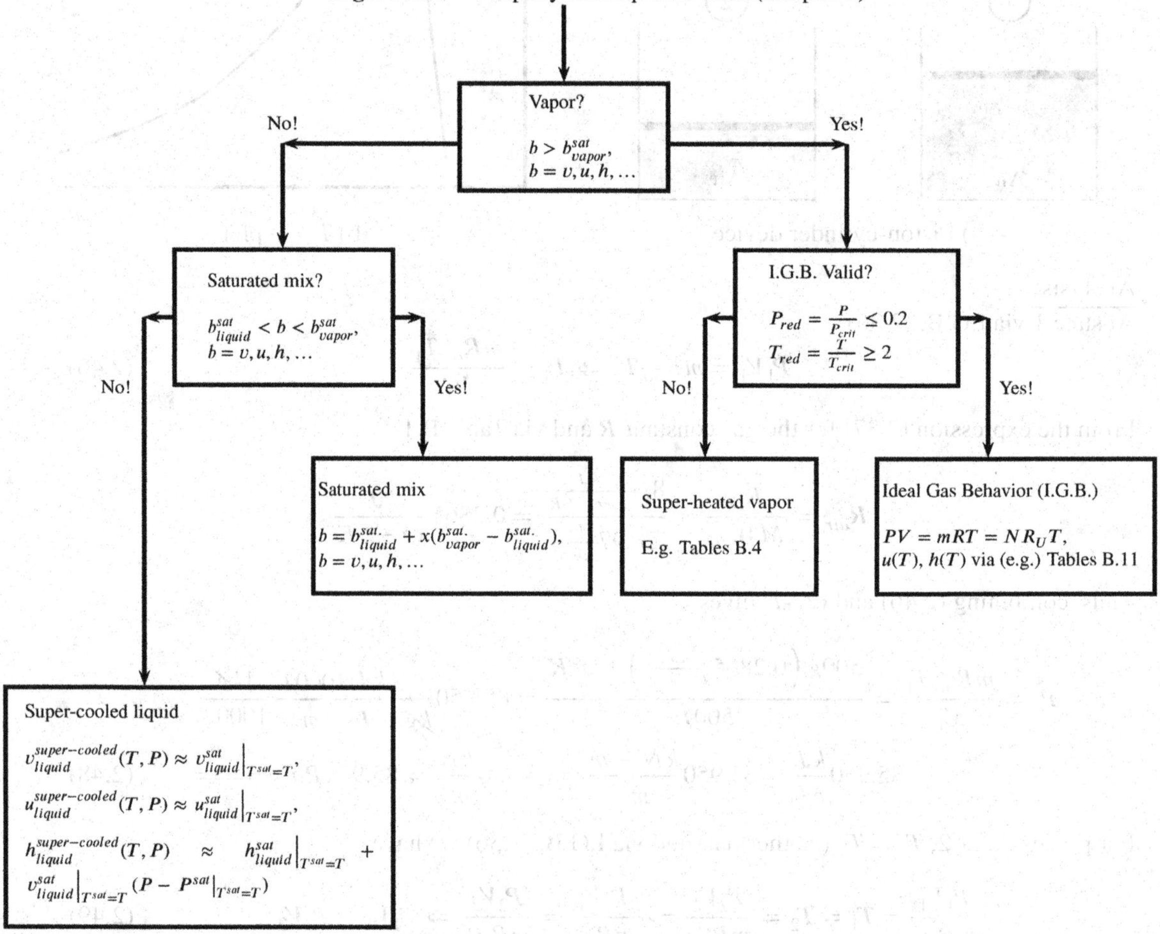

CHAPTER 3

HEAT AND WORK INTER-ACTIONS

In this chapter we will be introducing two energy transfer mechanisms: Work and Heat. While both these mechanisms represent flows of energy in or out of a system, the modes by which the transfer is affected are distinct. This distinction between heat and work, the rationale of which appears unclear at the moment, will nevertheless be crucial in the coming chapters, in particular during our investigation of entropy and the Second-Law of Thermodynamics (Chapters 5 and 6).

3.1 POINT-FUNCTIONS

We begin this chapter by investigating the nature of some of the thermodynamic quantities introduced up to this point (e.g. temperature T, pressure P, internal energy U, etc.) with respect to their mathematical characteristics. Let us begin with the "State Postulate":

Definition 3.1 ***State Postulate:*** *The state of a compressible system in thermodynamic equilibrium can be quantified by two independent thermodynamic properties.*

For example, if we have a container of CO_2 gas with temperature T and pressure P, we can specify density ρ or specific volume v. Or, if density ρ and pressure P are given, we can find the gas temperature T. Now, properties in general should be a function of where the system is (i.e. the state of the system), not how it got there! Or mathematically, if L is any property of a system, then for any path between states 1 and 2

$$\int_{1\to 2} dL = \int_{L=L_1}^{L_2} dL = \Delta L_{1\to 2} = L_2 - L_1 \qquad \text{Regardless of path!} \tag{3.1}$$

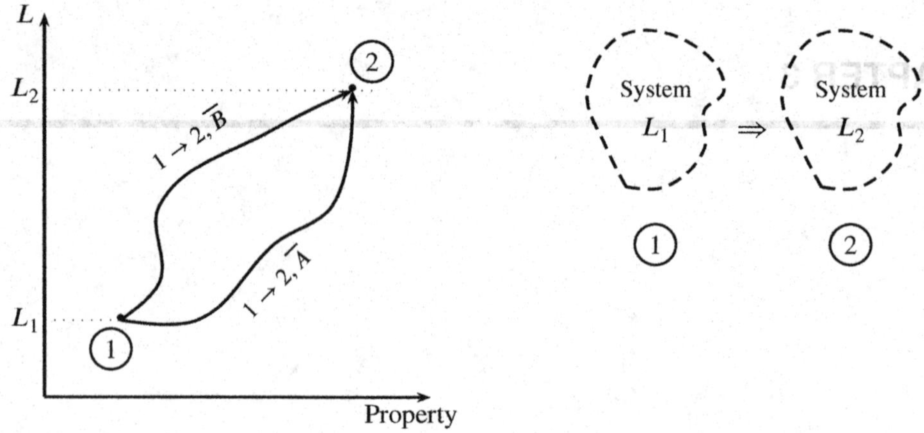

Figure 3.1 Pictorial illustration of process paths $\overline{A}$ and $\overline{B}$ between the same end-states 1 and 2.

as shown in Figure 3.1. Note that since $\Delta L_{1\to 2} = -\Delta L_{2\to 1}$ by definition, we have a null value for the close path integral

$$\oint dL = \left[\int_{1\to 2} dL\right]_{\overline{B}} - \left[\int_{2\to 1} dL\right]_{\overline{A}} = \Delta L_{1\to 2} - \Delta L_{2\to 1} = (\cancel{L_2 - L_1}) - (\cancel{L_1 - L_2}) = 0 \quad (3.2)$$

Thus, we formally define the term "point-function":

Definition 3.2 ***Point-Function:*** *If the value of the closed path integral* $\oint dL = 0$ *along any path, the quantity L is a point-function.*

System properties such as KE, P, T, pe, u, h, v and ρ which describe the state of the system are necessarily point-functions.

3.2 PATH-FUNCTIONS

Definition 3.3 ***Path-Function:*** *If the value of the closed path integral* $\oint d\phi \neq 0$ *along any arbitrary path, the quantity* ϕ *is a path-function.*

Let's say that we find the following integral $\int_{1\to 2} d\phi$ is not path independent, then from to the definition of point-function, ϕ is not a point-function. In other words from Figure 3.1

$$\left[\int_{1\to 2} d\phi\right]_{\overline{A}} \neq \left[\int_{1\to 2} d\phi\right]_{\overline{B}} \quad (3.3)$$

and ϕ is path dependent or a path-function.[1]

[1]The extent to which you master thermodynamics is a point-function, while the effort required to do so is a path-function!

3.2.1 NOTATION FOR POINT AND PATH-FUNCTIONS

The distinction between point and path-functions will permeate our future investigations of thermodynamics, thus it is important to symbolically differentiate between point and path-functions mathematically.

Let us return to the point-function L which exhibits the following behavior:

$$\Delta L_{1\to2} = L_2 - L_1 = \left[\int_{1\to2} dL\right]_{\overline{A}} = \left[\int_{1\to2} dL\right]_{\overline{B}} \tag{3.4}$$

or $\Delta L_{1\to2} = L_2 - L_1$regardless of path where Δ is the mathematical symbol representing a change in a point-function or a property. Because L is a property (i.e. the function L exists and has a value for any state of the system), we use the following notation to indicate changes in L over processes:

$$L : \begin{cases} \Delta L_{1\to2} = L_2 - L_1 = \displaystyle\int_{1\to2} dL, & \text{Overall process } 1 \to 2 \\ dL, & \text{Infinitesimal/differential process} \end{cases} \tag{3.5}$$

If L does not exist, (in other words $\int_{1\to2} dL$ is path dependent), then we have a path-function ϕ characterized by the notation:

$$\phi : \begin{cases} \phi_{1\to2} = \displaystyle\int_{1\to2} \delta\phi, & \text{Overall process } 1 \to 2 \\ \delta\phi, & \text{Infinitesimal/differential process} \end{cases} \tag{3.6}$$

Comparing (3.5) and (3.6) you will notice two differences: First, in the case of a point-function or property, it makes sense to speak of a change Δ of the property L, after all, the system state is described by the property and the system may change state. If the process is an infinitesimal or differential one (infinitely small changes in properties) then the finite property change becomes a differential change $\Delta L \to dL$. Secondly, if we are dealing with a path-function ϕ it makes no sense to speak of a change in ϕ because ϕ does not depend only on the state of the system. In the case of an infinitesimal (or differential) process, we do not write $d\phi$, but instead write $\delta\phi$ (lower-case delta δ) to signify explicitly that we are referring of a path-function: Hence, $\phi_{1\to2} = \int_1^2 \delta\phi$. By the end of this chapter, the student will have become intimately familiar with two path-functions, Work W and Heat Q, which represent *energy transfers* into or out of a system.

Conclusion: When encountering $\phi_{1\to2}$ or $\delta\phi$ we are dealing with a path-function. When $\Delta L_{1\to2}$ or dL, we have a point-function or a property.

3.3 WORK INTER-ACTIONS *W*

The first (of two) path functions to be introduced is Work W defined from elementary physics:

Definition 3.4 ***Work*** W: *A mechanical transfer of energy to or from a system:*

$$\boxed{\delta W = \vec{F} \cdot \vec{dx}} \tag{3.7}$$

where work is positive when performed by the system, negative when done on the system.

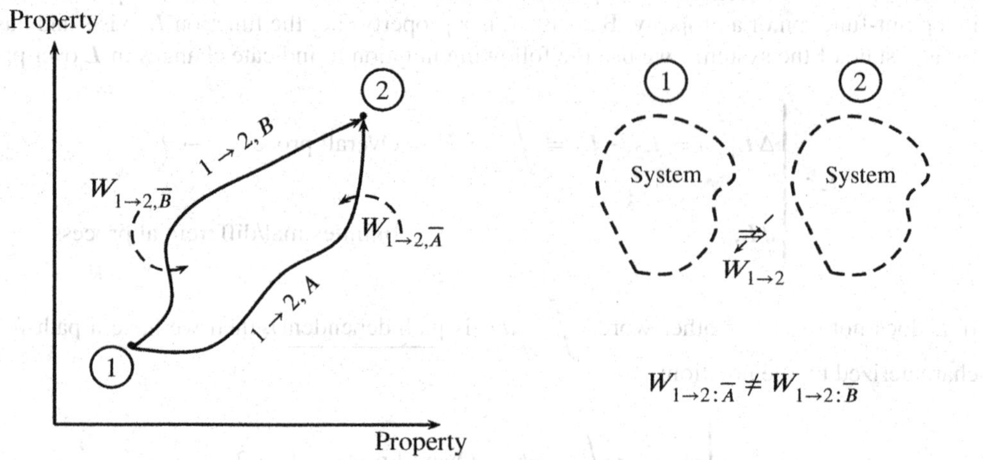

Figure 3.2 Pictorial illustration of path dependence of work interactions

Thus, as a system moves from state 1 to state 2, energy may flow out of the system in the form of work ($W > 0$) or conversely, work may be done on the system as it moves from state 1 to 2 ($W < 0$): Work is a *path function* and may be positive, negative or zero depending on the path as shown in Figure 3.2. Given that work represents an energy-transfer, it can be represented on various bases as given in Table 3.3: Work for an overall process, specific overall process, differential process or on a rate-basis.

There are several forms of work, each of which we will now turn our attention.

Table 3.1 Various bases of mechanical energy transfer (work) with dimensions and units

Basis	Symbol	Dimension	Unit (metric)	Unit (Imperial)
Overall process	W	Energy	J	BTU
Specific overall process	w	$\frac{\text{Energy}}{\text{mass}}$	$\frac{J}{kg}$	$\frac{BTU}{slug}$
Infinitesimal process	δW	Energy	J	BTU
Rate (Power)	$\dot{W} = \frac{\delta W}{dt}$	$\frac{\text{Energy}}{\text{time}}$	$\frac{J}{s} = Watt$	$\frac{BTU}{s}$

3.3.1 FORMS OF WORK: BOUNDARY WORK W_{bound}

Figure 3.3 shows a piston-cylinder device containing a system composed, let's say, of a gas. This piston with cross-section area A_{piston}, slides frictionlessly up or down within the cylinder and responds to the expansion or compression of the system. If the system is in equilibrium at pressure P, the force exerted on the piston is PA_{piston}. Let us now say that the gas expands (as does the system boundary) over some slow differential (and quasi-equilibrium) process resulting in an infinitesimal displacement dx (upwards is positive). From the definition of work (3.7) the differential boundary-work associated with this infinitesimal quasi-equilibrium expansion of the system boundary is

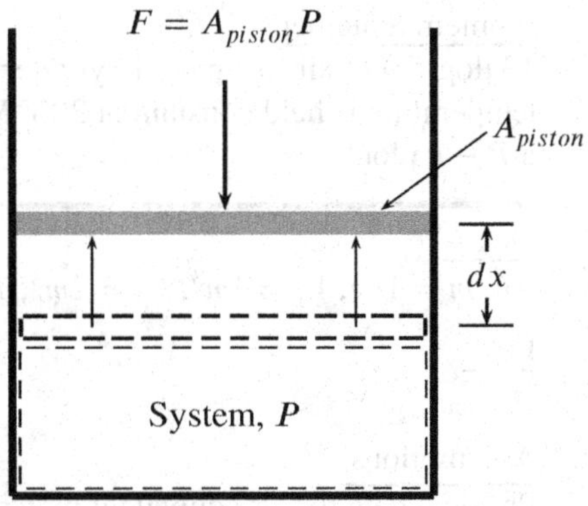

Figure 3.3 Differential boundary-work δW_{bound} for a gas in piston-cylinder device.

$$\delta W_{bound} = \vec{F} \cdot \vec{dx} = (PA_{piston})|\vec{dx}| = P(A_{piston}dx) = PdV \tag{3.8}$$

or

$$\boxed{\delta W_{bound} = PdV \text{ (By the system)/(On surroundings)}} \tag{3.9}$$

where we note that $A_{piston}dx = dV$. Thus, boundary-work is the work done by the system due to a change in volume (i.e. movement of the system boundary). Hence, if the system does not change volume, boundary-work is zero! How about the boundary-work associated with some (slow and hence quasi-equilibrium) *overall process* $1 \rightarrow 2$? Simply integrate (3.9) from V_1 to V_2 where pressure must be a known function of volume $P(V)$:

$$W_{bound_{1\rightarrow 2}} = \int_1^2 \delta W_{bound} = \int_{V=V_1}^{V_2} PdV \text{ (By the system)} \tag{3.10}$$

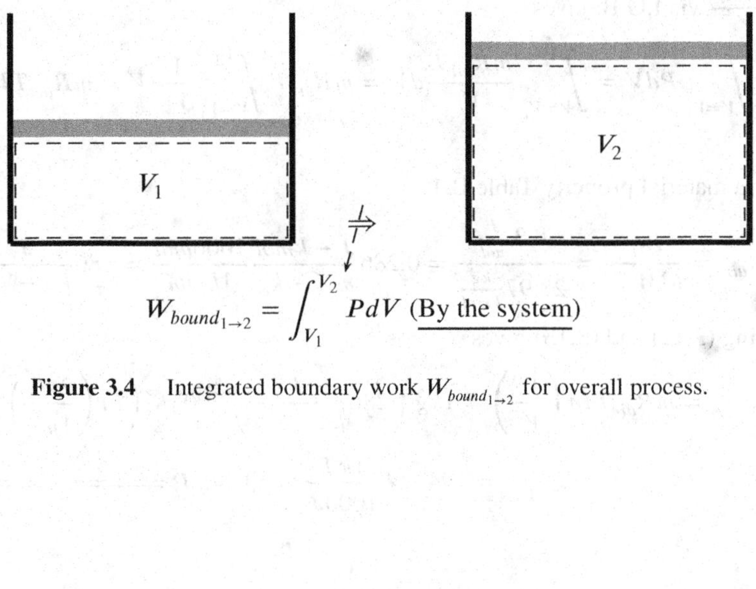

Figure 3.4 Integrated boundary work $W_{bound_{1\rightarrow 2}}$ for overall process.

EXAMPLE 3.1 Calculation of boundary work W_{bound}

Problem Statement:
1 kilogram of air in a piston-cylinder device expands from a volume of 1 to 2 cubic meters while temperature is held constant at 300° K. Determine the boundary-work in kJ. Show the process on a $P - v$ plot.

Given:
Air, $m = 1kg$, $V_1 = 1m^3$, $V_2 = 2m^3$, $T_1 = T_2 = 300°K$, isothermal process

Find:
$W_{bound_{1\rightarrow 2}}$

Assumptions:
Physical data for air equivalent material from Tables B.1,
I.G.B. (subject to validation)

Schematic:

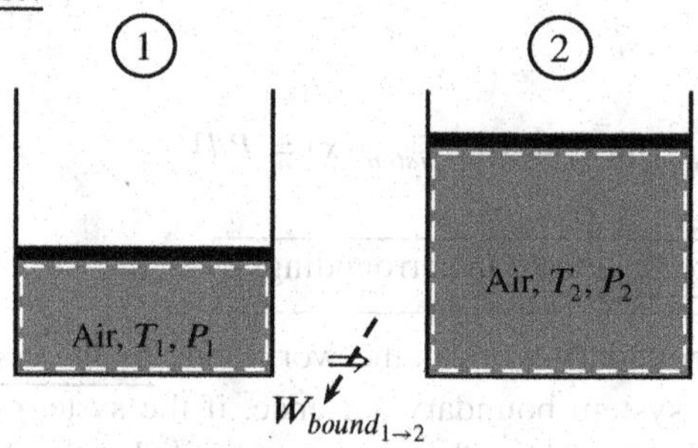

a) Piston-cylinder device

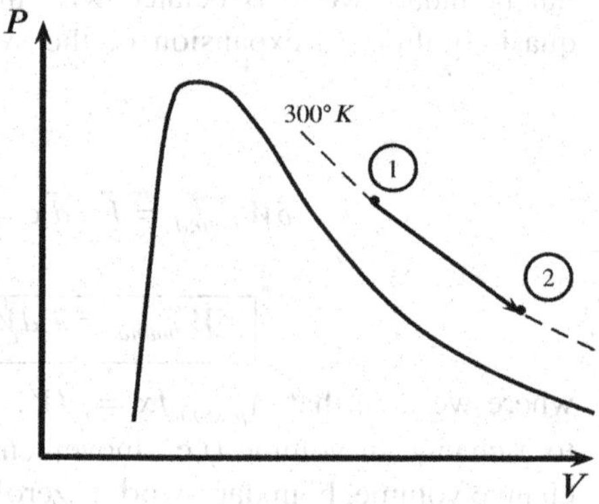

b) $P - V$ plot

Analysis:
From the definition of boundary work (3.9)

$$\delta W_{bound} = PdV \Rightarrow W_{bound_{1\rightarrow 2}} = \int_{V=V_1}^{V_2} PdV \tag{3.11}$$

where $P = \frac{mR_{air}T}{V}$ via I.G.B. gives

$$W_{bound_{1\rightarrow 2}} = \int_{V=V_1}^{V_2} PdV = \int_{V=V_1}^{V_2} \frac{mR_{air}T}{V} dV = mR_{air}T \int_{V=V_1}^{V_2} \frac{1}{V} dV = mR_{air}T ln\left(\frac{V_2}{V_1}\right) \tag{3.12}$$

Now for air, via material property Table B.1

$$R_{air} = \frac{R_U}{MW_{air}} = \frac{8.3\frac{J}{mol}}{28.97\frac{kg}{kmol}} = 0.286\frac{J - \cancel{kmol}}{\cancel{mol} - kg}\frac{1000\cancel{mol}}{1\cancel{kmol}} = 286\frac{J}{kg -° K} \tag{3.13}$$

Thus, combining (3.12) and (3.13) gives

$$W_{bound_{1\rightarrow 2}} = mR_{air}T ln\left(\frac{V_2}{V_1}\right) = 1\cancel{kg}\left(286\frac{J}{\cancel{kg} -° K}\right) 300°K\left(ln\left(\frac{2\cancel{m^3}}{1\cancel{m^3}}\right)\right)$$

$$= 59472\cancel{J}\frac{1kJ}{1000\cancel{J}} = 59.47kJ \Longleftarrow W_{bound_{1\rightarrow 2}} \tag{3.14}$$

Check validity of I.G.B.: $T_{red}|_1 = T_{red}|_2 = \frac{T_2}{T_{crit}} = \frac{300°K}{133°K} = 2.255 \gtrsim 2.$ $\surd$ Now, $P_{red}|_1 = \frac{P_1}{P_{crit}}$
$= \frac{\frac{mR_{air}T_1}{V_1}}{P_{crit}} = \frac{\frac{1kg(286\frac{J}{kg-°K})300°K}{1m^3}}{3770kPa} = \frac{85800Pa}{3770kPa}\frac{1kPa}{1000Pa} = 0.022 \lesssim 0.2$ and since $P_2 = P_1\frac{V_1}{V_2} = \frac{P_1}{2} < P_1$
$\Rightarrow P_{red}|_2 < P_{red}|_1 \lesssim 0.2.$ $\surd \Rightarrow$ I.G.B. valid over process!

$$\boxed{W_{bound_{1\to2}} = 59.47kJ}$$

□

3.3.2 GEOMETRIC INTERPRETATION OF BOUNDARY WORK W_{bound}

Noting that the expression for boundary work (3.10) for some overall process involves an integral, we have the following geometric interpretation for boundary-work given in Figure 3.5. Thus, graphical calculation of work for any given process is often relative straight forward, just determine the area under the $P-V$ curve from V_1 to V_2.

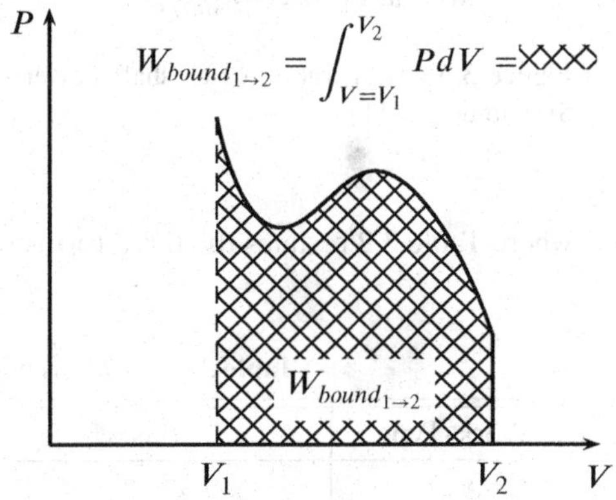

Figure 3.5 Integrated boundary-work W_{bound} (by the system) for overall process $1 \to 2$.

3.3.3 FORMS OF WORK: SHAFT-POWER $\dot{W}_{shaft}$

Work can be done on or by a system via a propeller or shaft as shown in Figure 3.6 a). Note that the natural expression for shaft work is actually *shaft power* $\dot{W}_{shaft}$ which, in terms of shaft torque τ and angular velocity ω is

$$\boxed{\dot{W}_{shaft} = \tau\omega} \tag{3.15}$$

Checking the physicality of this expression we note the dimensions of shaft power $\dot{W}_{shaft} = \tau\omega \sim (F-\ell)\frac{1}{t} = \frac{E}{t} = power$. Thus, shaft power has units of energy per-unit time or power.

3.3.4 FORMS OF WORK: ELECTRIC-POWER $\dot{W}_{electric}$

Finally, work can be done on or by a system via the flow of electric current as shown in Figure 3.6 b). The expression for electric power in terms of electric current I, resistance R and voltage V is

$$\boxed{\dot{W}_{electric} = VI = RI^2} \tag{3.16}$$

Checking the physicality of this expression we note the dimensions of electric power $\dot{W}_{electric} = VI \sim \frac{E}{coulomb}\frac{coulomb}{t} = \frac{E}{t} = power$. Thus, electric power has units of energy per-unit time or power as it should.

3.3.5 REVIEW FORMS OF WORK

Noting that a system may experience a combination of work inter-actions we define the total work W_{total}:

$$\boxed{W_{total} = W_{bound} + W_{shaft} + W_{electric}} \tag{3.17}$$

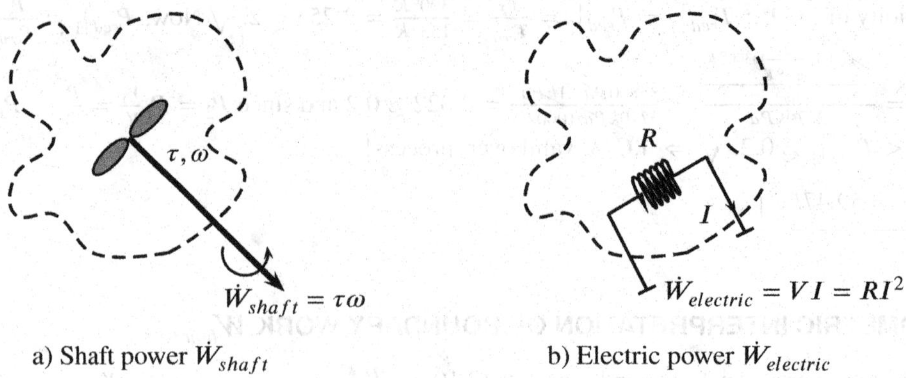

Figure 3.6 Illustration of a) shaft power (work/time) $\dot{W}_{shaft}$ and b) electric power (work/time) $\dot{W}_{electric}$. System denoted by − − −−.

where Table 3.2 restates the three forms of work or power using various basis.

Table 3.2 Various bases of mechanical energy transfer (work).

Work-Type	Basis	Expression
Boundary	Differential process	$\delta W_{bound} = PdV$ (By the system)
Boundary	Overall process	$W_{bound_{1\to2}} = \int_{V=V_1}^{V_2} PdV$ (By the system)
Boundary	Specific overall process	$w_{bound_{1\to2}} = \frac{W_{bound_{1\to2}}}{m} = \int_{v=v_1}^{v_2} Pdv$ (By the system)
Boundary	Rate based (Power)	$\dot{W}_{bound} = \frac{\delta W_{bound}}{dt} = P\frac{dV}{dt} = P\dot{V}$ (By the system)
Shaft-power	Rate based (Power)	$\dot{W}_{shaft} = \tau\omega$
Electric-power	Rate based (Power)	$\dot{W}_{electric} = VI = RI^2$

EXAMPLE 3.2 Calculation of total work W_{total}

Problem Statement:
A gas contained within a piston-cylinder device is heated via a $10Ohm$ electric resistor connected to a $12Volt$ battery. At the initial state the gas has a volume and pressure of $1m^3$ and $1bar$ respectively. Then, the resistor is connected the battery and current begins to flow through the resistor. After $10min$ the battery is discharged (the voltage reaches zero) and the gas volume is

found to be $1.001m^3$. If the voltage varies linearly from the initial to the final state and the expansion of the gas is isobaric, determine the total work done by the system.

Given:
Gas in piston-cylinder device, $V_1 = 1m^3$, $V_2 = 1.01m^3$,
$R = 10Ohm$,
$Voltage|_{t_1=0min} = 12Volt$, $Voltage|_{t_2=10min} = 0Volt$ (linearly varying voltage),
$P_1 = 1bar$,
isobaric process (expansion)

Find:
$W_{total_{1\to2}}$

Assumptions:
$\dot{W}_{shaft} = 0$

Schematic:

Isobaric process

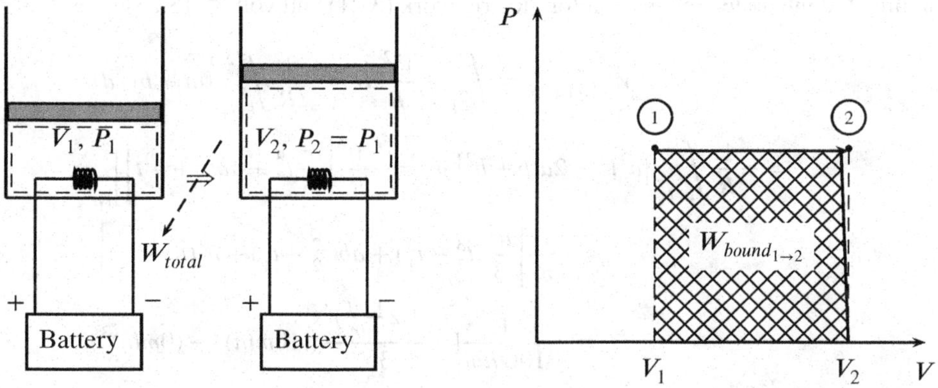

a) Piston-cylinder device b) $P - V$ plot

Analysis:
From the definition of total work done by the system (3.17)

$$W_{total_{1\to2}} = W_{bound_{1\to2}} + W_{elect._{1\to2}} + \cancelto{0}{W_{shaft_{1\to2}}} = W_{bound_{1\to2}} + W_{elect._{1\to2}} \tag{3.18}$$

Now via the integrated expression for boundary-work (3.10)

$$W_{bound_{1\to2}} = \int_{V=V_1}^{V_2} PdV = P(V_2 - V_2) = 1(bar)(1.01m^3 - 1m^3) =$$

$$1(bar)(0.01m^3) = 0.01(\cancel{bar} - m^3)\frac{100kPa}{1\cancel{bar}} = 1\cancel{kPa} - \cancel{m^3}\frac{kN/\cancel{m^2}}{} = 1kJ \tag{3.19}$$

From the definition of electric power (3.16) where V = Voltage

$$\dot{W}_{elect.} = \frac{\delta W_{elect.}}{dt} = IV = \left(\frac{V}{R}\right)V = \frac{V^2}{R} \tag{3.20}$$

Integrating electric power (3.20) from state 1 → 2 we have (again V = Voltage)

$$W_{elect._{1\to 2}} = \int_{1\to 2} \delta W_{elect.} = \int_{t=t_1}^{t_2} \frac{\delta W_{elect.}}{dt} dt = \int_{t=t_1}^{t_2} \dot{W}_{elect.} dt = \int_{t=t_1}^{t_2} -\frac{V^2}{R} dt \qquad (3.21)$$

Note, the negative sign indicates that $W_{elect.}$ is done on the system. Voltage varies linearly in time and thus has the functional form $Voltage(t) = at + b$. To determine constants a and b we apply initial and final conditions

$$\begin{cases} Voltage(t_1) = a(0) + b = 12Volt \Rightarrow b = 12Volt \\ Voltage(t_2) = a(10min) + b = 0Volt \Rightarrow a = \frac{-b}{t_2} = \frac{-12Volt}{10min} = -1.2\frac{Volt}{min} \end{cases} \qquad (3.22)$$

which yields our resulting expression for voltage as a function of time

$$Voltage(t) = -1.2\frac{Volt}{min}t + 12Volt \qquad (3.23)$$

Combining the integrated expression for electric work (3.21) and voltage (3.23) gives

$$\begin{aligned} W_{elect._{1\to 2}} &= \int_{t=t_1}^{t_2} -\frac{V^2}{R} dt = -\frac{1}{R}\int_{t=t_1}^{t_2} [at+b]^2 dt \\ &= -\frac{1}{R}\int_{t=t_1}^{t_2} \left[a^2t^2 + 2atb + b^2\right] dt = -\frac{1}{R}\left[\frac{a^2}{3}t^3 + abt^2 + b^2t\right]\Bigg|_{t_1}^{t_2} \\ &= -\frac{1}{R}\left[\frac{a^2}{3}(t_2^3 - t_1^3) + ab(t_2^2 - t_1^2) + b^2(t_2 - t_1)\right] \\ &= -\frac{1}{10Ohm}\left[\frac{(-1.2\frac{Volt}{min})^2}{3}((10min)^3 - (0min)^3)\right. \\ &\quad \left. +(-1.2\frac{Volt}{min})12Volt((10min)^2 - (0min)^2) + (12Volt)^2(10min - 0min)\right] \\ &= -48(\frac{Volt^{\not{2}\,1}}{\not{Ohm}} - min)\frac{1A - \not{Ohm}}{1\not{Volt}} = -48(Volt - A - \not{min})\frac{60s}{1\not{min}} \\ &= -2880(\frac{J}{\not{Coulomb}} - \frac{\not{Coulomb}}{\not{s}} - \not{s}) = -2880\not{J}\frac{1kJ}{1000\not{J}} = -2.88kJ \end{aligned} \qquad (3.24)$$

Finally, returning to (3.18) gives

$$W_{total_{1\to 2}} = W_{bound_{1\to 2}} + W_{elect._{1\to 2}} = 1kJ - 2.88kJ = -1.88kJ \xLeftarrow{W_{net_{1\to 2}}} \qquad (3.25)$$

$$\boxed{W_{total_{1\to 2}} = -1.88kJ}$$

Comment: Since $W_{total} < 0$, $1.88kJ$ of work has been done on the system. Also, it should be pointed out that the gas is doing boundary-work on the atmosphere equal to $P_{atm}(\Delta V_{1\to 2})$.

□

3.4 HEAT INTER-ACTIONS/HEAT TRANSFER Q

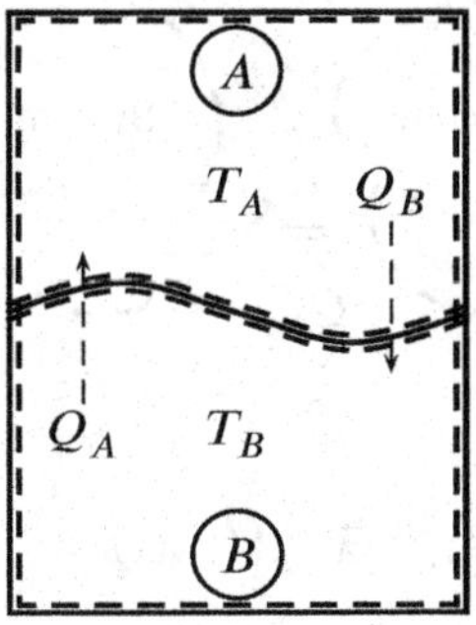

Temp. Relationship	Q_A	Q_B
$T_A > T_B$	$Q_A < 0$	$Q_B > 0$
$T_A = T_B$	$Q_A = 0$	$Q_B = 0$
$T_A < T_B$	$Q_A > 0$	$Q_B < 0$

Figure 3.7 Two systems A and B in thermal contact with associated direction of heat transfer.

We begin with a simple and intuitive definition of heat Q:

Definition 3.5 ***Heat Transfer Q:*** *Energy transfer between systems by virtue of a temperature difference. Heat transfer is positive when energy flows into the system.*

A pictorial representation of heat transfer between system A and B is given in Figure 3.7 as is the direction of heat transfer (from hot to cold). Like work, heat is an energy transfer and can be represented on various bases as given in Table 3.3. In addition, like work, heat is a path function

Table 3.3 Various bases of heat transfer with dimensions and units

Basis	Symbol	Dimension	Unit (metric)	Unit (Imperial)
Overall process	Q	Energy	J	BTU
Specific overall process	q	$\frac{\text{Energy}}{\text{mass}}$	$\frac{J}{kg}$	$\frac{BTU}{slug}$
Infinitesimal process	δQ	Energy	J	BTU
Rate (Power)	$\dot{Q} = \frac{\delta Q}{dt}$	$\frac{\text{Energy}}{\text{time}}$	$\frac{J}{s} = Watt$	$\frac{BTU}{s}$

and is therefore path dependent as illustrated in Figure 3.8. Hence, the movement of a system from state 1 to 2 under alternate paths $\overline{A}$ and $\overline{B}$ results in different heat inter-actions: $Q_{1\to 2,\overline{A}} \neq Q_{1\to 2,\overline{B}}$.

Processes which exclude heat transfer are referred to as *adiabatic processes*:

Definition 3.6 ***Adiabatic process:*** *No heat transfer into or out of the system.*

3.5 SPECIFIC HEAT C_P AND C_v

We are all familiar with the fact that adding heat to a system, for example a gas or liquid, results in an increase in temperature. Given that heat (and work) are path-functions, the logical point to make regarding the heat addition process is “what path do we take to raise the system temperature”? This question yields the following definition of “specific heat” or “heat capacity”:

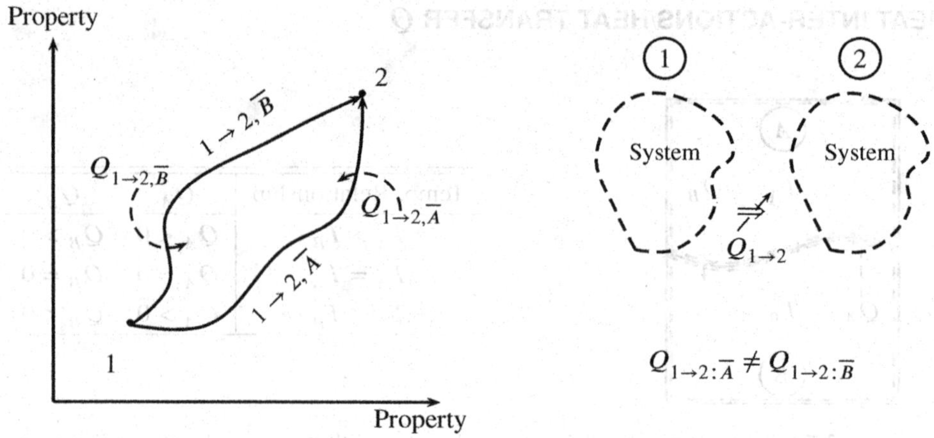

Figure 3.8 Pictorial illustration of path dependence of heat interactions

Definition 3.7 ***Specific heat or heat capacity*** *C: The amount of heat required to raise the temperature of one unit of mass of material by* 1 *degree.*

This quantity is necessarily material and path dependent. Thus, we define the pair of heat capacities: Specific heat at constant volume

$$C_v = \frac{1}{m}\frac{\delta Q}{dT}\bigg|_v \tag{3.26}$$

and specific heat at constant pressure

$$C_p = \frac{1}{m}\frac{\delta Q}{dT}\bigg|_p \tag{3.27}$$

where the dimensions of specific heat $\delta Q/(mdT)$ are $E/(mass{-}^{\circ})$. Note, these definitions include what looks like a derivative $\delta Q/dT$ (which would seem to imply the existence of a Q vs. T relationship). But, recall that heat is not a point-function, hence, Q is not a property and it is correct to write $\delta Q/dT$ not dQ/dT! A pictorial representation of specific heat at constant pressure C_p and constant volume C_v is given in Figure 3.9 a) and b) respectively.

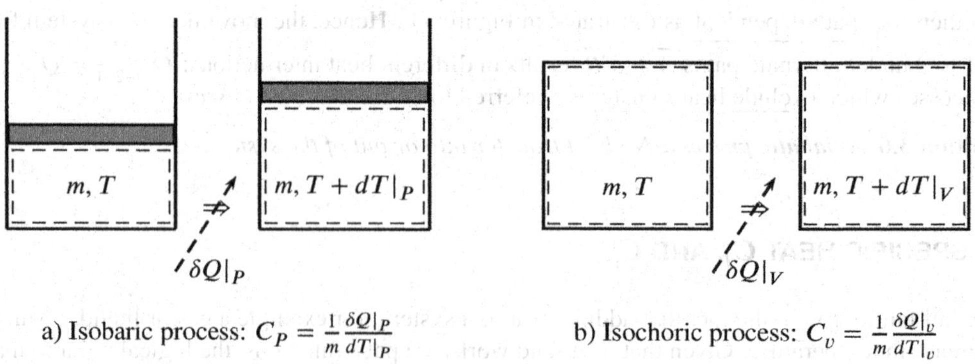

Figure 3.9 Illustration of experimental determination of a) C_P and b) C_v.

Now, intuitively, one would expect that if the system is allowed to expand (i.e. do boundary work) when adding heat, then temperature is liable to increase less than if the system volume were held constant. In other words, for a given amount of heat added to a system (a gas let's say), an isochoric heat addition process results in a greater temperature increase than the isobaric process: $dT|_v > dT|_p$ hence, $C_v < C_P$ for a compressible system (i.e. a gas). What about incompressible systems (solids or liquids)? Well, for incompressible systems, all processes are isochoric, thus $C_v = C_P = C$, the path is irrelevant and specific heat is denoted simply as C. Finally, from the State Postulate (Definition 3.1), two independent properties are required to determine the state of a compressible system. Thus, in general, specific heats are a function of two independent properties: $C_v(T, P)$ and $C_P(T, v)$.

Now, if the system behaves as an ideal gas, we have a functional relationship between T and P. For example, if the process is isochoric $P|_v = (R/v)T = f(T)$. Likewise, if the process is isobaric $v|_P = (R/P)T = g(T)$. Thus, we can substitute for the functional dependence of P in $C_v(T, P)$ and v in $C_P(T, v)$ yielding, for an ideal gas, $C_v(T)$ and $C_P(T)$ which are functions of temperature alone! For some process the variation in specific heat is sufficiently small (due to small variations in

Table 3.4 Summary of functionality of specific heat for various materials

Material behavior	Functionality
Incompressible (solid/liquid)	$C(T) = C_v(T) = C_P(T)$
Compressible (gas/vapor)	$C_v(T, P), C_P(T, v), C_P(T, P) > C_v(T, v)$
I.G.B.	$C_v(T), C_P(T), C_P(T) = C_v(T) + R$
Perfect gas (I.G.B const. C's)	$C_P, C_v, C_P = C_v + R$

system temperature) such that constant specific heats may be assumed for computational purposes. If the assumption of constant specific heats is made and I.G.B. is valid, the gas behavior is referred to as "perfect gas behavior ". In addition, it will be shown in §4.3.2 that C_P and C_v are related for ideal and perfect gas behavior via the expression $C_P = C_v + R$.

With respect to incompressible substances (liquids and solids), the heat addition path is by definition isochoric, thus specific heat is a function of temperature alone $C(T)$. Table 3.4 summarizes the relevant conclusions of this section and should be memorized.

EXAMPLE 3.3 Calculation of heat transfer Q using specific heats

Problem Statement:
A rigid container of $0.2kg$ of ice-tea (with properties equivalent to liquid H_2O) initially at $0°C$ is set in the sun until the temperature increases to $30°C$. Assuming some appropriate average value of specific heat, determine the approximate heat transfer to the ice tea.

Given:
$m = 0.2kg\ (H_2O)$,
$T_1 = 0°C$,
$T_2 = 30°C$

Schematic:

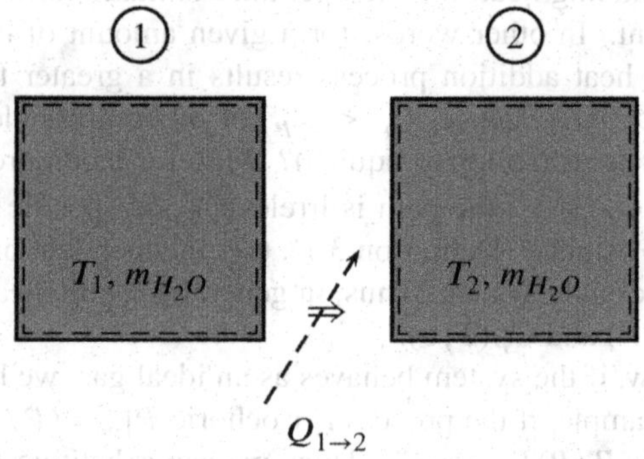

Find:
$Q_{1\to2}$

Assumptions:
Properties for H_2O from Table B.8,
incompressible substance,
$C_{av} = C(T)|_{T=15°C}$

Analysis:
Incompressible substance ($v = const$). Thus, from Table (3.4) and definition of specific heat (3.27)

$$C_P(T) = C_v(T) = C(T) = \frac{\delta Q}{mdT} \Rightarrow \delta Q = mC(T)dT \tag{3.28}$$

or

$$\delta Q = mC(T)dT \tag{3.29}$$

Integrating (3.29) from state $1 \to 2$ gives

$$Q_{1\to2} = \int_{1\to2} \delta Q = \int_{T=T_1}^{T_2} mC(T)dT = m\int_{T=T_1}^{T_2} C(T)dT$$
$$\approx m\int_{T=T_1}^{T_2} C_{av}dT = mC_{av}\int_{T=T_1}^{T_2} dT = mC_{av}(T_2 - T_1) \tag{3.30}$$

or

$$Q_{1\to2} \approx mC_{av}(T_2 - T_1) \tag{3.31}$$

Now, we choose some appropriate constant (or average) value of specific heat $C_{av} \approx C(T)|_{T=15°C}$. From a linear interpolation between $T = 27°C$ and $T = 2°C$ using data from Table B.8 we get

$$C(T)|_{T=15°C} = C(T)|_{T=2°C} + \frac{C(T)|_{T=27°C} - C(T)|_{T=2°C}}{27°C - 2°C}(15°C - 2°C) =$$
$$4.211\left(\frac{kJ}{kg - °K}\right) + \frac{4.179\left(\frac{kJ}{kg-°K}\right) - 4.211\left(\frac{kJ}{kg-°K}\right)}{27°\cancel{C} - 2°\cancel{C}}(15°\cancel{C} - 2°\cancel{C}) = 4.194\left(\frac{kJ}{kg - °K}\right) \tag{3.32}$$

Thus, evaluating (3.31) using the result from (3.32)

$$Q_{1\to2} \approx mC_{av}(T_2 - T_1) = 0.2(\cancel{kg})\left(4.194\left(\frac{kJ}{\cancel{kg} - °K}\right)\right)(30°\cancel{C} - 0°\cancel{C}) = 25.16kJ \xleftarrow{Q_{1\to2}} \tag{3.33}$$

$$\boxed{Q_{1\to2} \approx 25.166kJ}$$

□

CHAPTER 4

THE ENERGY-BALANCE

4.1 CLOSED SYSTEM ENERGY-BALANCE

We now are in the position to apply the energy-balance (1.7), also known as the First-Law of thermodynamics, for a closed system (with slight modification) as developed in §1:

$$\overbrace{Q}^{\text{Heat transfer \underline{to} system}} - \overbrace{W}^{\text{Work done \underline{by} system}} = \Delta E_{total}$$

$$= \overbrace{\Delta U}^{\text{Change in ``Internal Energy'' of system}} + \overbrace{\Delta KE + \Delta PE}^{\text{Change in ``Mech. Energy'' of system}} \tag{4.1}$$

or, in differential form

$$\boxed{\delta Q - \delta W = dE = dU + dKE + dPE, \text{ (Valid for closed systems)}} \tag{4.2}$$

where the work term represents work done by the system over the differential process, hence the negative sign. Like heat and work, the First-Law of Thermodynamics (the energy-balance) can be expressed in various forms as given in Table 4.1, all of which are equivalent!

For example, time-integration of the rate-based equation from time $t_1 \rightarrow t_2$ would yield the overall process balance from state $1 \rightarrow 2$. An even simpler approach is to integrate the differential process equation from state $1 \rightarrow 2$ again yielding the overall process energy-balance.

Table 4.1 Various bases for the First-Law (energy-balance) for a closed system with dimensions and units

Basis	Expression	Dimension	Unit (metric)	Unit (Imperi
Overall	$Q - W = \Delta E = \Delta U + \Delta KE + \Delta PE$	Energy	J	BTU
Specific overall	$q - w = \Delta e = \Delta u + \Delta ke + \Delta pe$	$\frac{\text{Energy}}{\text{mass}}$	$\frac{J}{kg}$	$\frac{BTU}{slug}$
Differential	$\delta Q - \delta W = dE = dU + dKE + dPE$	Energy	J	BTU
Rate (Power)	$\dot{Q} - \dot{W} = \dot{E} = \dot{U} + \dot{KE} + \dot{PE}$	$\frac{\text{Energy}}{\text{time}}$	$\frac{J}{s} = W$	$\frac{BTU}{s}$

EXAMPLE 4.1 Calculation of net work W_{net}: Energy-balance

Problem Statement:
A closed system undergoes a process by which $5kJ$ of heat is added resulting in an increase in system kinetic energy of $1kJ$, a decrease in potential energy of $2kJ$ and no change in internal energy. Determine the work done by the system.

Given:
Closed system,
$Q_{1\to2} = 5kJ$,
$\Delta KE_{1\to2} = 1kJ$,
$\Delta PE_{1\to2} = -2kJ$,
$\Delta U_{1\to2} = 0kJ$

Schematic:

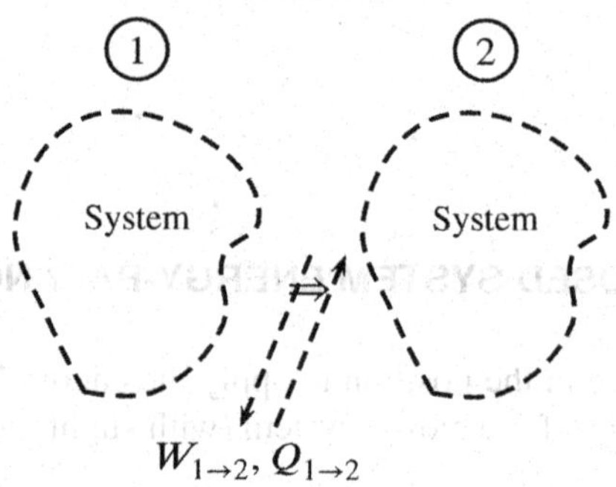

Find:
$W_{1\to2}$

Assumptions:
None

Analysis:
Application of the energy-balance for an overall process from Table (4.1)

$$Q_{1\to2} - W_{1\to2} = \Delta U_{1\to2} + \Delta KE_{1\to2} + \Delta PE_{1\to2} \tag{4.3}$$

Thus, solving for the work

$$W_{1\to2} = Q_{1\to2} - \left[\Delta U_{1\to2} + \Delta KE_{1\to2} + \Delta PE_{1\to2}\right]$$
$$= 5kJ - [0kJ + 1kJ + (-2kJ)] = 6kJ \xLeftarrow{W_{1\to2}} \tag{4.4}$$

$W_{1\to2} = 6kJ$ (by the system)

Comment: We need not have started with the energy-balance for an overall process, but could have instead integrated either the differential or rate-based equations over the process 1 → 2.

□

EXAMPLE 4.2 Rate-based application: Energy-balance

Problem Statement:
A closed system undergoes a process $1 \to 2$ where heat is added to the system at a constant rate of $3kW$, while work is done on the system at a rate of $1kW$. If the system initial and final total energy are $10kJ$ and $30kJ$ respectively, determine the amount of time elapsed over the process $1 \to 2$.

Given:
Closed system,
$\dot{Q} = 3kW$,
$\dot{W} = -1kW$,
$E_1 = 10kJ$,
$E_2 = 30kJ$

Find:
$\Delta t_{1 \to 2}$

Assumptions:
None

Schematic:

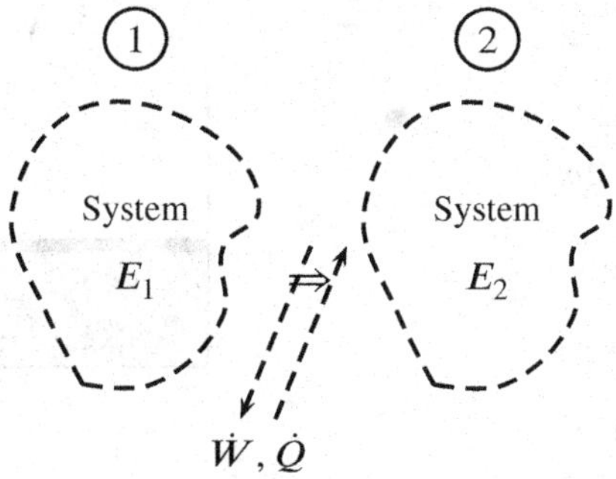

Analysis:
Application of the rate-based energy-balance from Table 4.1

$$\dot{Q} - \dot{W} = \dot{E} \Rightarrow \frac{\delta Q}{dt} - \frac{\delta W}{dt} = \frac{dE}{dt} \tag{4.5}$$

Time integrating (4.5) from $1 \to 2$

$$\int_{t=t_1}^{t_2} \underbrace{\frac{\delta Q}{dt}}_{\dot{Q}} dt - \int_{t=t_1}^{t_2} \underbrace{\frac{\delta W}{dt}}_{\dot{W}} dt = \int_{t=t_1}^{t_2} \frac{dE}{\cancel{dt}} \cancel{dt} = \int_{1\to 2} dE \Rightarrow \dot{Q} \int_{t_1}^{t_2} dt - \dot{W} \int_{t_1}^{t_2} dt = \int_{1\to 2} dE \tag{4.6}$$

Thus,

$$\dot{Q}(\Delta t_{1\to 2}) - \dot{W}(\Delta t_{1\to 2}) = \Delta E_{1\to 2} = E_2 - E_1 \tag{4.7}$$

Finally, solving (4.7) for $\Delta t_{1\to 2}$

$$\Delta t_{1\to 2} = \frac{E_2 - E_1}{\dot{Q} - \dot{W}} = \frac{30kJ - 10kJ}{3kW - (-1kW)} = \frac{20}{4}\frac{kJ}{kW} = 5\frac{\cancel{kJ}}{(\frac{\cancel{kJ}}{s})} = 5(s) \xleftarrow{\Delta t_{1\to 2}} \tag{4.8}$$

$\boxed{\Delta t_{1\to 2} = 5(s)}$

□

EXAMPLE 4.3 Energy-balance: Overall process with boundary-work

Problem Statement:
A piston-cylinder device containing air exhibiting ideal gas behavior, undergoes an isothermal expansion process whereby $69.3kJ$ of heat is added. If the initial and final volumes are $1m^3$ and $2m^3$ respectively with an initial pressure of $1bar$, determine the change in gas internal energy over the process.

Given:
Closed system,
gas (I.G.B.),
$Q_{1\to2} = 69.3kJ$,
$V_1 = 1m^3$, $V_2 = 2m^3$,
$P_1 = 1bar$,
isothermal process

Find:
$\Delta U_{1\to2}$, $W_{bound_{1\to2}}$

Schematic:

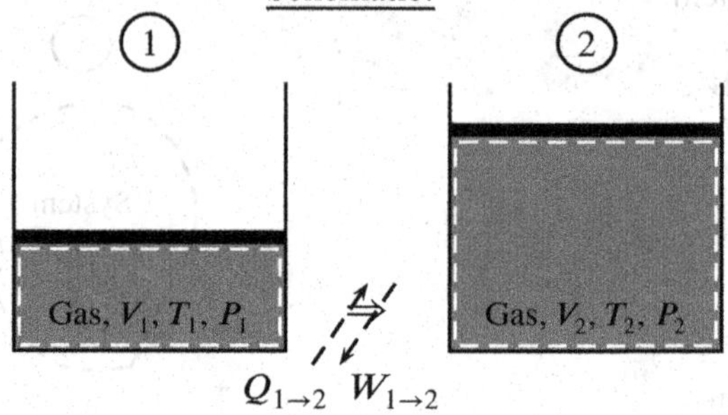

Assumptions:
$\Delta KE_{1\to2} = \Delta PE_{1\to2} = 0$,
$\dot{W}_{shaft} = \dot{W}_{electric} = 0$

Analysis:
Application of the energy-balance via Table 4.1 to the gas system

$$Q_{1\to2} - W_{1\to2} = \Delta E_{1\to2} \tag{4.9}$$

Now, from the definition of net work (3.17)

$$W_{1\to2} = W_{bound_{1\to2}} + \cancelto{0}{W_{shaft_{1\to2}}} + \cancelto{0}{W_{electric_{1\to2}}} = W_{bound_{1\to2}} \tag{4.10}$$

Combining this result (4.10) with the energy-balance (4.9) gives

$$Q_{1\to2} - \cancelto{W_{bound_{1\to2}}}{W_{1\to2}} = \Delta E_{1\to2} = \Delta U_{1\to2} + \cancelto{0}{\Delta KE_{1\to2}} + \cancelto{0}{\Delta PE_{1\to2}} \Rightarrow \Delta U_{1\to2} = Q_{1\to2} - W_{bound_{1\to2}} \tag{4.11}$$

Now via (3.10)

$$W_{bound_{1\to2}} = \int_{V=V_1}^{V_2} PdV \tag{4.12}$$

and $P = \frac{mRT}{V}$ (via I.G.B.). Thus, for an isothermal process $T = T_1 = T_2 = const$ and the integrated boundary-work is

$$W_{bound_{1\to2}} = \int_{V=V_1}^{V_2} PdV = \int_{V=V_1}^{V_2} \frac{mRT}{V} dV = mRT_1 \int_{V=V_1}^{V_2} \frac{1}{V} dV = mRT_1 ln\left[\frac{V_2}{V_1}\right] \tag{4.13}$$

Also, from I.G.B. $P_1V_1 = mRT_1$, thus evaluating (4.13)

$$W_{bound_{1\to2}} = mRT_1 ln\left[\frac{V_2}{V_1}\right] = P_1V_1 ln\left[\frac{V_2}{V_1}\right] = (1bar)(1m^3)ln\left[\frac{2m^3}{1m^3}\right] =$$

$$0.693(bar - m^3)\frac{100kPa}{1bar} = 69.3(kPa - m^3) = 69.3\left(\frac{kN}{m^2} - m^3\right) = 69.3kJ \xleftarrow{W_{bound_{1\to2}}} \quad (4.14)$$

Returning to (4.11)

$$\Delta U_{1\to2} = Q_{1\to2} - W_{bound_{1\to2}} = 69.3kJ - 69.3kJ = 0kJ \xleftarrow{\Delta U_{1\to2}} \quad (4.15)$$

$$\boxed{\Delta U_{1\to2} = 0kJ, W_{bound_{1\to2}} = 69.3kJ}$$

Comment: Recall that for I.G.B. $u(T)$. Thus, for isothermal processes $\Delta u(T) = 0$ as indicated by our previous result.

□

EXAMPLE 4.4 Energy-balance: Two-phased mixture application

Problem Statement:
$1kg$ of a liquid-vapor mixture of H_2O in a rigid container at a temperature of $100°C$ with initial quality 0.8 is heated till the mixture is a saturated vapor. Determine the heat transfer.

Given:
Closed system,
rigid container,
H_2O,
$m = 1kg$,
$T_1 = 100°C$,
$x_1 = 0.8$,
$x_2 = 1.0$

Find:
$Q_{1\to2}$

Assumptions:
Isochoric (rigid container),
$W_{shaft} = W_{electric} = 0$,
$\Delta PE = \Delta KE = 0$,
saturation temperature Tables B.2

Schematic:

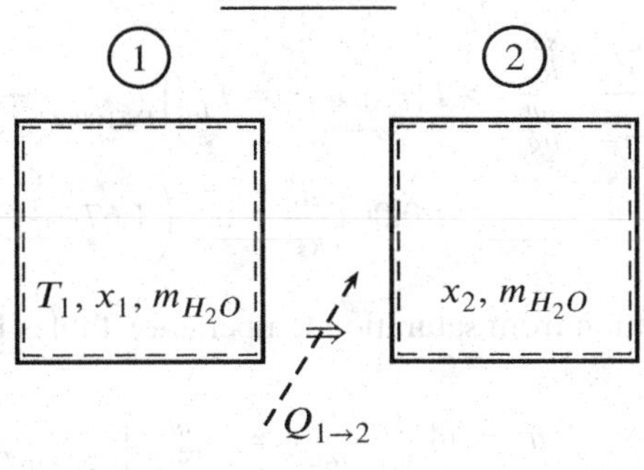

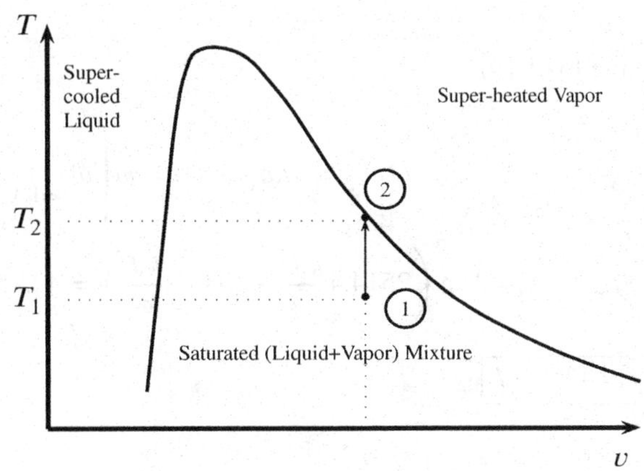

Analysis:
From the closed system energy-balance for an overall process (Table 4.1)

$$Q_{1\to2} - W_{1\to2} = \Delta E_{1\to2} \tag{4.16}$$

From the definition of total work (3.17)

$$W_{1\to2} = W_{total_{1\to2}} = \overset{0,\, isochoric}{\cancel{W}_{bound_{1\to2}}} + \overset{0}{\cancel{W}_{shaft_{1\to2}}} + \overset{0}{\cancel{W}_{electric_{1\to2}}} = 0 \tag{4.17}$$

Thus, combining (4.16) and (4.17) gives

$$Q_{1\to2} - \overset{0}{\cancel{W}_{1\to2}} = \Delta E_{1\to2} = \Delta U_{1\to2} + \overset{0}{\cancel{\Delta KE}_{1\to2}} + \overset{0}{\cancel{\Delta PE}_{1\to2}} = \Delta U_{1\to2} \tag{4.18}$$

or

$$Q_{1\to2} = \Delta U_{1\to2} = m\Delta u_{1\to2} = m(u_2 - u_1) = m\left(u|_{\substack{v=v_1,\\ x=1}} - u|_{\substack{T=100°C,\\ x=0.8}} \right) \tag{4.19}$$

Now, from the definition of quality (2.11) using saturation temperature Tables B.2

$$x = \frac{u^{sat}_{mix} - u^{sat}_{liq}}{u^{sat}_{vap} - u^{sat}_{liq}} \Rightarrow u|_{\substack{T=100°C,\\ x=0.8}} = u^{sat}_{liq}\Big|_{T=100°C} + x\,(u^{sat}_{vap} - u^{sat}_{liq})\Big|_{T=100°C}$$

$$= 418.9\frac{kJ}{kg} + 0.8\left(2506.5\frac{kJ}{kg} - 418.9\frac{kJ}{kg}\right) = 2088.9\frac{kJ}{kg} \tag{4.20}$$

Also, via saturation temperature Tables B.2 we have the initial (and final) specific volume

$$x = \frac{v^{sat}_{mix} - v^{sat}_{liq}}{v^{sat}_{vap} - v^{sat}_{liq}} \Rightarrow v|_{\substack{T=100°C,\\ x=0.8}} = v^{sat}_{liq}\Big|_{T=100°C} + x\,(v^{sat}_{vap} - v^{sat}_{liq})\Big|_{T=100°C}$$

$$= 0.001\frac{m^3}{kg} + 0.8\left(1.67\frac{m^3}{kg} - 0.001\frac{m^3}{kg}\right) = 1.336\frac{m^3}{kg} \tag{4.21}$$

Thus, via interpolation from saturation temperature Tables B.2, we have the final specific internal energy

$$u_2 = u|_{\substack{v=1.336\frac{m^3}{kg},\\ x=1}} = u^{sat}_{vapor}\Big|_{v=1.336\frac{m^3}{kg}} = 2514.4\frac{kJ}{kg} \tag{4.22}$$

Returning to (4.19)

$$Q_{1\to2} = m(u_2 - u_1) = m\left(u|_{\substack{v=1.336\frac{m^3}{kg},\\ x=1}} - u|_{\substack{T=100°C,\\ x=0.8}} \right)$$

$$= 1\cancel{kg}\left(2514.4\frac{kJ}{\cancel{kg}} - 2088.9\frac{kJ}{\cancel{kg}}\right) = 425.5kJ \xLeftarrow{Q_{1\to2}} \tag{4.23}$$

$$\boxed{Q_{1\to2} = 425.5kJ}$$

□

4.2 PROPERTY ENTHALPY H

Noting that U, V and P are properties of the system (i.e. point-functions), from §2.2.1 we re-state the definition of the property enthalpy H which tends to appear in situations involving open systems in general (see §4.4.3) or closed systems undergoing isobaric processes.

$$\boxed{H = U + PV} \tag{4.24}$$

Needless to say, enthalpy is an extensive property which, on a per-unit mass basis, is given by $h = \frac{H}{m} = u + Pv$.

EXAMPLE 4.5 Two-phase mixture (isobaric process): Energy-balance

Problem Statement:
$2.5kg$ of a saturated liquid H_2O at a temperature of $100°C$ is heated isobarically till all the liquid is vaporized. Determine the required heat transfer.

Schematic:

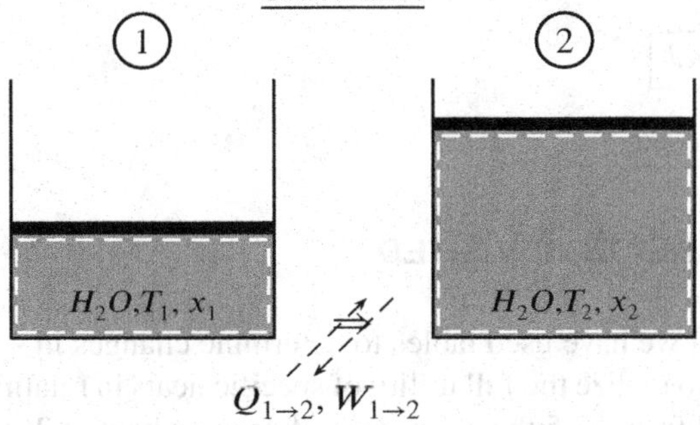

Given:
Closed system,
isobaric heating,
H_2O,
$m = 2.5kg$,
$T_1 = 100°C$,
$x_1 = 0.0$,
$x_2 = 1.0$

Find:
$Q_{1\to2}$

Assumptions:
$W_{shaft} = W_{electric} = 0$,
$\Delta PE = \Delta KE = 0$,
saturation Tables B.2

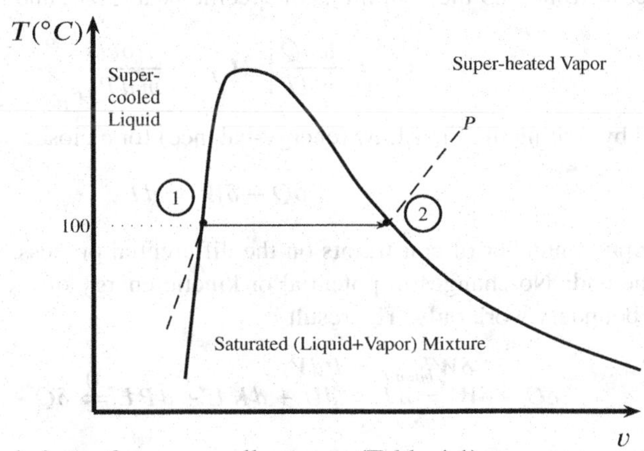

Analysis:
From the closed system energy-balance for an overall process (Table 4.1)

$$Q_{1\to2} - W_{1\to2} = \Delta E_{1\to2} \tag{4.25}$$

From the definition of total work (3.17) (note, the process is isobaric)

$$W_{1\to2} = W_{bound_{1\to2}} + \cancelto{0}{W_{shaft_{1\to2}}} + \cancelto{0}{W_{electric_{1\to2}}} = W_{bound_{1\to2}}$$
$$= \int_{V_1}^{V_2} PdV = P_1 \int_{V_1}^{V_2} dV = P_1(V_2 - V_1) \tag{4.26}$$

Thus, combining the energy-balance (4.25) and our expression for work (4.26)

$$Q_{1\to 2} - \overset{P_1(V_2-V_1)}{W_{1\to 2}} = \Delta E_{1\to 2} = \Delta U_{1\to 2} + \overset{0}{\cancel{\Delta KE_{1\to 2}}} + \overset{0}{\cancel{\Delta PE_{1\to 2}}} = \Delta U_{1\to 2} \quad (4.27)$$

or

$$Q_{1\to 2} = \Delta U_{1\to 2} + P_1(V_2 - V_1) = \Delta U_{1\to 2} + \Delta(PV)_{1\to 2}$$

$$= \Delta(U + PV)_{1\to 2} = \Delta H_{1\to 2} = m\Delta h_{1\to 2} = m(h_2 - h_1) \quad (4.28)$$

Note, process is isobaric, within saturation dome and <u>hence</u> isothermal. Thus, from (4.28) and the saturation temperature Tables B.2

$$Q_{1\to 2} = m(h_2 - h_1) = m\left(h|_{\substack{T=100°C,\\ x=1}} - h|_{\substack{T=100°C,\\ x=0}}\right) =$$

$$m\left(h^{sat}_{vapor}\Big|_{T=100°C} - h^{sat}_{liq}\Big|_{T=100°C}\right) = 2.5\cancel{kg}\left(2676\frac{kJ}{\cancel{kg}} - 419\frac{kJ}{\cancel{kg}}\right) = 5642kJ \xleftarrow{Q_{1\to 2}} \quad (4.29)$$

$$\boxed{Q_{1\to 2} = 5642kJ}$$

□

4.3 SPECIFIC HEATS REVISITED

Up to this point we have used tables to determine changes in system properties. However, the time has now come to realize the full utility of specific heats in relating changes in properties over a given process. Recall from §3.5 the definitions of specific heat (3.26) and (3.27)

$$C_v = \left.\frac{1}{m}\frac{\delta Q}{dT}\right|_v, C_P = \left.\frac{1}{m}\frac{\delta Q}{dT}\right|_P$$

We proceed by writing the First-Law (energy-balance) for a closed system in differential form

$$\delta Q - \delta W = dE$$

Let us now put a number of constraints on the differential process, none of which will turn out to matter in the end: No changes in potential or kinetic energy of the system where the work interactions are boundary work only. The result is

$$\delta Q - \overset{\delta W_{bound} = PdV}{\delta W} = dE = dU + \overset{0}{\cancel{dKE}} + \overset{0}{\cancel{dPE}} \Rightarrow \delta Q - PdV = dU \quad (4.30)$$

Finally, we constrain the system to an isochoric differential process as shown in Figure 4.1. Thus, the differential boundary work vanishes yielding

$$\delta Q|_v - \overset{0}{\cancel{PdV}} = dU|_v \Rightarrow \delta\, Q|_v = dU|_v \quad (4.31)$$

Solving (3.26) for $\delta Q|_v$ and substituting into the First-Law (4.31) yields

$$\delta Q|_v = mC_v(T,P)dT = dU|_v = mdu|_v \Rightarrow C_v(T,P)\,dT|_v = du|_v \quad (4.32)$$

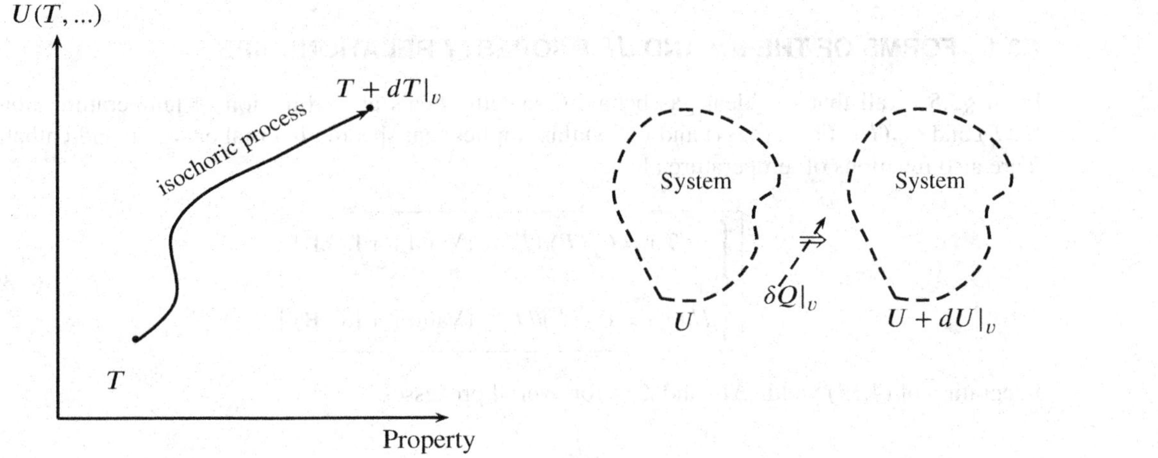

Figure 4.1 Pictorial illustration of isochoric differential process.

or

$$du|_v = C_v(T, P)\, dT|_v \tag{4.33}$$

Now, specific internal energy u is a property of the system and thus a function of the state of the system not how the system got there! This means that (4.33) is valid regardless of path! Thus, we drop the constraints on the path and simply write the property relationship:

$$\boxed{du = C_v(T, P)dT} \tag{4.34}$$

Hence (4.34) is true for all systems and processes regardless of path except within the saturation dome where specific heats are not applicable!

Question: What if we had instead chosen an isobaric process? In this case, boundary work no longer vanishes, yielding

$$\delta Q|_P - (\delta W_{bound})\Big|_P^{PdV} = dU|_P \Rightarrow \delta\, Q|_P - d(PV)|_P = dU|_P$$

$$\Rightarrow \delta Q|_P = dU|_P + d(PV)|_P = d(U + PV)|_P = dH|_P \tag{4.35}$$

Solving (3.27) for $\delta Q|_P$ and substituting into the First-Law (4.35) yields

$$\delta\, Q|_P = mC_P(T, v)\, dT|_P = dH|_P = mdh|_P \Rightarrow C_P(T, v)\, dT|_P = dh|_P \tag{4.36}$$

or

$$dh|_P = C_P(T, v)\, dT|_P \tag{4.37}$$

As before, specific enthalpy h $(u + Pv)$ is a property of the system and thus a function of the state of the system not how the system got there! Hence, (4.37) is valid regardless of path! Thus, we drop the constraint on the path and simply write our second property relationship:

$$\boxed{dh = C_P(T, v)dT} \tag{4.38}$$

which again is valid for any process except within the saturation dome where specific heats are not applicable!

4.3.1 FORMS OF THE du AND dh PROPERTY RELATIONSHIPS

From §3.5 recall that for ideal gas behavior, specific heats are a function of temperature alone: $C_v(T)$ and $C_P(T)$. From (4.34) and (4.38) this implies that specific internal energy u and enthalpy h are also functions of temperature alone:

$$\begin{cases} du(T) = C_v(T)dT, & \text{(Valid for I.G.B)} \\ dh(T) = C_P(T)dT, & \text{(Valid for I.G.B)} \end{cases} \tag{4.39}$$

Integration of (4.39) yields ΔU and ΔH for overall processes

$$\begin{cases} \Delta U(T)_{1\to 2} = m \displaystyle\int_{T=T_1}^{T_2} C_v(T)dT, & \text{(Valid for I.G.B)} \\ \Delta H(T)_{1\to 2} = m \displaystyle\int_{T=T_1}^{T_2} C_P(T)dT, & \text{(Valid for I.G.B)} \end{cases} \tag{4.40}$$

Specific heats for various substances under ideal gas conditions (as a function of temperature T) are given in either tabular form in Tables B.10 or polynomial curve-fits in Table B.9. Note, over-bared specific heats are on a per-kmol basis, hence $\overline{C}_P, \overline{C}_V \sim kJ/kmol - ^\circ K$.

If perfect gas behavior is assumed (i.e. constant specific heats), C_v and C_P are evaluated at some *representative or average* temperature T_{av} and (4.40) becomes

$$\begin{cases} \Delta U_{1\to 2} \approx m \displaystyle\int_{T=T_1}^{T_2} C_v(T)\big|_{T_{av}} dT = m C_v\big|_{T_{av}} \Delta T_{1\to 2}, & \text{(Perfect Gas)} \\ \Delta H_{1\to 2} \approx m \displaystyle\int_{T=T_1}^{T_2} C_P(T)\big|_{T_{av}} dT = m C_P\big|_{T_{av}} \Delta T_{1\to 2}, & \text{(Perfect Gas)} \end{cases} \tag{4.41}$$

4.3.2 RELATIONSHIP BETWEEN C_P AND C_v FOR I.G.B.

For an ideal gas (I.G.B.) there is a simple and precise relationship between C_P and C_v. Taking the *total derivative* of specific enthalpy h gives

$$h = u + Pv \Rightarrow dh = d(u + Pv) = du + d(Pv) \tag{4.42}$$

Combining this result with the differential property relationships (4.39) and the ideal gas law (2.34) (where, taking the total derivative, $d(Pv) = d(RT)$) we have

$$dh = du + d(Pv) = C_v(T)dT + d(RT) = C_v(T)dT + RdT$$

$$\Rightarrow C_P(T)dT = C_v(T)dT + RdT \Rightarrow C_P(T) = C_v(T) + R \tag{4.43}$$

or

$$\boxed{C_P(T) = C_v(T) + R, \text{ (Valid for I.G.B.)}} \tag{4.44}$$

Finally, it is often convenient in our analysis to specify a ratio of specific heats k

$$\boxed{k = \frac{C_P}{C_v}} \tag{4.45}$$

Combining the definition of k and the relationship between specific heats for an ideal gas (4.44)

$$\begin{cases} k = \frac{C_P}{C_v} = \frac{C_v + R}{C_v} = 1 + \frac{R}{C_v} \Rightarrow \boxed{C_v = \frac{R}{k-1}\text{, (Valid for I.G.B.)}} \\ k = \frac{C_P}{C_v} = \frac{C_P}{C_P - R} \Rightarrow C_P(k-1) = kR \Rightarrow \boxed{C_P = \frac{k}{k-1}R\text{, (Valid for I.G.B.)}} \end{cases} \tag{4.46}$$

4.3.3 SUPER-COOLED LIQUIDS-A SPECIAL CASE

The previous section provided the relationship between property variations and specific heats. However, recall from §3.5 that $C_P = C_v$ for incompressible systems yielding, from the definition of enthalpy $h = u + Pv$

$$\overset{C_v}{\overset{C_P dt}{dh}} = \overset{C_v dT}{du} + d(Pv) \Rightarrow C_v dT = C_v dT + d(Pv) \Rightarrow d(Pv) = v dP = 0 \Rightarrow dP = 0 \tag{4.47}$$

which would seem to imply that *all processes for an incompressible substance are isobaric*. But this is not true! Thus, for a super-cooled liquids $dh \neq C_P dT$ and variations in specific enthalpy are *not* a function of variations in temperature alone. Instead, changes in enthalpy for a super-cooled liquid are evaluated via

$$dh = du + d(Pv) \Rightarrow \Delta h = \Delta u + \Delta(Pv) = \Delta u + v\Delta(P) \tag{4.48}$$

or

$$\boxed{\Delta h = \Delta u + v\Delta(P) \quad \text{(Valid for super-cooled liquid (incompressible))}} \tag{4.49}$$

Finally, (2.22) derived in §2.3 is used to evaluate $h_{liq}^{sup.-cooled}$ from saturation Tables.

EXAMPLE 4.6 Application of specific heats for I.G.B.: Energy-balance

Problem Statement:

$1kg$ of air is heated in a rigid $1m^3$ container from $300°K$ to $500°K$. Find the heat transfer using:
a) Ideal gas property Tables B.11
b) Polynomial curve-fits for specific heat Table B.9

Given:
Rigid container of air,
$T_1 = 300°K, T_2 = 500°K$,
$V_1 = 1m^3$,
$m = 1kg$

Find:
$Q_{1\to2}$ via:
a) I.G.B. property Tables B.11
b) curve-fits for specific heat (Table B.9)

Schematic:

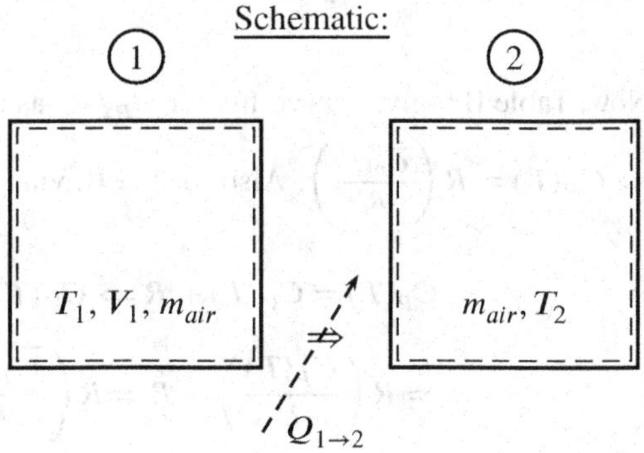

Assumptions:
Isochoric process (rigid container),
$W_{shaft_{1\to2}} = W_{electric_{1\to2}} = 0$,
$\Delta PE_{1\to2} = \Delta KE_{1\to2} = 0$,
I.G.B. (subject to validation)

Analysis:
From the closed system energy-balance for an overall process (Table 4.1)

$$Q_{1\to2} - W_{1\to2} = \Delta E_{1\to2} \tag{4.50}$$

From the definition of total work (3.17) for an isochoric process (thus, $W_{bound} = 0$)

$$W_{1\to2} = W_{bound_{1\to2}} + \cancelto{0}{W_{shaft_{1\to2}}} + \cancelto{0}{W_{electric_{1\to2}}} = \cancel{W_{bound_{1\to2}}} \int_{V_1}^{\cancelto{V_1}{V_2}} PdV = 0 \tag{4.51}$$

Hence

$$Q_{1\to2} - \cancelto{0}{W_{1\to2}} = \Delta E_{1\to2} = \Delta U_{1\to2} + \cancelto{0}{\Delta KE_{1\to2}} + \cancelto{0}{\Delta PE_{1\to2}} = \Delta U_{1\to2} \tag{4.52}$$

or

$$Q_{1\to2} = \Delta U_{1\to2} \tag{4.53}$$

a) Ideal gas property Tables B.11

Writing the energy-balance (4.53) in terms of changes in specific internal energy

$$Q_{1\to2} = \Delta U_{1\to2} = m\Delta u_{1\to2} = m(u_2 - u_1) \tag{4.54}$$

From I.G.B. property Table B.11 $u_2 = u|_{T=500^\circ K} = 359\frac{kJ}{kg}$ and $u_1 = u|_{T=300^\circ K} = 214\frac{kJ}{kg}$. Thus returning to (4.54)

$$Q_{1\to2} = m(u_2 - u_1) = 1\cancel{kg}\left(359\frac{kJ}{\cancel{kg}} - 214\frac{kJ}{\cancel{kg}}\right) = 145kJ \xLeftarrow{Q_{1\to2}} \tag{4.55}$$

b) Via I.G.B. curve-fits for specific heat (Table B.9)
Returning to the energy-balance (4.53) we have, noting the property relationships (4.39)

$$Q_{1\to2} = \Delta U_{1\to2} = m\Delta u_{1\to2} = m\int_{1\to2} du = m\int_{T=T_1}^{T_2} C_v(T)dT \tag{4.56}$$

Now, Table B.9 gives curve-fits for $\overline{C_P}/R_u$ as a function of temperature T where $\frac{\overline{C}_P(T)}{R_u} = \frac{C_P(T)}{R}$ $\Rightarrow C_P(T) = R\left(\frac{\overline{C}_P(T)}{R_u}\right)$. Also for I.G.B. via (4.44)

$$\begin{aligned} C_P(T) &= C_V(T) + R \Rightarrow C_V(T) = C_P(T) - R = \frac{R}{R}C_P(T) - R \\ &= R\left(\frac{C_P(T)}{R}\right) - R = R\left(\frac{\overline{C}_P(T)}{R_u}\right) - R = R\left(\frac{\overline{C}_P(T)}{R_u} - 1\right) \end{aligned} \tag{4.57}$$

where from Table B.9

$$\frac{\overline{C}_P(T)}{R_U} = \alpha + \beta T + \gamma T^2 + \delta T^3 + \epsilon T^4 \tag{4.58}$$

with best-fit coefficients α, β, γ, δ and ϵ listed. Substitution of (4.58) into (4.56) using (4.57) gives

$$\begin{aligned} Q_{1\to 2} &= m\int_{T=T_1}^{T_2} C_v(T)dT = m\int_{T=T_1}^{T_2} R\left(\frac{\overline{C}_P(T)}{R_u} - 1\right)dT \\ &= mR\int_{T=T_1}^{T_2} (\alpha + \beta T + \gamma T^2 + \delta T^3 + \epsilon T^4 - 1)\,dT \\ &= mR\left[\alpha T + \frac{\beta}{2}T^2 + \frac{\gamma}{3}T^3 + \frac{\delta}{4}T^4 + \frac{\epsilon}{5}T^5 - T\right]_{T_1}^{T_2} \end{aligned} \tag{4.59}$$

or

$$Q_{1\to 2} = mR\left[\alpha T + \frac{\beta}{2}T^2 + \frac{\gamma}{3}T^3 + \frac{\delta}{4}T^4 + \frac{\epsilon}{5}T^5 - T\right]_{T_1}^{T_2} \tag{4.60}$$

Coefficients are provided from Table B.9 giving

$$Q_{1\to 2} = mR\left[\alpha T + \frac{\beta}{2}T^2 + \frac{\gamma}{3}T^3 + \frac{\delta}{4}T^4 + \frac{\epsilon}{5}T^5 - T\right]_{T_1}^{T_2} =$$

$$\begin{aligned} &1kg\left(0.286\frac{kJ}{kg-{}^\circ \cancel{K}}\right)[3.653\,(500^\circ \cancel{K} - 300^\circ \cancel{K}) + \frac{-1.337(10^{-3})\frac{1}{{}^\circ \cancel{K}}}{2}\left((500^\circ \cancel{K})^2 - (300^\circ \cancel{K})^2\right) \\ &+\frac{3.294(10^{-6})\frac{1}{{}^\circ \cancel{K}^2}}{3}\left((500^\circ \cancel{K})^3 - (300^\circ \cancel{K})^3\right) + \frac{-1.913(10^{-9})\frac{1}{{}^\circ \cancel{K}^3}}{4}\left((500^\circ \cancel{K})^4 - (300^\circ \cancel{K})^4\right) + \\ &\frac{0.2763(10^{-12})\frac{1}{{}^\circ \cancel{K}^4}}{5}\left((500^\circ \cancel{K})^5 - (300^\circ \cancel{K})^5\right) - (500^\circ \cancel{K}) - (300^\circ \cancel{K})] = 139.7kJ \stackrel{Q_{1\to 2}}{\Longleftarrow} \end{aligned} \tag{4.61}$$

Assumption of I.G.B. is subject to validation: Via Table B.1

$$T_{red_1} = \frac{T_1}{T_{crit}} = \frac{300^\circ \cancel{K}}{133^\circ \cancel{K}} = 2.25 > 2 \ \surd,$$

$$P_{red_1} = \frac{P_1}{p_{crit}} = \frac{\frac{mRT_1}{V_1}}{p_{crit}} = \frac{\frac{1\cancel{kg}(0.286\frac{kJ}{\cancel{kg}-{}^\circ \cancel{K}})300^\circ \cancel{K}}{1m^3}}{37.7\cancel{bar}}\frac{1\cancel{bar}}{100kPa} = 0.022 < 0.2 \ \surd,$$

$$P_{red_2} = \frac{P_2}{p_{crit}} = \frac{\frac{mRT_2}{V_2}}{p_{crit}} = \frac{\frac{1\cancel{kg}(0.286\frac{kJ}{\cancel{kg}-{}^\circ \cancel{K}})500^\circ \cancel{K}}{1m^3}}{37.7\cancel{bar}}\frac{1\cancel{bar}}{100kPa} = 0.036 < 0.2 \ \surd$$

$\Rightarrow$ I.G.B is valid.

$$Q_{1\to 2} = \begin{cases} 145kJ & \text{I.G.B. properties Table B.11} \\ 139.7kJ & \text{Variable specific heat via polynomial expression in Table B.9} \end{cases}$$

Comment: The solution of highest accuracy corresponds to using property Table B.11. The polynomial expression for $\overline{C}_P/\overline{R}$, while producing a less accurate result, is used in various computational applications such as computational fluid dynamics codes (CFD).

□

4.4 ENERGY CONSERVATION: OPEN SYSTEMS

4.4.1 THE REYNOLDS TRANSPORT THEOREM (RTT)

Thus far we have dealt only with energy conservation (the First-Law or energy-balance) for closed systems. For open systems, where mass flows across the system boundary, we need to develop a methodology for transforming conservation equations (for energy *and* mass) for closed systems to their open system equivalent expressions. Such a methodology, the Reynolds Transport Theorem (RTT), is shown in Figure 4.2. To proceed, recall that any extensive property (e.g. m, E, H, etc.), represented by the Greek letter β, depends on the extent of the system, while intensive properties (e.g. e, h, v, etc.), represented by the letter b, do not. In addition, recall that an extensive property β can be converted to an intensive property b via normalization (division) by the system mass:

$$b = \frac{\beta}{m} \tag{4.62}$$

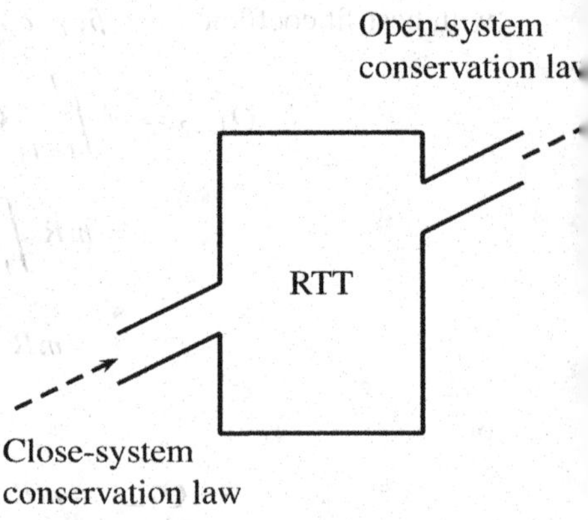

Figure 4.2 Pictorial illustration of Reynolds Transport Theorem (RTT).

Table 4.2 lists the various rate-based conservation laws for closed systems in terms of β and b.

Table 4.2 Closed system conservation equations

Extensive property	Conservation Equation	β	b
Mass	$0 = \frac{d}{dt} m_{\substack{closed \\ system}} = \frac{d}{dt}\beta_{\substack{closed \\ system}}$	$\beta = m$	$b = \frac{\beta}{m} = \frac{m}{m} = 1$
Momentum ($i = x, y, z$)	$\sum F_{i:\substack{closed \\ system}} = \frac{d}{dt}(mV)_{i:\substack{closed \\ system}} = \frac{d}{dt}\beta_{\substack{closed \\ system}}$	$\beta = (mV)_i$	$b = \frac{\beta}{m} = \frac{(mV)_i}{m} = V_i$
Energy	$(\dot{Q} - \dot{W})_{\substack{closed \\ system}} = \dot{E} = \frac{d}{dt}\beta_{\substack{closed \\ system}}$	$\beta = E$	$b = \frac{\beta}{m} = \frac{E}{m} = e$

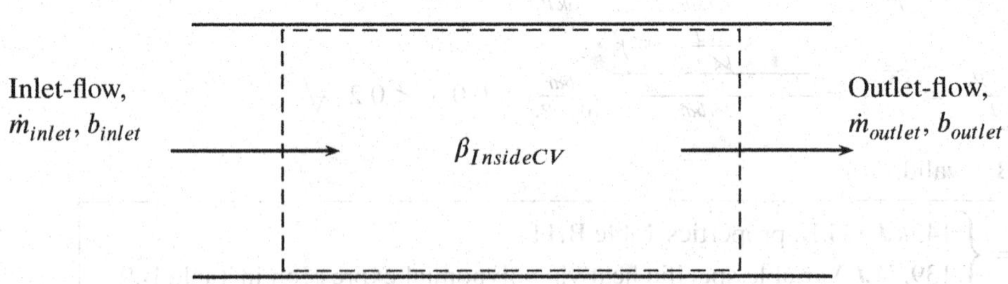

Figure 4.3 Illustration of open system control volume – – –– with a single inlet/outlet.

Next, we envision some fluid passing through a duct as shown in Figure 4.3. Note that the duct system – – –– is an open system, thus mass is flowing into and out of the open system (duct) control volume over time. Next, let us tag *all* of the mass encapsulated within the open system

CV – – –– at time t. This fixed mass we refer as the closed system which contains the amount of extensive property β equal to $\beta_{\substack{closed\\system}}\Big|_t$ as shown in Figure 4.4 a). Of course, at time t the amount of β within the open system control volume $\beta_{CV}\big|_t$ and closed system control volume $\beta_{\substack{closed\\system}}\Big|_t$ are equal

$$\beta_{CV}\big|_t = \beta_{\substack{closed\\system}}\Big|_t \tag{4.63}$$

The next question to ask is "how much β is in the open system CV at time $t + \Delta t$"? Well, at time

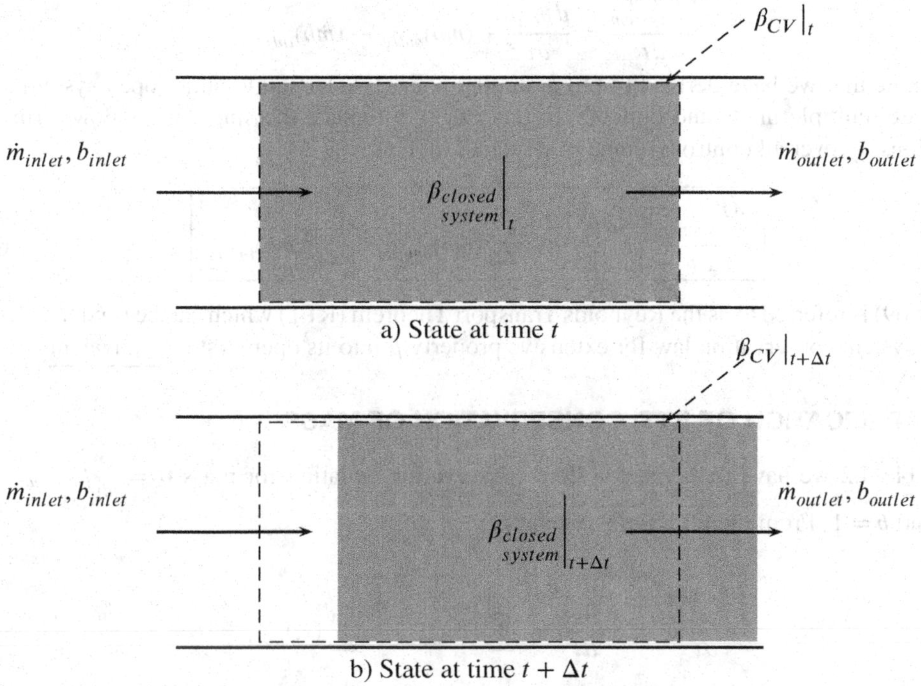

Figure 4.4 Illustration of fixed mass system moving through open system control volume – – –– with a single inlet/outlet.

$t + \Delta t$, $\beta_{CV}\big|_{t+\Delta t}$ is approximated as β within the closed system at time $t + \Delta t$ (i.e. $\beta_{\substack{closed\\system}}\Big|_{t+\Delta t}$), plus the inflow of β into the control volume over time Δt minus the outflow of β from the control volume over time Δt. Or

$$\beta_{CV}\big|_{t+\Delta t} \approx \beta_{\substack{closed\\system}}\Big|_{t+\Delta t} + \overbrace{(\dot{m}b)_{inlet}\Delta t}^{\text{Inflow of }\beta} - \overbrace{(\dot{m}b)_{outlet}\Delta t}^{\text{Outflow of }\beta} \tag{4.64}$$

Subtracting (4.63) from (4.64) and dividing by Δt gives

$$\frac{\beta_{CV}\big|_{t+\Delta t} - \beta_{CV}\big|_t}{\Delta t} \approx \frac{\beta_{\substack{closed\\system}}\Big|_{t+\Delta t} - \beta_{\substack{closed\\system}}\Big|_t}{\Delta t} + \frac{(\dot{m}b)_{inlet}\cancel{\Delta t} - (\dot{m}b)_{outlet}\cancel{\Delta t}}{\cancel{\Delta t}} \tag{4.65}$$

or

$$\frac{\Delta \beta_{CV}}{\Delta t} \approx \frac{\Delta \beta_{\substack{closed\\system}}}{\Delta t} + (\dot{m}b)_{inlet} - (\dot{m}b)_{outlet} \tag{4.66}$$

Taking the limit as $\Delta t \to 0$ the approximate relationship becomes exact and we have

$$\frac{d\beta_{CV}}{dt} = \frac{d\beta_{\substack{closed\\system}}}{dt} + (\dot{m}b)_{inlet} - (\dot{m}b)_{outlet} \tag{4.67}$$

Rearranging (4.67) by placing the $d\beta_{\substack{closed\\system}}/dt$ on the left-hand and $d\beta_{CV}/dt$ on the right-hand side we have

$$\frac{d\beta_{\substack{closed\\system}}}{dt} = \frac{d\beta_{CV}}{dt} + (\dot{m}b)_{outlet} - (\dot{m}b)_{inlet} \tag{4.68}$$

Finally, note that we have derived (4.68) assuming a single inlet/single outlet open system. What if there are multiple inlets and outlets? In this case we replace the single mass-flow terms with summations $\sum$ over all control volume inlets and outlets:

$$\boxed{\frac{d\beta_{\substack{closed\\system}}}{dt} = \frac{d\beta_{CV}}{dt} + \sum (\dot{m}b)_{outlets} - \sum (\dot{m}b)_{inlets}} \tag{4.69}$$

Thus, (4.69) is referred to as the Reynolds Transport Theorem (RTT) which can be used to transform a closed system conservation law for extensive property β into its open system equivalent.

4.4.2 APPLICATION OF RTT: CONSERVATION OF MASS

From Table 4.2 we have the closed system conservation equation for mass $0 = \frac{d}{dt}\beta_{\substack{closed\\system}}$ where $\beta = m$ and $b = 1$. From the RTT (4.69) we have

$$\cancelto{0}{\frac{dm_{\substack{closed\\system}}}{dt}} = \frac{dm_{CV}}{dt} + \sum (\dot{m})_{outlet} - \sum (\dot{m})_{inlet} = 0 \tag{4.70}$$

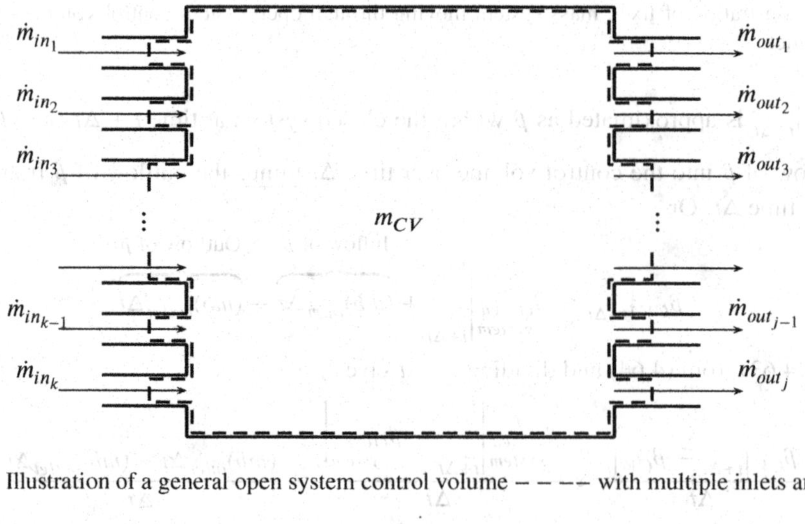

Figure 4.5 Illustration of a general open system control volume – – –– with multiple inlets and outlets.

or, rearranging terms we get our mass-conservation equation for an open system

$$\overbrace{\frac{dm_{CV}}{dt}}^{\text{Change of mass in CV}} = \overbrace{\sum (\dot{m})_{inlets} - \sum (\dot{m})_{outlets}}^{\text{Net inflow of mass}} \tag{4.71}$$

Figure 4.5 shows a general open system control volume with k inlets and j outlets. Finally, it is often the case that the particular open-flow device operates in such a way that properties and flow-rates do not change with time. Thus we define the terms "steady-state" and "steady-flow"

Definition 4.1 ***Steady-state operation:*** *No variation in system (control-volume) properties with respect to time:* $\frac{d}{dt} = 0$

Definition 4.2 ***Steady-flow operation:*** *Constant inlet/outlet mass-flow rates:* $\frac{d\dot{m}}{dt} = 0$

As far a mass-conservation is concerned, steady-state operation implies $dm_{CV}/dt = 0$ while steady-flow implies that the mass-flow ($\dot{m}$'s) at the inlets and outlets are not a function of time.

EXAMPLE 4.7 Steady-state, steady-flow through a nozzle: Mass-balance

Problem Statement:
A circular nozzle (flow acceleration device) has an inlet velocity and radius of $1m/s$ and $0.25m$ respectively. The outlet velocity is $10m/s$. Find the outlet radius of the nozzle. Assume steady-state/steady-flow.

Given:
Open system,
$Vel_1 = 1m/s$,
$r_1 = 0.25m$,
$Vel_2 = 10m/s$

Find:
r_2

Assumptions:
Steady-state (S.S.), steady-flow,
Incompressible flow (v is constant)

Schematic:

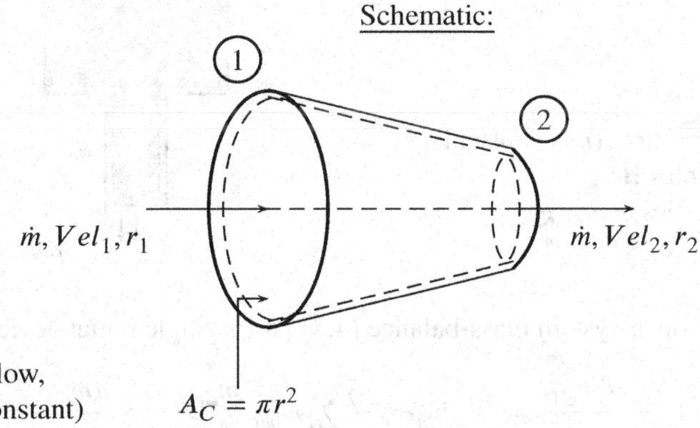

Analysis:
Applying the open-system mass-balance (4.71) for a steady-state, single inlet/single outlet device

$$\frac{dm_{CV}}{dt}\overset{0,\,S.S.}{} = \sum \dot{m}_{inlets} + \sum \dot{m}_{outlets} = 0 \Rightarrow \dot{m}_2 = \dot{m}_1 = \dot{m} \tag{4.72}$$

Note, at each inlet/outlet the cross-section area takes the form $A_C = \pi r^2$, hence

$$\dot{m} = A_c(Vel)\rho = \pi r^2(Vel)\rho \Rightarrow \dot{m}_1 = \pi r_1^2(Vel_1)\rho_1, \dot{m}_2 = \pi r_2^2(Vel_2)\rho_2 \tag{4.73}$$

The flow is assumed incompressible ($v_1 = v_2 \Rightarrow \rho_1 = \rho_2$), and combining (4.73) with the mass-balance (4.72) gives

$$\dot{m}_1 = \dot{m}_2 \Rightarrow \cancel{\pi} r_1^2 (Vel_1)\cancel{\rho_1} = \cancel{\pi} r_2^2 (Vel_2)\cancel{\rho_2}$$

$$\Rightarrow r_2 = r_1\sqrt{\frac{Vel_1}{Vel_2}} = 0.25m\sqrt{\frac{1\frac{\cancel{m}}{\cancel{s}}}{10\frac{\cancel{m}}{\cancel{s}}}} = 0.079m \xleftarrow{\quad r_2 \quad} \tag{4.74}$$

$\boxed{r_2 = 0.079m}$

□

EXAMPLE 4.8 Draining tank: Unsteady mass-balance

Problem Statement:
A water tank with a drain at the bottom initially contains $4L$ of water. Water flows at a constant rate of $0.1\frac{kg}{s}$ out through the drain. Find the time required to empty the tank.

Given:
Open system,
water (H_2O liquid),
$V_1 = 4L$,
$V_2 = 0L$,
$\dot{m}_{drain} = 0.1\frac{kg}{s}$

Find:
$\Delta t_{1\rightarrow 2}$

Assumptions:
Incompressible flow (v is constant),
saturation Tables B.2,
Liquid properties$|_{T=300°K, P=1atm}$

Schematic:

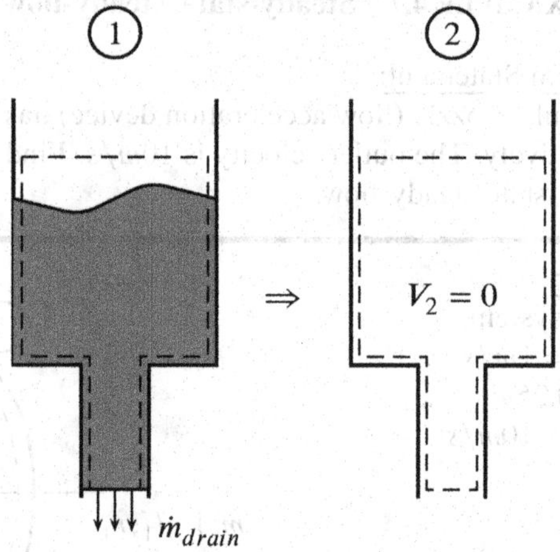

Analysis:
Applying the open-system mass-balance (4.71) for a single outlet device

$$\frac{dm_{CV}}{dt} = \sum \cancelto{0}{\dot{m}_{inlets}} - \sum \cancelto{\dot{m}_{drain}}{\dot{m}_{outlets}} \Rightarrow \frac{dm_{CV}}{dt} = -\dot{m}_{drain} \tag{4.75}$$

Multiplying the mass-conservation equation (4.75) by dt and integrating from state $1 \rightarrow 2$ gives

$$\cancelto{\Delta m_{1\rightarrow 2}}{\int_{t=t_1}^{t_2} \frac{dm_{CV}}{dt} dt} = -\int_{t=t_1}^{t_2} \dot{m}_{drain} dt = -\dot{m}_{drain}\int_{t=t_1}^{t_2} dt = -\dot{m}_{drain}\Delta t_{1\rightarrow 2} \tag{4.76}$$

or

$$\Delta m_{1\rightarrow 2} = -\dot{m}_{drain}\Delta t_{1\rightarrow 2} \Rightarrow \Delta t_{1\rightarrow 2} = -\frac{\Delta m_{1\rightarrow 2}}{\dot{m}_{drain}} \tag{4.77}$$

For H_2O we have via (2.19) and Table B.2 $\upsilon_{H_2O} \approx \upsilon_{liq}^{sat}\Big|_{300° K} = 1.0035(10^{-3})\frac{m^3}{kg}$ and $\rho = \frac{1}{\upsilon} = 996.5\frac{kg}{m^3}$. For the tank $m = \rho V$ and thus via (4.77)

$$\Delta t_{1\to 2} = -\frac{\Delta m_{1\to 2}}{\dot{m}_{drain}} = -\frac{\Delta(\rho V)_{1\to 2}}{\dot{m}_{drain}} = -\frac{\rho \Delta V_{1\to 2}}{\dot{m}_{drain}} = -\frac{\rho(V_2 - V_1)}{\dot{m}_{drain}}$$

$$= -\frac{996.5\frac{kg}{m^3}(0L - 4L)}{0.1\frac{kg}{s}}\frac{1m^3}{1000L} = 40sec \overset{\Delta t_{1\to 2}}{\Longleftarrow\!=\!=\!=} \tag{4.78}$$

$\boxed{\Delta t_{1\to 2} = 40sec}$

Comment: If the flow is via gravity feed, do you expect the outlet mass-flow to remain constant or decrease with decreasing water height h?

□

4.4.3 APPLICATION OF THE RTT: ENERGY CONSERVATION

From Table 4.2 we have the energy conservation equation for a closed system $(\dot{Q} - \dot{W})_{\substack{closed\\system}} = \frac{d}{dt}\beta_{\substack{closed\\system}}$ where $\beta = E$ and $b = e$. Applying the RTT (4.69) we have

$$(\dot{Q} - \dot{W})_{\substack{closed\\system}} = \frac{dE_{CV}}{dt} + \sum(\dot{m}e)_{outlets} - \sum(\dot{m}e)_{inlets} \tag{4.79}$$

Now it would seem that we are ready to apply the above expression to any open system such as that shown in Figure 4.6. However, a new and subtle form of power has been introduced: The work over time required to push the fluid in and/or out of the open system control volume (CV).

4.4.4 OPEN SYSTEMS: THE ROLE OF PRESSURE WORK

To examine the work over time (or power) required to push fluid into or out of the open system control volume (CV) we return to the pipe-flow investigated in deriving the RTT. Noting that work corresponds to $force \cdot distance$ and thus power to $force \cdot velocity$ we have, on the left in Figure 4.7, pressure-power at the inlet given by $(\vec{F}_{pres} \cdot \vec{Vel})_{inlet} > 0$. At the outlet, on the right in Figure 4.7, we have $(\vec{F}_{pres} \cdot \vec{Vel})_{outlet} < 0$. Hence, the pressure force opposes the fluid motion at the outlet and facilitates the motion at the inlet. Thus, the closed system receives work due to pressure power at the control volume inlet where fluid is forced into the system, while at the outlet the system is doing pressure work to expel the fluid. Noting that $P = Force/Area$, the net pressure power on the close system which spatially coincides with the control volume CV is

$$\text{Net Pressure-power (on system)} = \dot{W}_{pressure}^{on\ system}$$

$$= (\vec{F}_{pres} \cdot \vec{Vel})_{inlet} + (\vec{F}_{pres} \cdot \vec{Vel})_{outlet} = [PA(Vel)]_{inlet} - [PA(Vel)]_{outlet}$$

$$= [P(A(Vel))]_{inlet} - [P(A(Vel))]_{outlet} = [P\dot{V}]_{inlet} - [P\dot{V}]_{outlet} \tag{4.80}$$

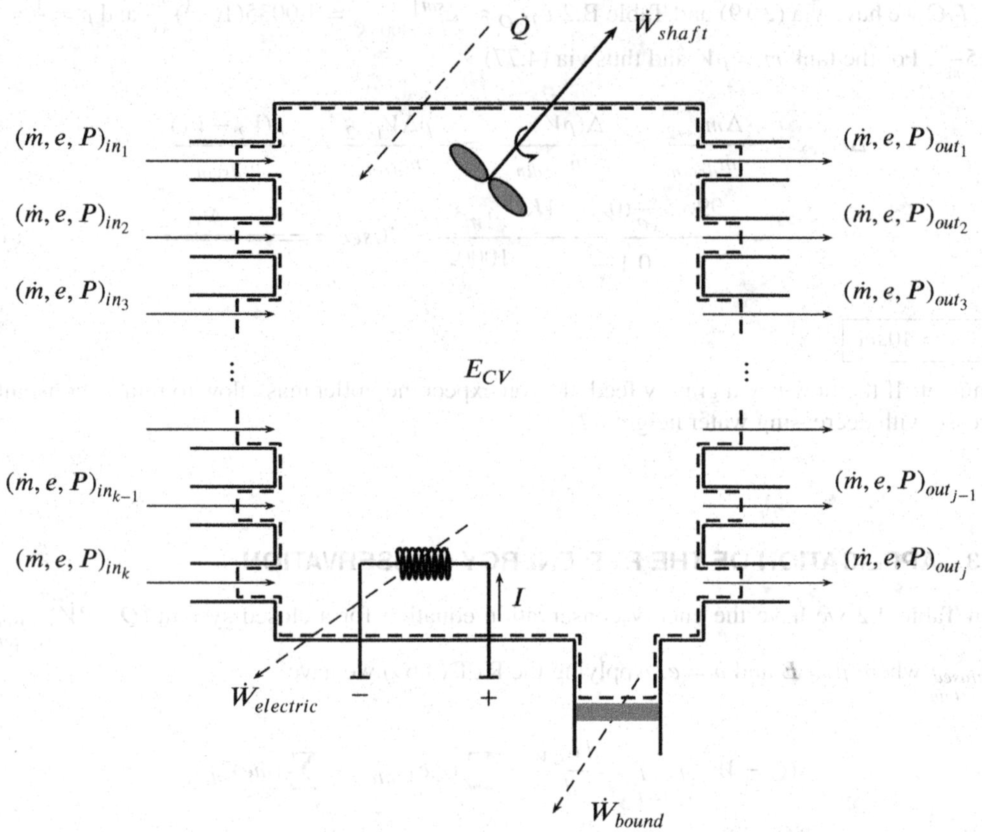

Figure 4.6 Illustration of open system control volume (CV) – – – – with a multiple inlets and outlets. Energy inter-actions $\dot{Q}$ and $\dot{W}$ shown.

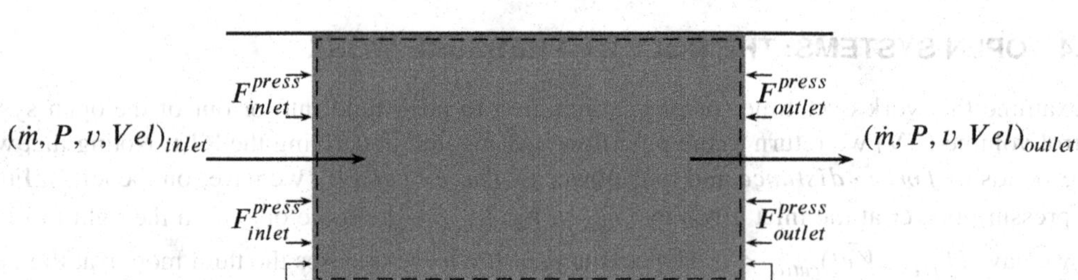

Figure 4.7 Illustration of pressure power for a single inlet/outlet duct.

where A is the flow area and $\dot{V}$ is the volume flow rate. Finally, given the definition of specific volume $\upsilon = V/m$ while noting that $\dot{W}^{by\ system}_{pressure} = -\dot{W}^{on\ system}_{pressure}$ we have

$$\dot{W}^{by\ system}_{pressure} = -\dot{W}^{on\ system}_{pressure} = -\left[(P\dot{V})_{inlet} - (P\dot{V})_{outlet}\right]$$

$$= -(P\dot{V})_{inlet} + (P\dot{V})_{outlet} = -(P\upsilon\dot{m})_{inlet} + (P\upsilon\dot{m})_{outlet} \tag{4.81}$$

Now, let us apply the energy-balance given in (4.79) to the duct in Figure 4.7 noting that the power term $\dot{W}_{\substack{closed\\system}}$ is now composed of *four* distinct types of power: shaft, electric, boundary and pressure power!

$$\overbrace{\dot{Q} - (\dot{W}_{bound} + \dot{W}_{shaft} + \dot{W}_{electric} + \dot{W}^{by\,system}_{pressure})}^{LHS} = \overbrace{\frac{dE_{CV}}{dt} + (\dot{m}e)_{outlet} - (\dot{m}e)_{inlet}}^{RHS} \tag{4.82}$$

Moving the pressure power term $\dot{W}^{by\,system}_{pressure}$ to the RHS of (4.82) and using the expression for pressure power (4.81) gives

$$\begin{aligned}\dot{Q} - (\dot{W}_{boundary} + \dot{W}_{shaft} + \dot{W}_{electric}) &= \frac{dE_{CV}}{dt} + (\dot{m}e)_{outlet} - (\dot{m}e)_{inlet} + \dot{W}^{by\,system}_{pressure} = \\ &\frac{dE_{CV}}{dt} + (\dot{m}e)_{outlet} - (\dot{m}e)_{inlet} + \left[-(Pv\dot{m})_{inlet} + (Pv\dot{m})_{outlet}\right] = \\ &\frac{dE_{CV}}{dt} + (\dot{m}(e + Pv))_{outlet} - (\dot{m}(e + Pv))_{inlet}\end{aligned} \tag{4.83}$$

Noting the definition of specific enthalpy $h = u + Pv$ as well as total specific energy $e = u + pe + ke$, (4.83) reduces to

$$\dot{Q} - \dot{W}_{\text{less pressure power}} = \frac{dE_{CV}}{dt} + (\dot{m}(h + ke + pe))_{outlets} - (\dot{m}(h + ke + pe))_{inlets} \tag{4.84}$$

or, for multiple inlet and outlet systems (the general case)

$$\boxed{\dot{Q} - \dot{W} = \frac{dE_{CV}}{dt} + \sum(\dot{m}(h + ke + pe))_{outlets} - \sum(\dot{m}(h + ke + pe))_{inlets}} \tag{4.85}$$

where, given that the control-volume CV coincides spatially with closed system in the limit, the heat and work transfer are on the open system, and the power term $\dot{W}$ conforms to the definition used previously for closed systems: $\dot{W} = \dot{W}_{shaft} + \dot{W}_{electric} + \dot{W}_{bound}$. While this might have seemed a lot of effort just to account for a new form of power, the formulation given by (4.85) saves us the effort of evaluating the pressure work explicitly for each application!

EXAMPLE 4.9 Fan driven flow through a heated duct: Energy-balance

Problem Statement:

A painter's heat gun contains a heating chamber composed of a duct with a fan and electric heating element through which air at $1atm$ and $27°C$ flows at a rate of $2L/s$. The fan power is $0.05kW$ while the heating element draws current at a rate of $2.25A$ from a power source of $120Volts$. Determine the outlet air-flow temperature.

Given:
Open system, air,
$P_1 = 1atm$, $T_1 = 27°C$,
$\dot{V}_1 = 2L/s$,
$\dot{W}_{shaft} = 0.05kW$,
$I_{Element} = 2.25A$,
$Voltage_{Element} = 120V$

Schematic:

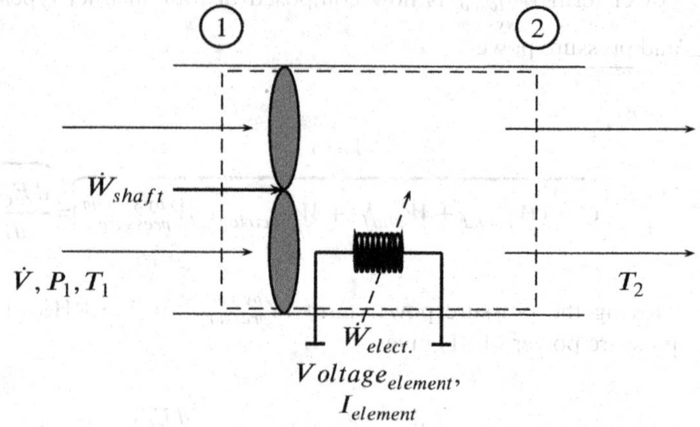

Find:
T_2

Assumptions:
Steady-state (S.S.), steady-flow,
$\Delta ke_{1\to2} = \Delta pe_{1\to2} = 0$,
$\dot{W}_{bound} = 0$, $\dot{Q} = 0$,
I.G.B. air Tables B.11

Analysis:
Applying the open system energy-balance (4.85)

$$\cancelto{0}{\dot{Q}} - \dot{W} = \cancelto{0,\,S.S.}{\frac{dE_{CV}}{dt}} + \sum \cancelto{(\dot{m}(h+ke+pe))_2^{\,0}}{(\dot{m}(h+ke+pe))_{outlets}} - \sum \cancelto{(\dot{m}(h+ke+pe))_1^{\,0}}{(\dot{m}(h+ke+pe))_{inlets}} \tag{4.86}$$

or

$$\dot{W} = (\dot{m}h)_2 - (\dot{m}h)_1 \tag{4.87}$$

Since $\dot{W} = \dot{W}_{shaft} + \dot{W}_{electric}$ (4.87) becomes

$$-\dot{W}_{electric} - \dot{W}_{shaft} = (\dot{m}h)_2 - (\dot{m}h)_1 \tag{4.88}$$

where h_2 will allow us to determine T_2. Now, applying the open system mass-balance (4.71) for a steady-state, single inlet single outlet device

$$\cancelto{0,\,S.S.}{\frac{dm_{CV}}{dt}} = -\sum \cancelto{\dot{m}_2}{(\dot{m})_{outlets}} + \sum \cancelto{\dot{m}_1}{(\dot{m})_{inlets}} \tag{4.89}$$

or

$$\dot{m}_1 = \dot{m}_2 = \dot{m} \tag{4.90}$$

Substituting this result into the energy-balance (4.88) gives

$$-\dot{W}_{electric} - \dot{W}_{shaft} = (\dot{m}h)_2 - (\dot{m}h)_1 = \dot{m}(h_2 - h_1)$$

$$\Rightarrow h_2 = \frac{-\dot{W}_{electric} - \dot{W}_{shaft}}{\dot{m}} + h_1 = \frac{-(-I_{Element}Voltage_{Element}) - \dot{W}_{shaft}}{\dot{m}} + h_1 \tag{4.91}$$

where we note that $\dot{W}_{electric} = -I_{Element}Voltage_{Element} < 0$ (element does work on the duct system). Now, from the definition of specific volume

$$\dot{m} = \dot{m}_1 = \dot{V}_1\frac{1}{v_1} = \dot{V}_1\rho_1 \tag{4.92}$$

where via I.G.B. $\rho_1 = \frac{P_1}{RT_1}$. Thus

$$\dot{m} = \dot{m}_1 = \dot{V}_1\rho_1 = \dot{V}_1\frac{P_1}{RT_1} = 2(\frac{L}{s})\frac{1atm}{0.286\frac{kJ}{kg-°K}300°K}$$

$$= 0.0232\frac{L-atm-kg}{kJ}\frac{10^3 m^3}{1L}\frac{1.01(10^5)Pa}{1atm}\frac{1kJ}{10^3 J} = 0.00234\frac{kg}{s} \tag{4.93}$$

Next, note that $h_1 = h|_{T=300°K} = 300.2\frac{kJ}{kg}$ (I.G.B. air Tables B.11). Hence, we have from (4.91) where both power terms are on the system and thus negative

$$h_2 = \frac{-(-I_{Element}Voltage_{Element} - \dot{W}_{shaft})}{\dot{m}} + h_1$$

$$= \frac{-(-(2.25\overset{\frac{Coulomb}{s}}{A})(120\overset{\frac{J}{Coulomb}}{V}) - 0.05\overset{\frac{kJ}{s}}{kW})}{0.00232\frac{kg}{s}} + 300.2\frac{kJ}{kg} =$$

$$\frac{-(-(2.25\frac{Coulomb}{s})(120\frac{J}{Coulomb})\frac{1kJ}{1000J} - 0.05\frac{kJ}{s})}{0.00234\frac{kg}{s}} + 300.2\frac{kJ}{kg} = 437\frac{kJ}{kg} \tag{4.94}$$

or $h_2 = 437\frac{kJ}{kg}$. Interpolating from I.G.B. air Table B.11 gives

$$T_2 = T|_{h=437\frac{kJ}{kg}} = 164°C \stackrel{T_2}{\Longleftarrow} \tag{4.95}$$

Check validity of I.G.B.: $T_{red_1} = \frac{T_1}{T_{crit}} = \frac{300°K}{133°K} = 2.25 > 2\surd$ Also, $P_{red} = \frac{P_1}{P_{crit}} = \frac{1atm}{33bar}\frac{1.01bar}{1atm}$ $= 0.03 < 0.2\surd \Rightarrow$ I.G.B. valid!

$\boxed{T_2 = 164°C}$

Comment: If one had instead chosen a control volume *excluding* the heating element, the electric work interaction would become a heat interaction where $\dot{Q}_{Element} = Voltage_{Element}I_{Element}$.

□

4.5 OPEN FLOW DEVICES

Various open flow devices (of which there are many) are classified according to function. In the following section we will group these flow devices into two broad categories: Those which accelerate or decelerate mass flow, and those which affect a heat or a work transfer. In terms of the First-Law, the former convert pressure power into internal and kinetic energy (or visa verse), while the latter convert pressure power and internal energy to heat or work (or visa verse).

4.5.1 NOZZLES AND DIFFUSERS

Often, for the purpose of generating thrust or producing a pressure rise (or fall), it is necessary to decelerate or accelerate fluid flow. Hence the definition of an adiabatic nozzle:

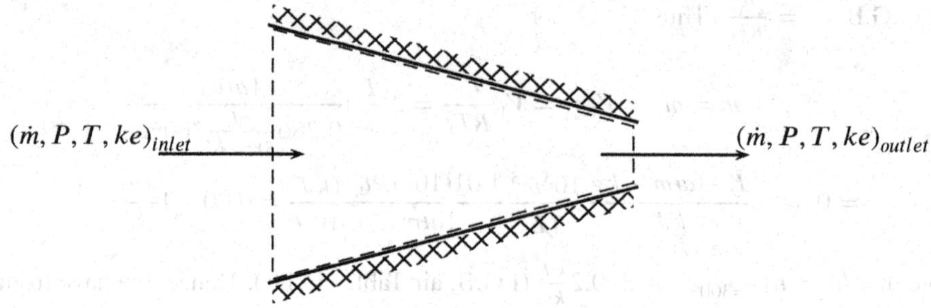

$$P_{outlet} < P_{inlet}$$
$$T_{outlet} \leq T_{inlet}$$
$$ke_{outlet} > ke_{inlet}$$

Figure 4.8 Illustration of Adiabatic Nozzle.

Definition 4.3 ***Adiabatic Nozzle:*** *A nozzle utilizes pressure-power and an internal energy change to affect an acceleration of the flow and a corresponding pressure drop. Such a device along with typical relationships between inlet and outlet properties are given in Figure 4.8:*

Note, as shown in Figure 4.8, nozzles are work-less and often adiabatic $\dot{Q} = 0$. In addition, it is often convenient for engineering purposes, to make the following approximation if $ke_{outlet} \gg ke_{inlet}$: $\Delta ke_{inlet \to outlet} = ke_{outlet} - ke_{inlet} \approx ke_{outlet}$.

Definition 4.4 ***Adiabatic Diffuser:*** *A diffuser converts flow kinetic energy into pressure-power and/or an increase in internal energy by decelerating the flow. Such a device along with typical relationships between inlet and outlet properties are given in Figure 4.9:*

As indicated in Figure 4.9, diffusers, like nozzles, are work-less and often adiabatic $\dot{Q} = 0$. Again, as in the case of nozzles, it is often convenient for engineering purposes to make the following approximation if $ke_{outlet} \ll ke_{inlet}$: $\Delta ke_{inlet \to outlet} = ke_{outlet} - ke_{inlet} \approx -ke_{inlet}$.

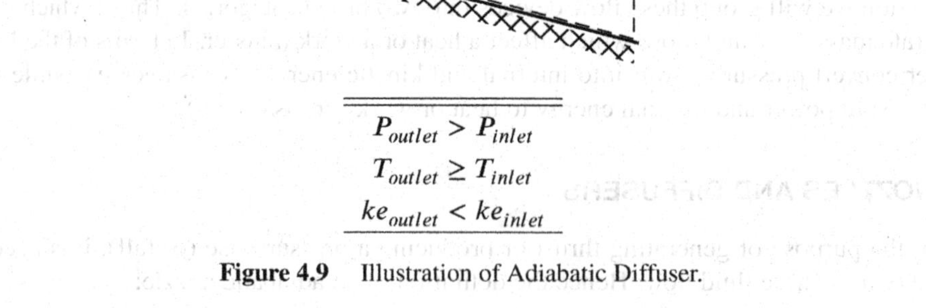

$$P_{outlet} > P_{inlet}$$
$$T_{outlet} \geq T_{inlet}$$
$$ke_{outlet} < ke_{inlet}$$

Figure 4.9 Illustration of Adiabatic Diffuser.

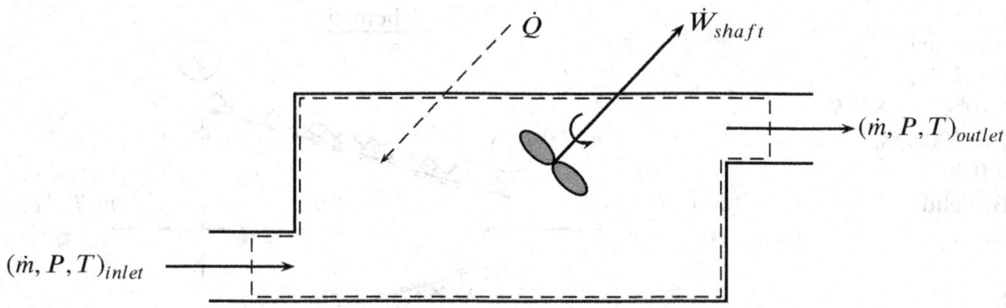

Device	$\dot{Q}$	$\dot{W}_{shaft}$
Adiabatic Turbine	0	+
Adiabatic Compressor/pump	0	−
Heat Exchanger	+, −	0

Figure 4.10 Illustration of an adiabatic turbine/compressor and heat-exchanger functionality.

4.5.2 TURBINES, COMPRESSORS AND HEAT EXCHANGERS

The second group of open-flow devices transform pressure work and flow energy into either work or heat (or visa verse). These devices are referred to as turbines, compressors/pumps and heat exchangers.

Definition 4.5 ***Compressor or pump:*** *Compressors and pumps convert shaft-power into flow energy and pressure-power. Such devices are work consuming devices which typically increase flow pressure and/or temperature as shown in Figure 4.10. Compressors as opposed to pumps are distinguished by the behavior of the working fluid: If the flow is incompressible, the device is a pump. If the flow is compressible, such a device is referred to as a compressor.*

Definition 4.6 ***Turbine:*** *A turbine converts flow energy and pressure-power into shaft-power. Such a device is thus a work producing device which typically decreases flow pressure and/or temperature as shown in Figure 4.10.*

Definition 4.7 ***Heat Exchanger:*** *A heat exchanger converts flow energy and pressure-power into heat (or visa verse). Such a device is typically work-less and decreases or increases flow temperature as shown in Figure 4.10.*

EXAMPLE 4.10 Flow through an adiabatic diffuser: Energy-balance

Problem Statement:
An (air) diffuser with inlet temperature, pressure, velocity and flow area of $10°C$, $80kPa$, $200\frac{m}{s}$ and $0.4m^2$ respectively, decelerates the flow such that the outlet flow kinetic energy is negligible. Find the mass-flow rate and outlet temperature. I.G.B. is valid.

Given:
Diffuser (air),
$P_1 = 80kPa$,
$T_1 = 10°C = 283°K$,
$Vel_1 = 200m/s$,
$A_1 = 0.4m^2$,
I.G.B. Valid,
$ke_1 \gg ke_2$

Schematic:

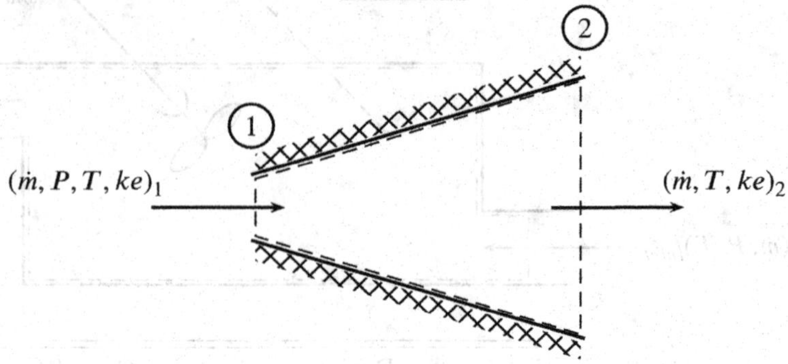

Find:
T_2,
$\dot{m}$

Assumptions:
Steady-state (S.S.), steady-flow,
$\Delta pe_{1\to 2} = 0$,
$\dot{W} = 0$,
$\dot{Q} = 0$ (adiabatic),
I.G.B. air Tables B.11

Analysis:
Applying the open system energy-balance (4.85) for a steady-state, single-inlet, single-outlet device

$$\cancelto{0}{\dot{Q}} - \cancelto{0}{\dot{W}} = \cancelto{0, S.S.}{\frac{dE_{cv}}{dt}} + \cancelto{(\dot{m}(h+ke+pe))_2}{\sum(\dot{m}(h+ke+pe))_{outlets}} - \cancelto{(\dot{m}(h+ke+pe))_1}{\sum(\dot{m}(h+ke+pe))_{inlets}} \tag{4.96}$$

or

$$0 = (\dot{m}(h+ke))_2 - (\dot{m}(h+ke))_1 \tag{4.97}$$

Hence, the resulting expression for h_2 which will allow for the determination of T_2

$$h_2 = \frac{(\dot{m}(h+ke))_1 - (\dot{m}ke)_2}{\dot{m}_2} = \frac{\dot{m}_1}{\dot{m}_2}(h+ke)_1 - ke_2 \tag{4.98}$$

Now, applying the open-system mass-balance (4.71)

$$\cancelto{0, S.S.}{\frac{dm_{cv}}{dt}} = -\sum \cancelto{\dot{m}_2}{(\dot{m})}_{outlets} + \sum \cancelto{\dot{m}_1}{(\dot{m})}_{inlets} \Rightarrow \dot{m}_1 = \dot{m}_2 = \dot{m} \tag{4.99}$$

Combining this result and energy-balance (4.98) gives

$$h_2 = \cancelto{\dot{m}_2}{\frac{\dot{m}_1}{\dot{m}_2}}(h+ke)_1 - ke_2 = (h+ke)_1 - ke_2 = h_1 + (ke_1 - ke_2) \tag{4.100}$$

Now since $ke_1 \gg ke_2$, (4.100) becomes

$$h_2 = h_1 + \cancelto{\approx ke_1}{(ke_1 - ke_2)} \approx h_1 + ke_1 = h_1 + \frac{Vel_1^2}{2} \tag{4.101}$$

From I.G.B. air Table B.11 $h_1 = h|_{T=10°C} = 283\frac{kJ}{kg}$ and (4.101) gives

$$h_2 \approx h_1 + \frac{Vel_1^2}{2} = 283\frac{kJ}{kg} + \frac{(200\frac{m}{s})^2}{2} = 283kJ + 2(10^4)\frac{m^2}{s^2}$$

$$= 283\frac{kJ}{kg} + 2(10^4)\frac{J}{kg}\frac{1kJ}{1000J} = 303\frac{kJ}{kg} \tag{4.102}$$

Interpolating from Table B.11 gives

$$T_2 \approx T|_{h=303kJ} = 303°K \xLeftarrow{T_2} \tag{4.103}$$

Finally, from the mass-balance (4.99) noting that I.G.B. is valid

$$\dot{m} = \dot{m}_1 = (A\rho Vel)_1 = A_1\frac{P_1}{RT_1}Vel_1 = 0.4m^2\frac{80kPa}{0.286\frac{kJ}{kg-°K}283°K}200\frac{m}{s}$$

$$= 79\frac{m^2 - kPa - kg - m}{kJ - s} = 79\frac{kN - m\ kg}{kJ\ \ s} = 79\frac{kg}{s} \xLeftarrow{\dot{m}} \tag{4.104}$$

$$\boxed{T_2 \approx 303°K, \dot{m} = 79\frac{kg}{s}}$$

□

EXAMPLE 4.11 Adiabatic steam turbine: Energy-balance

Problem Statement:
An adiabatic steam-turbine with an inlet mass-flow rate of $3kg$ of steam per second produces power at a rate of $2.35MW$. The steam at the inlet has a pressure and temperature of $10MPa$ and $500°C$. The outlet steam pressure is $35kPa$. Determine the outlet steam temperature.

Given:
Adiabatic turbine (H_2O),
$T_1 = 500°C$,
$P_1 = 10MPa$,
$P_2 = 35kPa$,
$\dot{m}_1 = 3\frac{kg}{s}$,
$\dot{W}_{shaft} = 2.35MW$

Find:
T_2

Assumptions:
Steady-state (S.S.), steady-flow,
$\dot{W} = \dot{W}_{shaft}$,
$\Delta pe_{1\to2} = \Delta ke_{1\to2} = 0$,
H_2O steam and saturation Tables B.4 and B.3

Schematic:

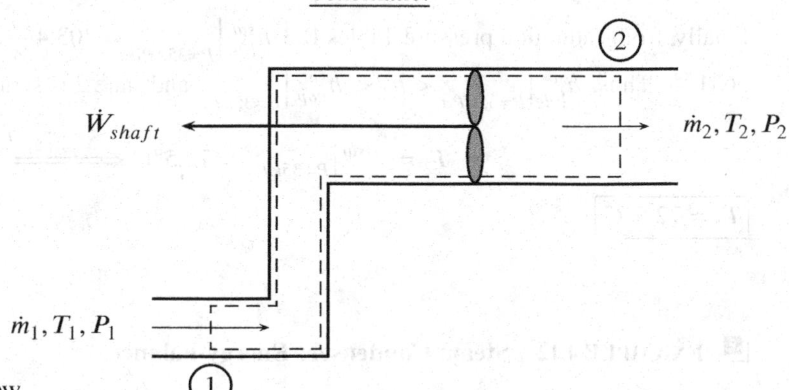

Analysis:
Applying the open-system energy-balance (4.85) for a steady-state, single inlet, single outlet device

$$\overset{0}{\cancel{\dot{Q}}} - \overset{\dot{W}_{shaft}}{\cancel{\dot{W}}} = \overset{0,S.S.}{\cancel{\frac{dE_{CV}}{dt}}} + \sum \overset{\dot{m}_2(h+ke+pe)_2}{\cancel{(\dot{m}(h+ke+pe))_{outlets}}} - \sum \overset{\dot{m}_1(h+ke+pe)_1}{\cancel{(\dot{m}(h+ke+pe))_{inlets}}} \tag{4.105}$$

or

$$-\dot{W}_{shaft} = \dot{m}_2(h+ke+pe)_2 - \dot{m}_1(h+ke+pe)_1 \tag{4.106}$$

where h_2 will allow for the determination of T_2. Applying the open system mass-balance (4.71)

$$\overset{0,S.S.}{\cancel{\frac{dm_{CV}}{dt}}} = \sum \overset{\dot{m}_1}{\cancel{\dot{m}_{inlets}}} - \sum \overset{\dot{m}_2}{\cancel{\dot{m}_{outlets}}} \Rightarrow \dot{m}_2 = \dot{m}_1 = \dot{m} \tag{4.107}$$

Combining the result of the energy (4.106) and mass-balance (4.107) gives

$$\begin{aligned} -\dot{W}_{shaft} &= \overset{\dot{m}}{\cancel{\dot{m}_2}}(h+ke+pe)_2 - \overset{\dot{m}}{\cancel{\dot{m}_1}}(h+ke+pe)_1 \\ &= \dot{m}[(h_2-h_1) + \overset{0}{\cancel{(ke_2-ke_1)}} + \overset{0}{\cancel{(pe_2-pe_1)}}] = \dot{m}(h_2-h_1) \end{aligned} \tag{4.108}$$

or solving for h_2 (again, which will allow for the calculation of T_2)

$$h_2 = h_1 - \frac{\dot{W}_{shaft}}{\dot{m}} \tag{4.109}$$

From super-heated vapor Tables B.4 $h_1 = h_{vapor}^{super-heated}\Big|_{\substack{T=500°C,\\ P=10MPa}} = 3373\frac{kJ}{kg}$. Thus from (4.109)

$$\begin{aligned} h_2 = h_1 - \frac{\dot{W}_{shaft}}{\dot{m}} &= 3373\frac{kJ}{kg} - \frac{2.35MW}{3\frac{kg}{s}} \\ = 3373\frac{kJ}{kg} - 0.783\frac{\cancel{MJ}}{kg}\frac{10^3kJ}{1\cancel{MJ}} &= 3373\frac{kJ}{kg} - 783\frac{kJ}{kg} = 2590\frac{kJ}{kg} \end{aligned} \tag{4.110}$$

Finally, from saturation pressure Tables B.3 $h_{liq.}^{sat.}\Big|_{P=35kPa} = 303.4\frac{kJ}{kg}$ and $h_{vap.}^{sat.}\Big|_{P=35kPa} = 2631\frac{kJ}{kg}$. Thus, $h_{liq.}^{sat.}\Big|_{P=35kPa} < h_2 < h_{vap.}^{sat.}\Big|_{P=35kPa}$ and state 2 is saturated mixture

$$T_2 = T^{sat}\big|_{P=35kPa} = 72.5°C \overset{T_2}{\Longleftarrow} \tag{4.111}$$

$\boxed{T_2 = 72.5°C}$

□

EXAMPLE 4.12 Steam Condenser: Energy-balance

Problem Statement:
An adiabatic condenser uses cooling water with temperature $18°C$ to cool and then condense steam at $240°C$ and $1bar$. The outflow of steam condensate from the condenser is a mixture of saturated liquid and vapor with quality 0.25. The cooling water outflow temperature is $27°C$ with a

mass-flow rate of $101\frac{kg}{s}$. If the cooling water is at a pressure of $1\,bar$, find the mass-flow rate for the steam.

Schematic:

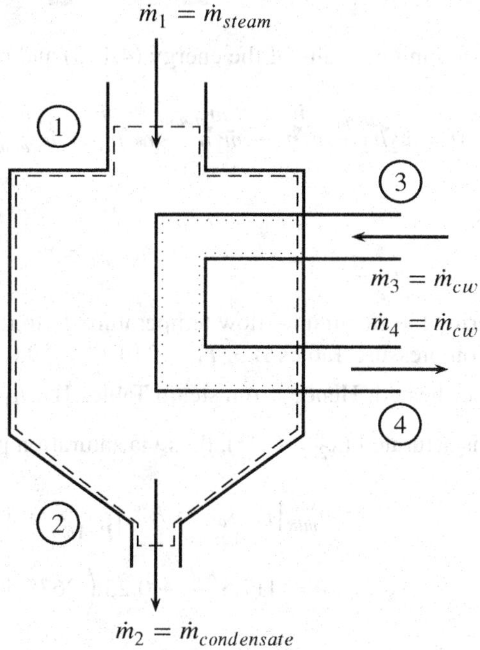

Given:
Adiabatic condenser (H_2O),
$T_1 = 240°C$,
$T_3 = 18°C$,
$T_4 = 27°C$,
$P_1 = 1\,bar$,
$P_{cw} = P_3 = P_4 = 1\,bar$,
$x_2 = 0.25$,
$\dot{m}_{cw} = 101\frac{kg}{s}$

Find:
$\dot{m}_{steam}$

Assumptions:
Steady-state (S.S.), steady-flow
$\dot{W} = 0$,
flow pe, ke negligible,
isobaric condensation and heating,
steam and saturation Tables B.4, B.3 and B.2

Analysis:
Applying the open system energy-balance (4.85) for CV – – – – (the overall condenser)

$$\overset{0,\,adia.}{\cancel{\dot{Q}}} - \overset{0}{\cancel{\dot{W}}} = \overset{0,\,S.S.}{\cancel{\frac{dE_{CV}}{dt}}} + \sum (\dot{m}(h + ke + pe))_{outlets} - \sum (\dot{m}(h + ke + pe))_{inlets} \tag{4.112}$$

or

$$0 = \sum (\dot{m}(h + ke + pe))_{outlets} - \sum (\dot{m}(h + ke + pe))_{inlets} \tag{4.113}$$

From the schematic the inlet locations are 1, 3 and outlets are 2 and 4. Thus, the open system energy balance (4.113) reduces to

$$0 = \sum (\dot{m}(h + ke + pe))_{outlets} - \sum (\dot{m}(h + ke + pe))_{inlets} =$$

$$\left(\dot{m}_2(h + \overset{0}{\cancel{ke}} + \overset{0}{\cancel{pe}})_2 + \dot{m}_4(h + \overset{0}{\cancel{ke}} + \overset{0}{\cancel{pe}})_4\right) - \left(\dot{m}_1(h + \overset{0}{\cancel{ke}} + \overset{0}{\cancel{pe}})_1 + \dot{m}_3(h + \overset{0}{\cancel{ke}} + \overset{0}{\cancel{pe}})_3\right)$$

$$= \dot{m}_2 h_2 + \dot{m}_4 h_4 - \dot{m}_1 h_1 - \dot{m}_3 h_3 \tag{4.114}$$

Hence

$$0 = \dot{m}_2 h_2 + \dot{m}_4 h_4 - \dot{m}_1 h_1 - \dot{m}_3 h_3 \tag{4.115}$$

Applying the open system mass-balance (4.71) for steady-state to the ...... CV (single inlet/single outlet cooling-water pipe)

$$\overset{0,\,S.S.}{\cancel{\frac{dm_{CV}}{dt}}} = \sum \overset{\dot{m}_3}{\cancel{\dot{m}_{inlets}}} - \sum \overset{\dot{m}_4}{\cancel{\dot{m}_{outlets}}} \Rightarrow \dot{m}_4 = \dot{m}_3 = \dot{m}_{cw} \tag{4.116}$$

Applying the open system mass-balance (4.71) for a steady-state, to the – – –– less ...... CV (single inlet/single outlet volume (condenser less coolant pipe))

$$\frac{dm_{CV}}{dt}\overset{0,\,S.S.}{=} \sum \dot{m}_{inlets} - \sum \dot{m}_{outlets} \Rightarrow \dot{m}_2 = \dot{m}_1 = \dot{m}_{steam} \tag{4.117}$$

Thus, combining results of the energy (4.115) and mass-balances, (4.116) and (4.117) gives

$$0 = \dot{m}_2 h_2 + \dot{m}_4 h_4 - \dot{m}_1 h_1 - \dot{m}_3 h_3 = \dot{m}_{steam} h_2 + \dot{m}_{cw} h_4 - \dot{m}_{steam} h_1 - \dot{m}_{cw} h_3 \tag{4.118}$$

or

$$\dot{m}_{steam} = \dot{m}_{cw}\left[\frac{h_4 - h_3}{h_1 - h_2}\right] \tag{4.119}$$

Comparing the steam inlet flow temperature T_1 to the saturation temperature at P_1 we find, via saturation pressure Tables B.3, $T_1 = 240°C > 99.63°C = T^{sat}\big|_{1bar}$ or $T_1 > T^{sat}\big|_{1bar}$. Thus, state 1 is super-heated. Hence, from steam Tables B.4 $h_1 = h_{vapor}^{super-heated}\Big|_{T=240°C,\; P=1bar} = 2954.5\frac{kJ}{kg}$. Now, state 2 is saturated ($x_2 = 0.25$), thus via saturation pressure Tables B.3

$$\begin{aligned} h_2 &= h_{mix}^{sat}\Big|_{P=1bar,\; x=0.25} = h_{liq}^{sat}\Big|_{P=1bar} + x\left(h_{vap}^{sat}\Big|_{P=1bar} - h_{liq}^{sat}\Big|_{P=1bar}\right) \\ &= 417.5\frac{kJ}{kg} + 0.25\left(2675.5\frac{kJ}{kg} - 417.5\frac{kJ}{kg}\right) = 982\frac{kJ}{kg} \end{aligned} \tag{4.120}$$

Also $T_3, T_4 < T^{sat}\big|_{P=1bar} = 99.63°$ hence, states 3 and 4 are super-cooled liquids. From the approximation for enthalpy of a super-cooled liquid (2.22) and saturation temperature Tables B.2 we have

$$h_3 = h_{liq}^{sup.-cooled}\Big|_{18°C,\; 1bar} \approx h_{liq}^{sat}\Big|_{T^{sat}=18°C} + v_{liq}^{sat}\Big|_{T^{sat}=18°C}\left(1bar - P^{sat}\big|_{T=18°C}\right)$$

$$= 75.58\frac{kJ}{kg} + 1.001(10^{-3})\frac{m^3}{kg}(1bar - 0.02bar)\frac{100kPa}{1bar} = 75.58\frac{kJ}{kg} + \overset{\approx 0}{0.1\frac{kJ}{kg}} \approx 75.58\frac{kJ}{kg} \tag{4.121}$$

or $h_3 = h_{liq}^{sup.-cooled}\Big|_{18°C,\; 1bar} \approx h_{liq}^{sat}\Big|_{T^{sat}=18°C} = 75.58\frac{kJ}{kg}$. Similarly $h_4 = h_{liq}^{sup.-cooled}\Big|_{27°C,\; 1bar} \approx h_{liq}^{sat}\Big|_{T^{sat}=27°C} = 113.25\frac{kJ}{kg}$. Returning to (4.119) gives

$$\dot{m}_{steam} = \dot{m}_{cw}\left[\frac{h_4 - h_3}{h_1 - h_2}\right] = 101\frac{kg}{s}\left[\frac{113.25\frac{kJ}{kg} - 75.58\frac{kJ}{kg}}{2954.5\frac{kJ}{kg} - 982\frac{kJ}{kg}}\right] = 1.93\frac{kg}{s} \overset{\dot{m}_{steam}}{\Longleftarrow} \tag{4.122}$$

$$\boxed{\dot{m}_{steam} = 1.93\frac{kg}{s}}$$

□

EXAMPLE 4.13 Steady-state adiabatic mixer: Energy-balance

Problem Statement:
An adiabatic mixing chamber is composed of two inlet elbows and a single outlet. Steam is fed into the mixer at a pressure of $300kPa$ and $300°C$ and mixed isobarically with liquid water at a temperature of $20°C$. The resulting mixture of pressurized hot water exits the mixer as a liquid at a temperature and mass-flow rate of $60°C$ and $1.8kg/s$. Determine the rate of steam and cold water injection.

Given:
Adiabatic isobaric mixer (H_2O),
$T_1 = 20°C$,
$T_2 = 300°C$,
$P_2 = 300kPa$,
$T_3 = 60°C$,
$\dot{m}_{hw} = 1.8\frac{kg}{s}$ (hot water)

Schematic:

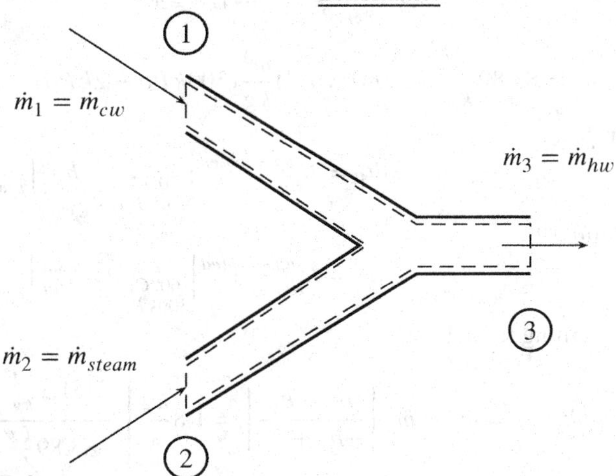

Find:
$\dot{m}_{steam}$, $\dot{m}_{cw}$ (cold water)

Assumptions:
Steady-state (S.S.), steady-flow,
$\dot{W} = 0$,
neglect pe and ke of flow,
steam and saturation Tables B.4, B.3 and B.2

Analysis:
Applying the open system energy-balance (4.85) for CV − − −−

$$\cancelto{0}{\dot{Q}} - \cancelto{0}{\dot{W}} = \cancelto{0, S.S.}{\frac{dE_{CV}}{dt}} + \sum(\dot{m}(h + ke + pe))_{outlets} - \sum(\dot{m}(h + ke + pe))_{inlets} \tag{4.123}$$

or

$$0 = \sum(\dot{m}(h + ke + pe))_{outlets} - \sum(\dot{m}(h + ke + pe))_{inlets} \tag{4.124}$$

From the schematic the inlets are at locations 1, 2 while the outlet is at 3. Thus

$$0 = \sum(\dot{m}(h + \cancelto{0}{ke + pe}))_{outlets} - \sum(\dot{m}(h + \cancelto{0}{ke + pe}))_{inlets} = \dot{m}_3h_3 - \dot{m}_1h_1 - \dot{m}_2h_2 \tag{4.125}$$

yielding

$$0 = \dot{m}_3h_3 - \dot{m}_1h_1 - \dot{m}_2h_2 \tag{4.126}$$

Applying the open system mass-balance (4.71) for steady-state/steady-flow, to CV − − −− and solving for the cold water mass-flow rate $\dot{m}_{cw} = \dot{m}_1$

$$\cancelto{0, S.S.}{\frac{dm_{CV}}{dt}} = \sum\cancelto{\dot{m}_2 + \dot{m}_1}{\dot{m}_{inlets}} - \sum\cancelto{\dot{m}_3}{\dot{m}_{outlets}} \Rightarrow \dot{m}_1 = \dot{m}_3 - \dot{m}_2 \tag{4.127}$$

returning to the the energy-balance (4.126)

$$0 = \dot{m}_3h_3 - \cancelto{\dot{m}_3 - \dot{m}_2}{\dot{m}_1}h_1 - \dot{m}_2h_2 = \dot{m}_3(h_3 - h_1) - \dot{m}_2(-h_1 + h_2) \tag{4.128}$$

or

$$\dot{m}_{steam} = \dot{m}_2 = \dot{m}_3 \left[\frac{h_3 - h_1}{-h_1 + h_2} \right] \tag{4.129}$$

From saturation pressure Tables B.3 $T^{sat}\big|_{P=300kPa} = 133° < T_2$ hence state 2 is super-heated vapor and $h_2 = h_{vapor}^{super-heated}\Big|_{T=300°C,\ P=300kPa} = 3069.5\frac{kJ}{kg}$ via steam Tables B.4. Now $T_1, T_3 < T^{sat}\big|_{P=300kPa} = 133°C$, thus state 1 and 3 are super-cooled liquids. From the approximation for enthalpy of a super-cooled liquid (2.22) and saturation temperature Table B.2

$$h_1 = h_{liq}^{sup.-cooled}\Big|_{20°C,\ 300kPa} \approx h_{liq}^{sat}\Big|_{T^{sat}=20°C} + v_{liq}^{sat}\Big|_{T^{sat}=20°C}(300kPa - P^{sat}|_{T=20°C})$$

$$= 83.89\frac{kJ}{kg} + 1.001(10^{-3})\frac{m^3}{kg}(300kPa - 2kPa) = 83.89\frac{kJ}{kg} + \underbrace{0.2\frac{kJ}{kg}}_{\approx 0} \approx 83.89\frac{kJ}{kg} \tag{4.130}$$

or

$$h_1 = h_{liq}^{sup.-cooled}\Big|_{20°C,\ 300kPa} \approx h_{liq}^{sat}\Big|_{T^{sat}=20°C} = 83.89\frac{kJ}{kg} \tag{4.131}$$

Similarly,

$$h_3 = h_{liq.}^{sup.-cooled}\Big|_{60°C,\ 300kPa} \approx h_{liq.}^{sat}\Big|_{T^{sat}=60°C} = 251.1\frac{kJ}{kg} \tag{4.132}$$

Returning to (4.129)

$$\dot{m}_{steam} = \dot{m}_2 = \dot{m}_3 \left[\frac{h_3 - h_1}{-h_1 + h_2} \right] = 1.8\frac{kg}{s} \left[\frac{251.1\frac{kJ}{kg} - 83.89\frac{kJ}{kg}}{-83.89\frac{kJ}{kg} + 3069.5\frac{kJ}{kg}} \right] = 0.10\frac{kg}{s} \xLeftarrow{\dot{m}_{steam}} \tag{4.133}$$

Finally, from the mass-balance (4.127)

$$\dot{m}_{cw} = \dot{m}_1 = \dot{m}_3 - \dot{m}_2 = \dot{m}_{hw} - \dot{m}_{steam} = 1.80\frac{kg}{s} - 0.10\frac{kg}{s} = 1.70\frac{kg}{s} \xLeftarrow{\dot{m}_{cw}} \tag{4.134}$$

$$\boxed{\dot{m}_{steam} = 0.10\frac{kg}{s},\ \dot{m}_{cw} = 1.70\frac{kg}{s}}$$

□

4.6 CYCLIC DEVICES: EFFICIENCY AND C.O.P.

Definition 4.8 ***Cyclic Device:*** *A mechanism or device which executes a cycle* $1 \to 2 \to \ldots \to 1$ *is referred to as cyclic device.*

Note, over a cycle, the change in all properties are equal to zero (i.e. $\Delta E_{cycle} = \Delta U_{cycle} = \Delta P_{cycle} = \Delta T_{cycle} = 0$, etc.). As in the case of open system devices, closed system *cyclic* devices can be categorized according to function: Work producing cyclic devices produce work over a cycle are referred to as "heat engines". Work consuming devices consume work over a cycle and are referred to as "heat pumps" or "refrigerators". In the following sections we shall be investigating such devices, but before we do, we must define the concept of efficiency.

Definition 4.9 ***Efficiency:*** *Efficiency refers to a ratio of benefit-to-cost: In everyday terms "what you get" over "what you put in". Mathematically*

$$\boxed{Efficiency = \frac{Benefit}{Cost}} \tag{4.135}$$

4.6.1 WORK PRODUCING DEVICES: HEAT ENGINES

Figure 4.11 a) illustrates the operation of a work producing cyclic device receiving heat Q_H from a high temperature heat source at a temperature T_H. Heat Q_L is then rejected to a low temperature heat sink at a temperature T_L yielding cycle work W_{cycle}. With respect to efficiency, the benefit is the work produce over the cycle W_{cycle}, while the cost is the heat input Q_H from the high temperature source required to run the device. Thus, applying (4.135) we have the efficiency of a heat engine

$$\eta = \frac{W_{cycle}}{Q_H}, \text{ (Heat-engine)} \tag{4.136}$$

Note, (4.136) can be re-written with the help of Figure 4.11 a) in conjunction with the energy-balance applied to the heat engine. Specifically, noting that the heat-engine (HE), a closed system, is subject to two heat inter-actions Q_L and Q_H, then $Q_{cycle} = Q_H - Q_L$. Application of the energy-balance over the cycle gives

$$\overset{Q_H - Q_L}{\cancel{Q}_{cycle}} - W_{cycle} = \overset{0,\,cycle}{\Delta \cancel{E}_{cycle}} \Rightarrow W_{cycle} = Q_H - Q_L \tag{4.137}$$

substitution of this result into (4.136) yields

$$\eta = \frac{W_{cycle}}{Q_H} = \frac{Q_H - Q_L}{Q_H} = 1 - \frac{Q_L}{Q_H} \tag{4.138}$$

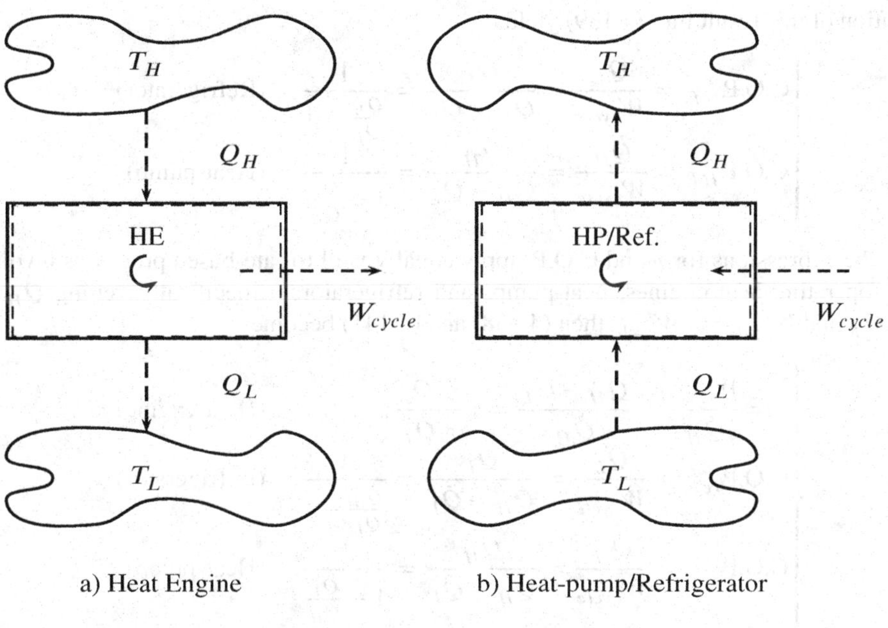

Figure 4.11 Illustration of cyclic devices: Heat engine a) and heat-pump/refrigerator b). Device control volume indicated by – – ––.

4.6.2 WORK CONSUMING DEVICES: HEAT PUMPS/REFRIGERATORS

Figure 4.11 b) illustrates the operation of a work consuming cyclic device receiving heat Q_L from a low temperature heat source at a temperature T_L while rejecting heat Q_H to a high temperature heat sink at a temperature T_H. Needless to say, such a heat transfer process runs counter to what spontaneously occurs in nature (where heat naturally flows from hot to cold). Therefore, such a device requires a work input W_{cycle} over the cycle.

With respect to efficiency, the benefit of a work consuming device is either, in the case for a refrigerator, the heat transfer from the low temperature source Q_L or, in the case of a heat pump, the heat transferred to the high temperature sink Q_H. In both cases, the cost is the work input W_{cycle}. Thus, applying (4.135), and renaming efficiency the "Coefficient of Performance" (C.O.P.), we have

$$\boxed{\begin{cases} \text{C.O.P.}_{ref.} = \dfrac{Q_L}{W_{cycle}}, \text{ (Refrigerator)} \\ \text{C.O.P.}_{HP} = \dfrac{Q_H}{W_{cycle}}, \text{ (Heat-pump)} \end{cases}} \tag{4.139}$$

As in the case of the heat engine, (4.139) can be re-written with the help of Figure 4.11 b) in conjunction with the energy-balance applied to the refrigerator/heat-pump. Again, noting that the closed system, (refrigerator or heat-pump) is subject to two heat inter-actions Q_L and Q_H, then $Q_{cycle} = -Q_H + Q_L$ and application of the energy-balance gives (note the cycle work in Figure 4.11 b) represents a work input)

$$\overset{-Q_H + Q_L}{\cancel{Q}_{cycle}} - (-W_{cycle}) = \Delta E_{cycle} \overset{0,\ cycle}{\Longrightarrow} W_{cycle} = Q_H - Q_L \tag{4.140}$$

substitution of this result into (4.139) yields

$$\begin{cases} \text{C.O.P.}_{ref.} = \dfrac{Q_L}{W_{cycle}} = \dfrac{Q_L}{Q_H - Q_L} = \dfrac{1}{\frac{Q_H}{Q_L} - 1}, & \text{(Refrigerator)} \\ \text{C.O.P.}_{HP} = \dfrac{Q_H}{W_{cycle}} = \dfrac{Q_H}{Q_H - Q_L} = \dfrac{1}{1 - \frac{Q_L}{Q_H}}, & \text{(Heat-pump)} \end{cases} \tag{4.141}$$

Finally, the expressions for η and C.O.P. apply equally well to rate-based processes involving continually operating heat engines, heat-pumps and refrigerators. Specifically, letting $Q_H \Rightarrow \dot{Q}_H$, $Q_L \Rightarrow \dot{Q}_L$ and $W_{cycle} \Rightarrow \dot{W}_{cycle}$ then (4.138) and (4.141) become

$$\begin{cases} \eta = \dfrac{\dot{W}_{cycle}}{\dot{Q}_H} = \dfrac{\dot{Q}_H - \dot{Q}_L}{\dot{Q}_H} = 1 - \dfrac{\dot{Q}_L}{\dot{Q}_H}, & \text{(Heat-engine)} \\ \text{C.O.P.}_{ref.} = \dfrac{\dot{Q}_L}{\dot{W}_{cycle}} = \dfrac{\dot{Q}_L}{\dot{Q}_H - \dot{Q}_L} = \dfrac{1}{\frac{\dot{Q}_H}{\dot{Q}_L} - 1}, & \text{(Refrigerator)} \\ \text{C.O.P.}_{HP} = \dfrac{\dot{Q}_H}{\dot{W}_{cycle}} = \dfrac{\dot{Q}_H}{\dot{Q}_H - \dot{Q}_L} = \dfrac{1}{1 - \frac{\dot{Q}_L}{\dot{Q}_H}}, & \text{(Heat-pump)} \end{cases} \tag{4.142}$$

EXAMPLE 4.14 Refrigeration Cycle: Calculation of *C.O.P.*

Problem Statement:

A refrigerator extracts heat from a low temperature source while rejecting $20kJ$ of heat to a high temperature sink. The coefficient of performance for the device is 2.5. Determine cycle work and the heat extracted from the low temperature source.

Schematic:

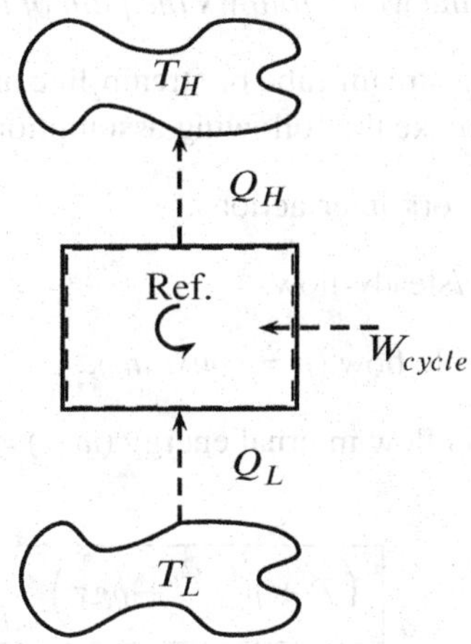

Given:
Refrigerator,
cyclic operation,
$Q_H = 20kJ$,
$C.O.P._{ref} = 2.5$

Find:
W_{cycle},
Q_L

Assumptions:
None

Analysis:
For a Refrigerator we have via (4.139)

$$\text{C.O.P.}_{ref.} = \frac{Q_L}{W_{cycle}} \tag{4.143}$$

Noting from the schematic that Q_L is into the system (positive), Q_H is out of the system (and contributes a negative heat transfer) while W_{cycle} is on the system (and contributes to a negative work transfer by the system), application of the energy-balance to the device over the cycle gives

$$\overset{Q_L - Q_H}{\not{Q}} - \overset{-W_{cycle}}{\not{W}} = \overset{0,\,cycle}{\not{\Delta E}} = 0 \Rightarrow W_{cycle} = Q_H - Q_L \tag{4.144}$$

Returning to (4.143) and combining with the energy-balance (4.144)

$$\begin{aligned} \text{C.O.P.}_{ref.} &= \frac{Q_L}{W_{cycle}} = \frac{Q_L}{Q_H - Q_L} \Rightarrow (Q_H - Q_L)\text{C.O.P.}_{ref.} = Q_L \\ \Rightarrow Q_L &= \frac{Q_H C.O.P._{ref.}}{1 + C.O.P._{ref.}} = \frac{20kJ(2.5)}{2.5+1} = 14.28kJ \xLeftarrow{Q_L} \end{aligned} \tag{4.145}$$

Returning to our result via the energy-balance (4.144)

$$W_{cycle} = Q_H - Q_L = 20kJ - 14.28kJ = 5.72kJ \text{ (work on the device)} \xLeftarrow{W_{cycle}} \tag{4.146}$$

$\boxed{Q_L = 14.28kJ, W_{cycle} = 5.72kJ \text{ (on Refrigerator)}}$

□

4.7 THE BERNOULLI EQUATION

Situations often arise whereby fluid flows from one location to another along some path we will referred to herein as a "stream-line" (an infinitely thin tube encapsulating a stream of fluid).

Definition 4.10 ***Stream-line:*** *A region consisting of an imaginary tube through which fluid flows from one location to another is referred to as a stream-tube. If the tube becomes infinitely thin, the tube becomes a line which follows the path of the fluid flow and is referred to as a* stream-line.

Needless to say, a stream-tube or stream-line are single inlet/single outlet open-system control volumes. Next, we make the following assumptions:

1. No heat or work inter-actions.
2. Steady-state/steady-flow.
3. Incompressible flow ($\rho = constant$).
4. No change in flow internal energy ($u(...) = constant$).

giving via (4.85)

$$\left(P + \rho\frac{Vel^2}{2} + \rho g z\right)_{outlet} = \left(P + \rho\frac{Vel^2}{2} + \rho g z\right)_{inlet} \tag{4.147}$$

Equation (4.147) is referred to as the Bernoulli Equation which relates pressures, velocities and elevations (z) under restricted conditions along a stream-line. Thus, provided we can draw line or tube along which the flow travels (the end-points being the inlet and outlet), where the region satisfies items #1-4, the Bernoulli equation (4.147) is valid. Figure 4.12 shows a gravity assisted pump driven flow through a water heater: Shaded areas indicate regions of applicability of the Bernoulli equation. The remaining regions do not satisfy (for example) assumption #1, hence, the Bernoulli equation is not applicable in these regions.

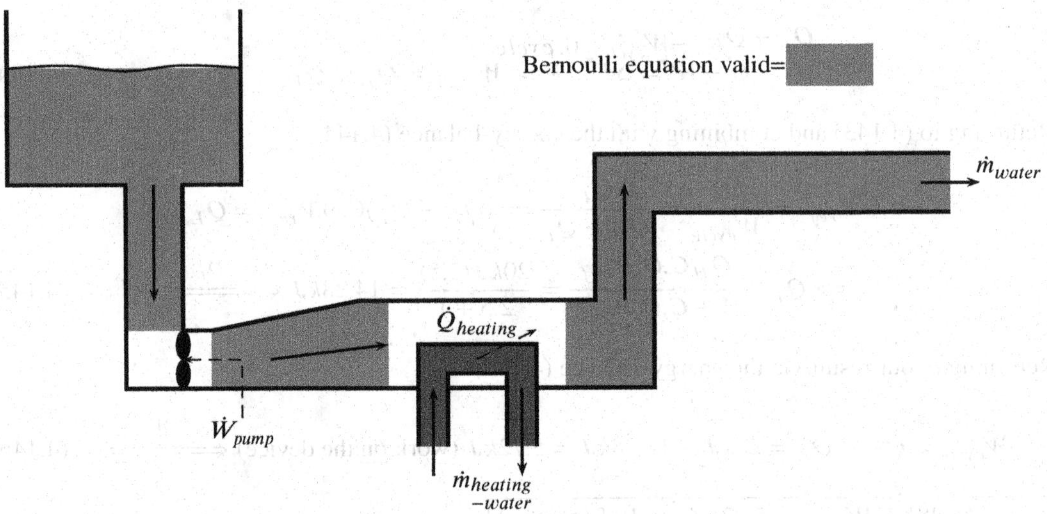

Figure 4.12 Gravity and pump-drive water heater with heat and work inter-actions: Shaded regions indicate validity of Bernoulli equation.

4.7.1 PRESSURES: TOTAL, LOCAL/STATIC & DYNAMIC

Note (4.147) implies that terms $\rho\frac{Vel^2}{2}$, $\rho g z$ and hence the sum $\left(P + \rho\frac{Vel^2}{2} + \rho g z\right)$ have unit of pressure. Hence we define the following:

Definition 4.11 ***Dynamic Pressure:*** *The pressure contribution due to the kinetic energy in the flow* $P_{dynamic} = \rho \frac{Vel^2}{2}$ *is termed the "Dynamic Pressure"* $P_{dynamic}$.

Definition 4.12 ***Total Pressure:*** *The total pressure contribution due the sum of the local or static pressure (the pressure) P, dynamic pressure* $P_{dynamic}$ *and the gravitational potential energy* $\rho g z$ *is termed the "Total Pressure"* P_{Total}. *Provided the Bernoulli equation is valid, total pressure* P_{Total} *is constant along a stream-line.*

The fact that total pressure is constant (provided the Bernoulli equation is valid) is gleaned from the fact that for a fixed stream-tube inlet location, we are free to choose arbitrary locations for the stream-line outlet: The left-hand side of (4.147) is of fixed value, thus the total pressure P_{total} is fixed and hence *any* outlet location along a stream-line has the same total pressure.

EXAMPLE 4.15 Application of the Bernoulli equation: Water jet

Problem Statement:

An isothermal, steady-flow column of water exits a pipe with diameter $0.01m$ as a circular jet. The jet flows vertically upward $0.2m$ and strikes a circular flat disk with diameter $0.1m$ and splays out radially. The thickness of the flow as it exits the disk edge is $0.4mm$. At the very center of the disk is a small vertical tube: The column of water in the tube is stationary with height H. If the pressure at the horizontal flow outlet (at the disk edge), top of the water column and pipe outlet are atmospheric, determine a) the water velocity at the pipe outlet and b) the column height H

Given:

Isothermal and steady-flow (H_2O),
$D_{pipe} = 0.01m$,
$H_{disk-pipe} = 0.2m$,
$D_{disk} = 0.1m$,
$t_{thickness} = 0.4mm$,
$P_1 = P_2 = P_3 = 1atm$

Find:

a) Vel_1
b) H

Assumptions:

$\dot{W} = \dot{Q} = 0$,
$g = 9.81\frac{m}{s^2}$,
Steady-state (S.S.)

Schematic:

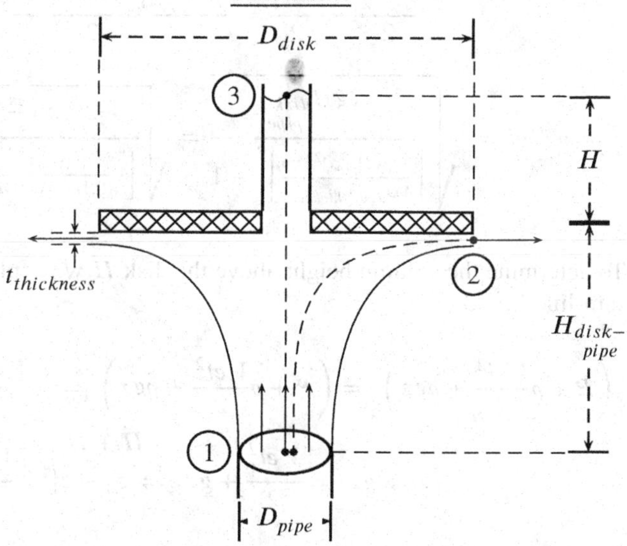

Analysis:

Note, the Bernoulli equation (4.147) is applicable along any stream-line within the flow (see §4.7).

a) Applying the Bernoulli equation (4.147) along the stream-line $1 \rightarrow 2$ we have

$$\left(\cancel{P} + \rho\frac{Vel^2}{2} + \rho g z\right)_2 = \left(\cancel{P} + \rho\frac{Vel^2}{2} + \rho g z\right)_1 \Rightarrow \cancel{\rho}\frac{Vel_2^2}{2} + \cancel{\rho} g z_2 = \cancel{\rho}\frac{Vel_1^2}{2} + \cancel{\rho} g z_1 \Rightarrow$$

$$\frac{Vel_2^2}{2} + g z_2 = \frac{Vel_1^2}{2} + g z_1 \Rightarrow \frac{Vel_2^2}{2} - \frac{Vel_1^2}{2} = g z_1 - g z_2 = g(z_1 - z_2) = -g H_{disk-pipe} \quad (4.148)$$

or

$$\frac{Vel_2^2}{2} - \frac{Vel_1^2}{2} = -gH_{disk-pipe} \tag{4.149}$$

Now, the disk edge velocity Vel_2 is unknown, hence we need another equation. Applying the open-system mass-balance (4.71) to the jet gives

$$\frac{dm_{CV}}{dt}^{0,\,S.S.} = \sum \dot{m}_{inlets} + \sum \dot{m}_{outlets} = 0 \Rightarrow \overbrace{\dot{m}_{pipe}}^{Vel_1\pi(D_{pipe}/2)^2} = \overbrace{\dot{m}_{disk}}^{Vel_2\pi D_{disk}t_{thickness}}$$

$$\Rightarrow Vel_1\pi(D_{pipe}/2)^2 = Vel_2\pi D_{disk}t_{thickness}$$

$$\Rightarrow Vel_2 = Vel_1\left[\frac{(D_{pipe}/2)^2}{D_{disk}t_{thickness}}\right] = Vel_1\left[\frac{(D_{pipe})^2}{4D_{disk}t_{thickness}}\right] \tag{4.150}$$

or

$$Vel_2 = Vel_1\left[\frac{(D_{pipe})^2}{4D_{disk}t_{thickness}}\right] \tag{4.151}$$

Substitution of (4.151) into (4.154) gives

$$\frac{Vel_2^2}{2} - \frac{Vel_1^2}{2} = \frac{\left(Vel_1\left[\frac{(D_{pipe})^2}{4D_{disk}t_{thickness}}\right]\right)^2}{2} - \frac{Vel_1^2}{2} = -gH_{disk-pipe}$$

$$\Rightarrow Vel_1 = \sqrt{\frac{-2gH_{disk-pipe}}{\left[\frac{(D_{pipe})^2}{4D_{disk}t_{thickness}}\right]^2 - 1}} = \sqrt{\frac{-2(9.81\frac{m}{s^2})0.2m}{\left[\frac{(0.01m)^2}{4(0.1m)0.0004m}\right]^2 - 1}} = 2.54\frac{m}{s} \overset{Vel_1}{\Longleftarrow} \tag{4.152}$$

b) To determine the column height above the disk H we apply the Bernoulli equation along the stream-line $1 \to 3$

$$\left(P + \rho\frac{Vel^2}{2} + \rho gz\right)_3 = \left(P + \rho\frac{Vel^2}{2} + \rho gz\right)_1 \Rightarrow \rho\frac{Vel_3^2}{2}^{0} + \rho gz_3 = \rho\frac{Vel_1^2}{2} + \rho gz_1$$

$$\Rightarrow gz_3 = \frac{Vel_1^2}{2} + gz_1 \Rightarrow \overbrace{z_3 - z_1}^{H + H_{disk-pipe}} = \frac{Vel_1^2}{g2} \Rightarrow H = \frac{Vel_1^2}{g2} - H_{disk-pipe} \tag{4.153}$$

or

$$H = \frac{Vel_1^2}{g2} - H_{disk-pipe} = \frac{\left(2.54\frac{m}{s}\right)^2}{\left(9.81\frac{m}{s^2}\right)2} - 0.2m = 0.129m \overset{H}{\Longleftarrow} \tag{4.154}$$

$Vel_1 = 2.54\frac{m}{s},\ H = 0.129m$

□

CHAPTER 5

THE SECOND-LAW OF THERMODYNAMICS

We begin this chapter by recalling the non-equivalence of work and heat as discussed in §1.1: Indeed, if heat and work were (equivalent), heat engine efficiencies of 100% and heat-pump/refrigerator C.O.P's of ∞ would be *theoretically* achievable. No such devices have ever been built!

5.1 THEORETICAL MOTIVATION

While the inability to construct perfect heat-pumps or refrigerators may be of practical importance to you as engineers in training, we will instead begin our investigation of the non-equivalence of work and heat by pointing out one severe theoretical short-coming in the First-Law or energy-balance: There is no indication of the direction of time! As far as the energy-balance is concerned, all the examples and problems solved thus far could easily have been reversed: The initial and final states could have been reversed (the signs of heat and work would have been reversed along with the change in all properties). The energy-balance makes no distinction between time running forward or in reverse! To the energy equation, all processes are reversible. This is in contrast to the natural world where most processes are uni-direction and irreversible (i.e. they proceed spontaneously in one direction). Examples of processes which are uni-direction and irreversible are:

- Processes involving *heat transfer between systems with a finite temperature differences.*
- Processes which *include friction.*
- Processes which *include un-restrained expansion.*

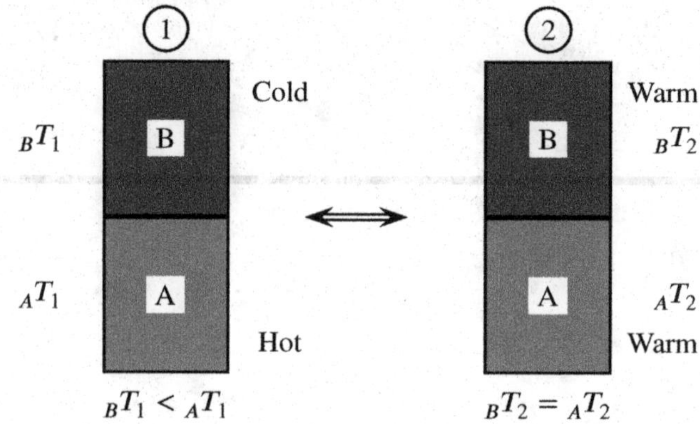

a) Heat transfer across a finite temperature difference: Process 1 → 2 spontaneous, 2 → 1 not spontaneous.

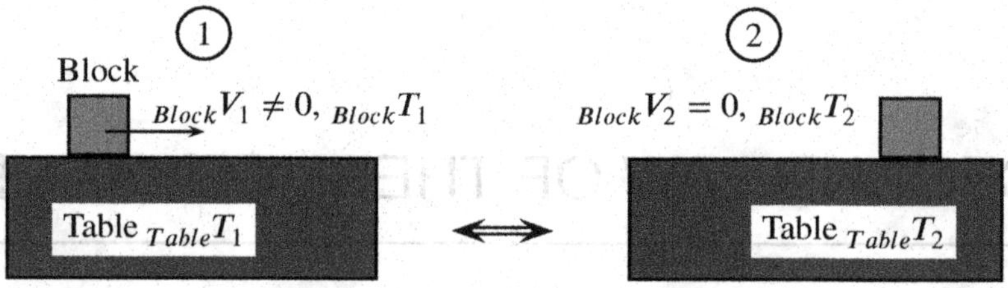

b) Frictional processes: Process 1 → 2 spontaneous ($T_2 > T_1$), 2 → 1 not spontaneous.

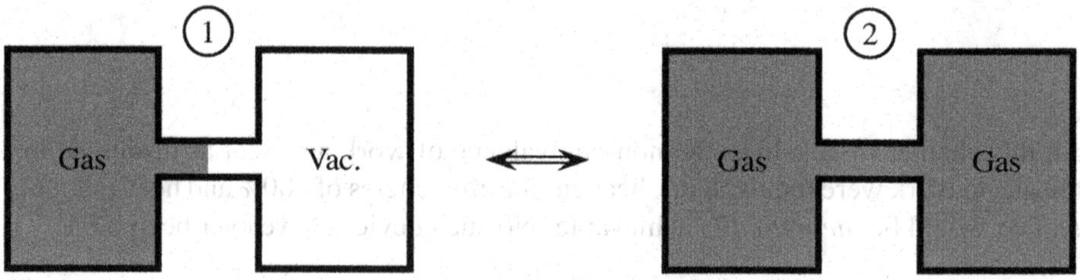

c) Unrestrained expansion: Process 1 → 2 spontaneous, process 2 → 1 not spontaneous.

Figure 5.1 Illustration of spontaneous or natural processes.

Figure 5.1 illustrates these three processes, and their reverse, which never spontaneously occur in nature. Amazingly, the previous restrictions on natural or spontaneous processes can be expressed as a single law illustrated by one of two pictorial representations: The Second-Law of Thermodynamics. However, prior to stating this law we must define the various terms used in the "language" associated with the Second-Law.

5.2 THE SECOND-LAW OF THERMODYNAMICS

5.2.1 TERMINOLOGY

Referring to the notion of reversible (non-spontaneous) vs. irreversible (spontaneous) process, we will officially define them here:

Definition 5.1 ***Irreversible/Spontaneous Processes:*** *Processes which, in nature, move an isolated system from an initial state to a final state, but not in the reverse.*

For students who want a more concrete definition: *An isolated system moving from an initial state* 1 *to a final state* 2 *is said to undergo a spontaneous process* $1 \rightarrow 2$, *if the reversed process* $2 \rightarrow 1$ *does not naturally/spontaneously occur.* One the other hand, for the student with a more "philosophical " disposition: *The arrow of time.* Thus, the progression of time in the natural world is a continual succession of irreversible processes such as those illustrated in Figure 5.1: *These processes can only be reversed via application of work by an external entity.* Conversely, *reversible processes only exist as theoretical concepts which can be approached, but not achieved in nature.*

Definition 5.2 ***Reversible Processes:*** *An isolated system moving from an initial state* 1 *to a final state* 2, *is said to undergo a reversible process* $1 \rightarrow 2$, *if the process excludes irreversibilities.*

Strictly speaking, reversible processes do not exist, but the concept of a reversible process (and indeed, a reversible device) are often of theoretical interest as you shall later see.

Now, before I move one, let's introduce a subtle distinction which will help to bypass much potential confusion later: The notion of internal vs. external reversibility defined as follows:

Definition 5.3 ***Internally reversible processes:*** *No irreversibilities occur within the system, over the process.*

Definition 5.4 ***Externally reversible processes:*** *No irreversibilities occur outside of the system, over the process.*

In most cases, the internal vs. external reversibility of a process will be clearly implied.

Definition 5.5 ***Thermal Reservoir:*** *Heat source/sink of essentially constant temperature.*

Or the slightly longer version: *A body/system of large enough extent such that any finite heat inter-action (i.e. addition or rejection of heat) results in a negligible temperature change.* Examples of thermal reservoirs include systems of very large mass such as the ocean (or some other large body of water), the atmosphere, etc. Needless to say, the notion of what is, or is not a thermal reservoir is dependent on the size of the heat inter-action: For small heat inter-actions, systems of relative modest size can be characterized as thermal reservoirs. In addition, the thermal properties of the reservoir material such as heat capacity C affect to what extent, the system behaves as a reservoir: High specific heats imply small changes in temperature per-unit specific heat transfer and visa verse.

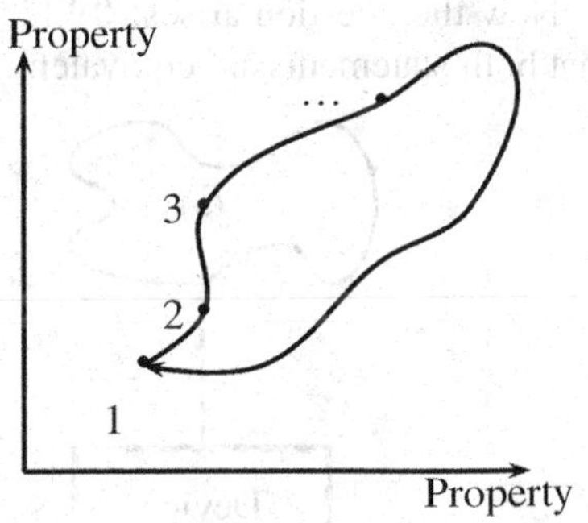

Figure 5.2 Illustration of cyclic process $1 \rightarrow 2 \rightarrow \ldots \rightarrow 1$.

Definition 5.6 ***Cyclic Process:*** *A system begins at an initial state* 1, *proceeds through a number of intermediate states* $2 \rightarrow 3 \rightarrow \ldots$ *and returns to the initial state* 1 *is said to undergo a cycle.*

The path illustrated in Figure 5.2 indicates a cyclic process. Note, that for any property L, $\Delta L_{cycle} = 0$.

Having extensive experience with systems both open and closed, we introduce two additional terms:

Definition 5.7 ***The Universe:*** *Everything everywhere.*

Definition 5.8 ***The Environment:*** *The universe less the system.*

An illustration of the universe relative to the system and environment is given in Figure 5.3. Note, sometimes we will shift focus from a system to the environment and thus may refer to the "environmental system" by which we mean a region outside of a some well defined system.

5.2.2 CLAUSIUS STATEMENT

The pictorial statement that processes involving *heat transfer between systems with finite temperature differences* are uni-directional or irreversible in nature is shown in Figure 5.4 a) and referred to as the **Clausius** Statement of the Second-Law of Thermodynamics. Thus, to move energy in the form of heat from a low to a high temperature region (thermal reservoir) requires work input! Heat will not spontaneously flow from cold to hot!

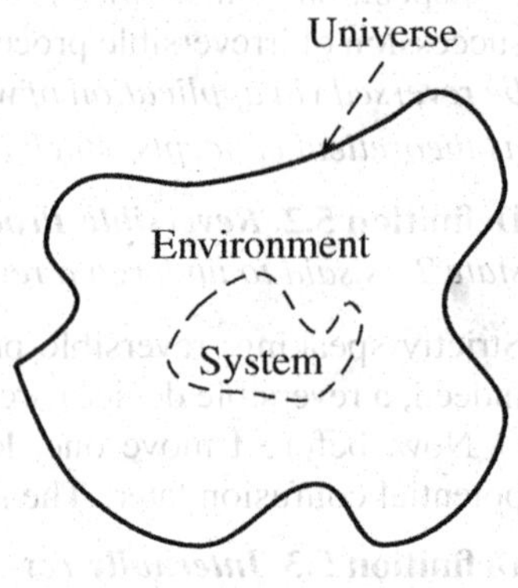

Figure 5.3 The system, environment and the universe.

5.2.3 KELVIN-PLANK STATEMENT

Another version of the Second-Law, the **Kelvin-Plank** statement, is more relevant to engineering applications and states "it is impossible to construct a work producing device which executes a cycle and inter-acts with only a single thermal reservoir". Like the Clausius Statement, the Kelvin-Plank Statement of the Second-Law yields a pictorial representation as shown in Figure 5.4 b). Thus, work and heat are not equivalent. In fact, only a fraction of the heat transferred from a high temperature source (thermal reservoir) can be converted into usable work by a cyclic device.

Now the question arises: "Which statement of the Second-Law is preferable?" The answer is that both statements are equivalent:

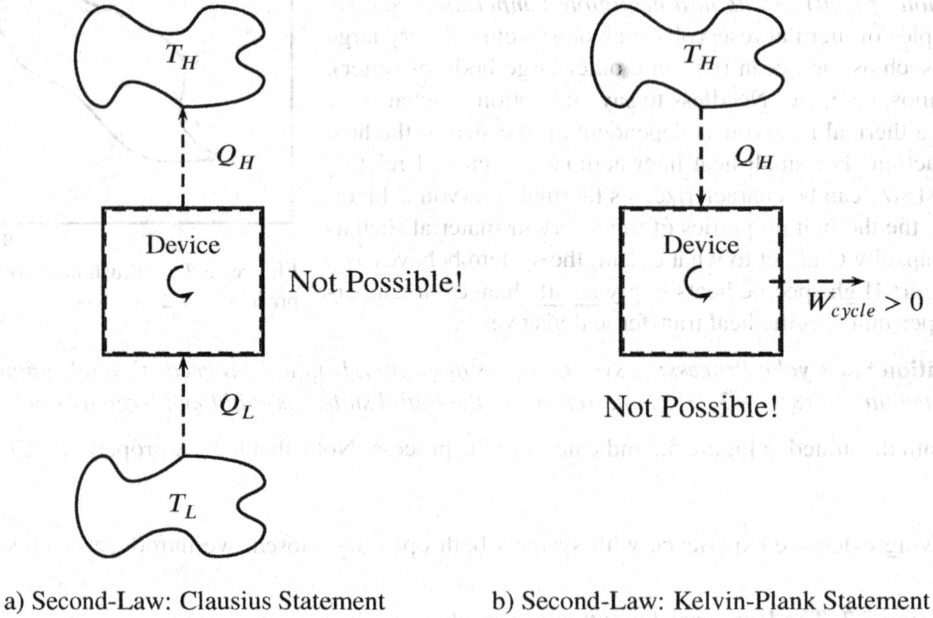

Figure 5.4 Pictorial illustration of the Second Law of Thermodynamics: Clausius Statement and Kelvin-Plank Statement.

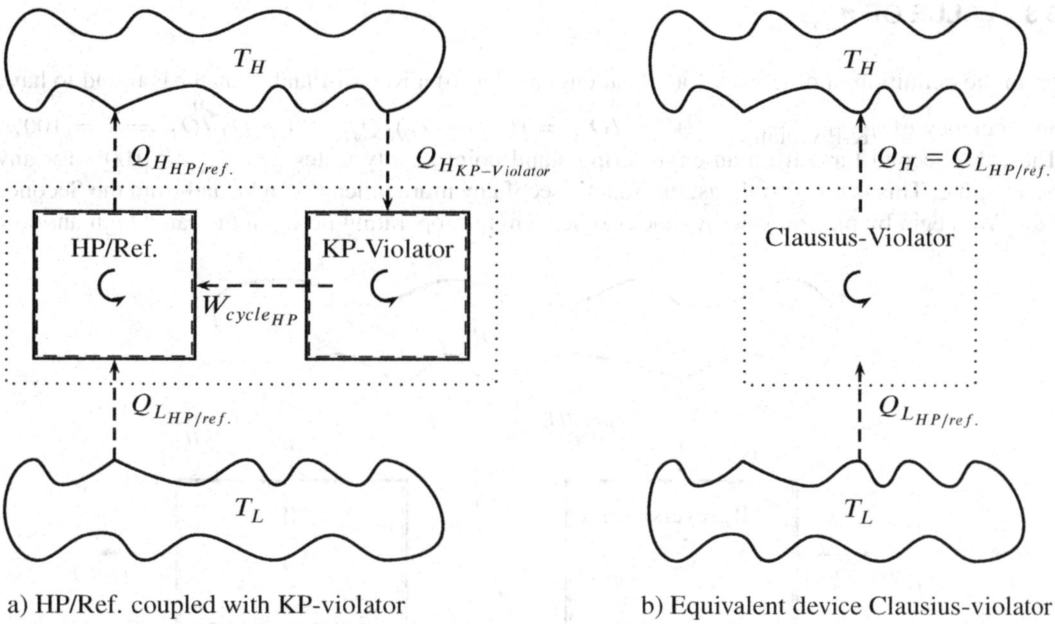

Figure 5.5 Equivalance of Kelvin-Plank and Clausius Statements of the Second Law

Proof: Let us pair two cyclic devices, a heat-pump and a heat engine, one beside the other, interacting between a high and low temperature thermal reservoir. Specifically, as shown in Figure 5.5 a), a heat-pump or refrigerator driven by a Kelvin-Plank violator (a heat-engine which inter-acts with only a single heat source (in this case, the high temperature reservoir)).

Drawing a control volume around the heat-pump/refrigerator and the Kelvin-Plank violator, we have the *combined* device shown in Figure 5.5 b) which extracts heat $Q_L = Q_{L_{HP/Ref}}$ from the low temperature thermal reservoir and rejects heat Q_H to the high temperature reservoir. To find the heat rejection Q_H of this combined device we apply the differential form of the energy-balance to the combined device control volume .... in Figure 5.5 b) and integrate over the cycle (recall, $\oint$ denotes a cyclic or close-path integral)

$$\overbrace{\oint \delta Q}^{Q_{L_{HP/Ref}}} - \overbrace{\oint \delta W}^{-Q_H} \;\;{}^{0} = \oint dE = \overbrace{\Delta E_{cycle}}^{0} \Rightarrow Q_{L_{HP/Ref}} - Q_H = 0 \tag{5.1}$$

or

$$Q_H = Q_{L_{HP/Ref}} \tag{5.2}$$

Hence, the combined device extracts heat $Q_{L_{HP/Ref}}$ from the low temperature thermal reservoir and rejects the *same* amount of heat $Q_H = Q_{L_{HP/Ref}}$ to the high temperature reservoir. Referring to Figure 5.5 b) we see that this equivalent composite device is a Clausius Statement violator. Thus, a Kelvin-Plank violator is equivalent to and implies a Clausius violator. Hence, both statements of the Second-Law are equivalent! ■

5.3 VALUE OF η_{max}

From the definition of efficiency for a heat engine (4.136) a Kelvin-Plank violator is found to have an efficiency of $\eta_{\text{K-P violator}} = W_{cycle}/Q_H = (Q_H - Q_L)/Q_H = 1 - \cancelto{0}{Q_L}/Q_H = 1 = 100\%$. Thus, the Second-Law, from an engineering stand-point simply states that $\eta_{max} < 100\%$ for any heat engine. This is not terribly useful, so let's see if any more mileage can be had from the Second-Law. We begin by placing side-by-side two heat engines operating between the same high and low

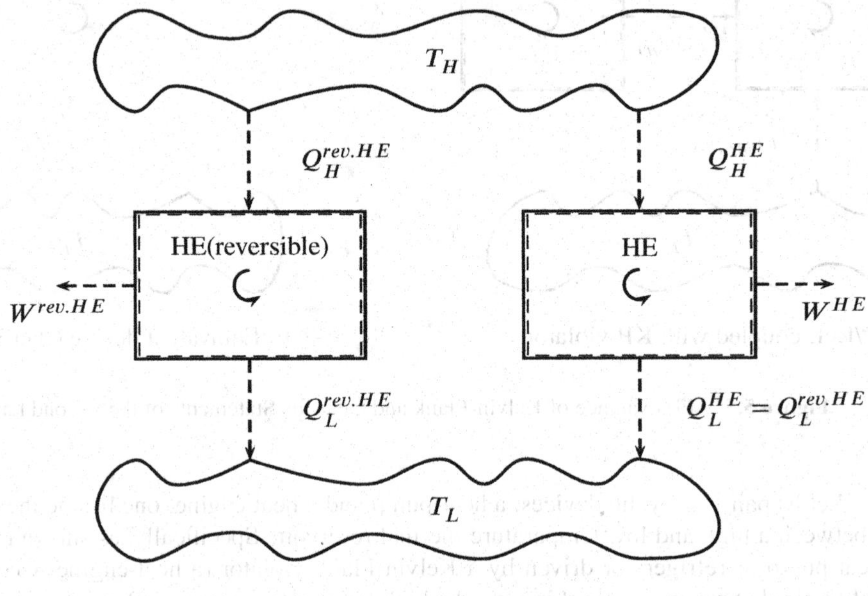

Figure 5.6 Reversible HE and HE composite device

temperature thermal reservoirs. The first heat engine (on the left) in Figure 5.6 is reversible, while the second (on the right) may be reversible <u>or</u> irreversible.[1] In addition, the second heat engine (on the right) has been scaled (increased or decreased in size) till $Q_L^{HE} = Q_L^{rev.HE}$. Note that the efficiency of the reversible heat engine (on the left in Figure 5.6) is given by

$$\eta^{rev.HE} = \frac{W^{rev.HE}}{Q_H^{rev.HE}} = 1 - \frac{\cancelto{Q_L^{HE}}{Q_L^{rev.HE}}}{Q_H^{rev.HE}} = 1 - \frac{Q_L^{HE}}{Q_H^{rev.HE}} \tag{5.3}$$

or

$$\eta^{rev.HE} = 1 - \frac{Q_L^{HE}}{Q_H^{rev.HE}} \tag{5.4}$$

In addition, the efficiency of the (reversible or irreversible) heat engine on the right in Figure 5.6 is given by

$$\eta^{HE} = \frac{W^{HE}}{Q_H^{HE}} = 1 - \frac{Q_L^{HE}}{Q_H^{HE}} \tag{5.5}$$

[1] In other words, the second heat engine (on the right) may, <u>or may not</u> exhibit friction, unrestrained expansion or heat transfer across a finite temperature difference.

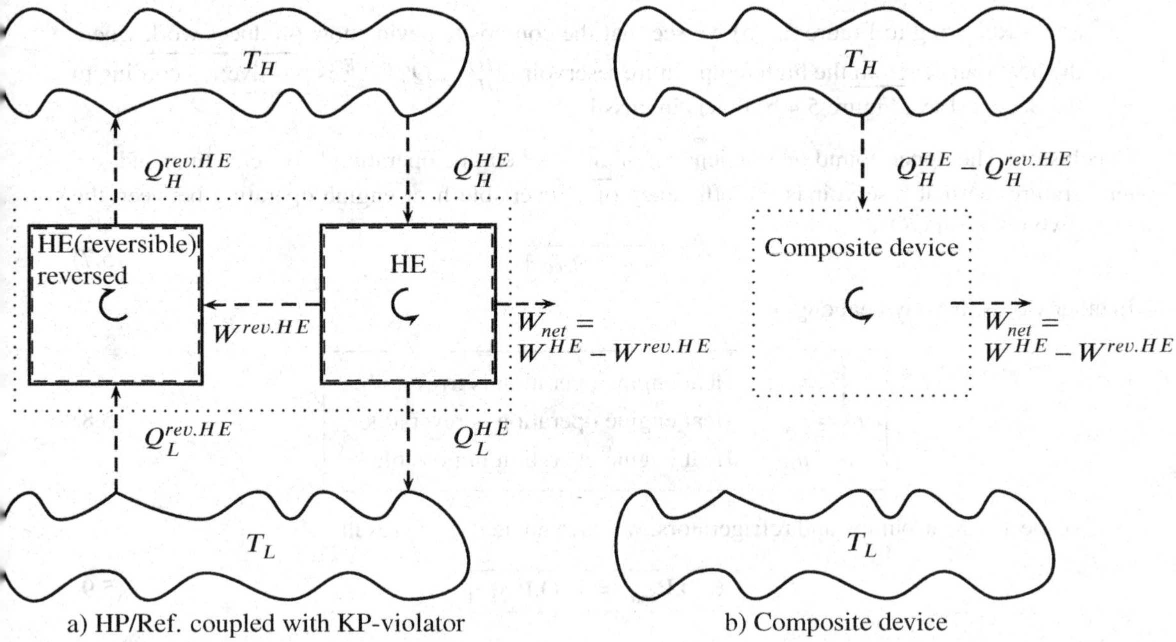

Figure 5.7 Reversible HE reversed and HE composite device

or

$$\eta^{HE} = 1 - \frac{Q_L^{HE}}{Q_H^{HE}} \tag{5.6}$$

We now reverse the reversible HE (on the left) using some of the work produce by the HE (on the right) to drive the, now reversed reversible HE as shown in Figure 5.7 a).

Now, let's look at three possible state of affairs:

1. In the first case, the reversible heat engine (i.e. prior to reversal) accepts a higher heat addition from the high temperature reservoir than the (irreversible or reversible) heat engine: $Q_H^{HE} < Q_H^{rev.HE}$ in Figure 5.6. Comparing (5.4) and (5.6) this implies that $\eta^{rev.HE} > \eta^{HE}$. Referring to Figure 5.7 b) we see that the composite device is now a work absorber with heat transfer from the high temperature reservoir of $Q_H^{HE} - Q_H^{Rev.HE}$ which is negative. In other words, work is done on the composite device in 5.7 b) and the energy is rejected to a single thermal reservoir: According to the Second-Law, Figure 5.4 b) this is possible!
2. In the second case, the reversible heat engine (prior to reversal) accepts an equal amount of heat from the high temperature thermal reservoir as the (irreversible or reversible) heat engine: $Q_H^{HE} = Q_H^{rev.HE}$ in Figure 5.6. Comparing (5.4) and (5.6) this implies that $\eta^{rev.HE} = \eta^{HE}$. Referring to Figure 5.7 b) we see that the composite device now produces no work where the heat transfer from the high temperature reservoir $Q_H^{HE} - Q_H^{Rev.HE}$ is zero: According to the Second-Law, Figure 5.4 b) this is possible!
3. In the third and final case, the reversible heat engine (prior to reversal) accepts a lower heat addition from the high temperature thermal reservoir than the (irreversible or reversible) heat engine: $Q_H^{HE} > Q_H^{rev.HE}$ in Figure 5.6. Comparing (5.4) and (5.6) this implies that $\eta^{rev.HE} <$

η^{HE}. Referring to Figure 5.7 b) we see that the composite device now produces work where the heat transfer from the high temperature reservoir $Q_H^{HE} - Q_H^{rev.HE}$ is positive: According to the Second-Law, Figure 5.4 b) this is impossible!

Conclusion: The upper bound on efficiency η of any heat engine operating between a high and low temperature thermal reservoir is the efficiency of a reversible heat engine operating between the same two reservoirs. Or

$$\boxed{\eta_{max} = \eta_{rev}} \tag{5.7}$$

In other words, for any heat engine

$$\eta \begin{cases} < \eta_{rev}, & \text{Heat engine operation is irreversible} \\ = \eta_{rev}, & \text{Heat engine operation is reversible} \\ > \eta_{rev}, & \text{Heat engine operation impossible} \end{cases} \tag{5.8}$$

With respect to heat-pumps and refrigerators, we have an analogous result:

$$\boxed{\text{C.O.P.}_{max} = \text{C.O.P.}_{rev}} \tag{5.9}$$

and thus

$$\text{C.O.P.} \begin{cases} < \text{C.O.P.}_{rev}, & \text{HP/Refrigerator operation is irreversible} \\ = \text{C.O.P.}_{rev}, & \text{HP/Refrigerator operation is reversible} \\ > \text{C.O.P.}_{rev}, & \text{HP/Refrigerator operation is impossible} \end{cases} \tag{5.10}$$

5.4 FUNCTIONAL FORMS OF η_{rev} AND C.O.P.$_{rev}$

As engineers "in training" we are interested in answering the question: "What is the exact expression for η_{rev}?" This question can be refined by recalling the definition of efficiency and applying the energy-balance

$$\eta_{rev} = \left[\frac{W_{cycle}}{Q_H}\right]_{rev} = \left[\frac{Q_H - Q_L}{Q_H}\right]_{rev} = 1 - \left[\frac{Q_L}{Q_H}\right]_{rev} \tag{5.11}$$

Thus, we are essentially interested in an expression for the ratio of the heat inter-actions for a reversible process $\left[\frac{Q_L}{Q_H}\right]_{rev}$. Now, (5.11) does not, in any way, describe or restrict the internal mechanism of the device apart for requiring reversibility. Thus, the efficiency of a reversible heat engine is not a function of the device internal mechanism! In fact, the only remaining parameters which might effect the operation (i.e. efficiency) of the reversible heat engine are the thermal reservoir temperatures T_H and T_L. Thus, we conclude

$$\left[\frac{Q_L}{Q_H}\right]_{rev} = f(T_L, T_H) \tag{5.12}$$

If we can determine the functional form of $f()$ we can calculate η_{rev}!

We begin by placing two work-producing devices, side-by-side, both of which operate reversibly between the same two reservoirs as shown in Figure 5.8. From (5.12) we have, for reversible HE #3,

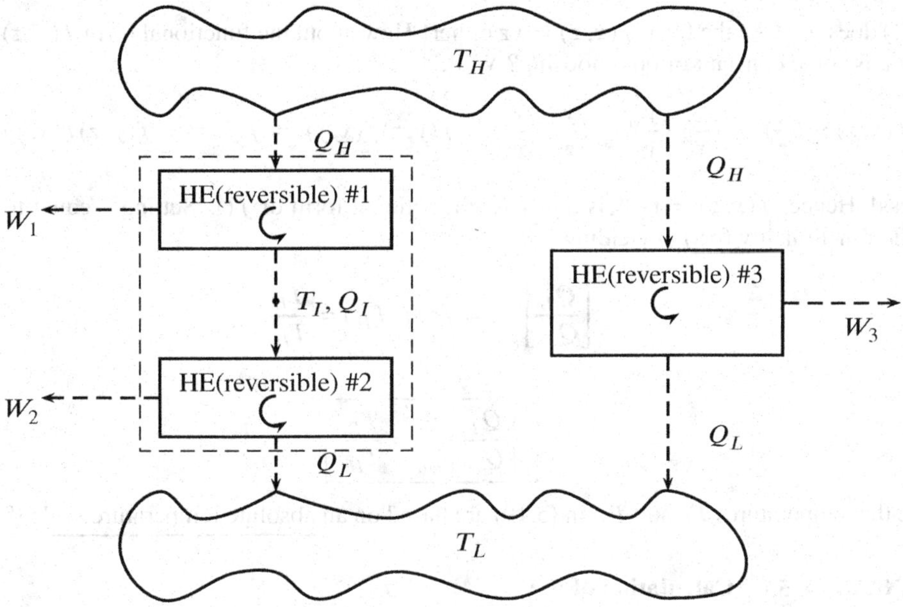

Figure 5.8 Pair of reversible devices inter-acting with the same reserviors. Composite device on left denoted by CV – – ––.

$(Q_L/Q_H)_{rev} = f(T_L, T_H)$. Likewise, for reversible HE #1 which rejects heat Q_I at an *intermediate* temperature T_I, $(Q_I/Q_H)_{rev} = f(T_I, T_H)$. Finally, for reversible HE #2 which accepts heat Q_I at an *intermediate* temperature T_I, $(Q_L/Q_I)_{rev} = f(T_L, T_I)$. Now the composite device containing HE #1 and #2 signified by – – – – – – is reversible and equivalent to reversible HE #3 (recall, the internal mechanism of the reversible heat engine is irrelevant). Thus,

$$f(T_L, T_H) = \left[\frac{Q_L}{Q_H}\right]_{rev} = \left[\frac{Q_L}{Q_H}\right]_{rev} \frac{Q_I}{Q_I} =$$

$$= \left[\frac{Q_L}{Q_H}\frac{Q_I}{Q_I}\right]_{rev} = \left[\frac{Q_I}{Q_H}\right]_{rev} \left[\frac{Q_L}{Q_I}\right]_{rev} = f(T_I, T_H) f(T_L, T_I) \tag{5.13}$$

or

$$f(T_L, T_H) = f(T_I, T_H) f(T_L, T_I) \tag{5.14}$$

We now ask ourselves the question: "What is the functional form of $f()$ which satisfies (5.14)?" Letting $T_L \to x$, $T_I \to y$ and $T_H \to z$, we ask "what function exhibits the following mathematical behavior: $f(x, z) = f(y, z) f(x, y)$?" How about $f(x, z) = x - z$? Well,

$$f(x, z) = x - z \neq yx - y^2 - zx + zy = (y - z)(x - y) = f(y, z) f(x, y)$$

$$\Rightarrow f(x, z) \neq f(y, z) f(x, y) \tag{5.15}$$

Thus, $f()$ does *not* take the form $f(x, z) = x - z$! How about the functional form $f(x, z) = xz$? Well,

$$f(x, z) = xz \neq y^2 xz = (yz)(xy) = f(y, z) f(x, y) \Rightarrow f(x, z) \neq f(y, z) f(x, y) \tag{5.16}$$

and $f()$ does *not* take the form $f(x,z) = xz$ either! How about the functional form $f(x,z) = (x/z)^\epsilon$ where ϵ is some constant of our choosing? Well,

$$f(x,z) = (\frac{x}{z})^\epsilon = (\frac{x}{z})^\epsilon(\frac{y}{y})^\epsilon = (\frac{y}{z})^\epsilon(\frac{x}{y})^\epsilon = f(y,z)f(x,y) \Rightarrow f(x,z) = f(y,z)f(x,y) \quad (5.17)$$

Success! Hence, $f(x,z) = (\frac{x}{z})^\epsilon$ is a *possible* functional form of $f()$. Setting ϵ equal to 1 gives a specific functionality for $f()$ yielding

$$\left[\frac{Q_L}{Q_H}\right]_{rev} = f(T_L, T_H) = \frac{T_L}{T_H} \quad (5.18)$$

or

$$\boxed{\left[\frac{Q_L}{Q_H}\right]_{rev} = \frac{T_L}{T_H}} \quad (5.19)$$

where the temperatures T_L and T_H in (5.19) are based on an absolute temperature scale ($^\circ K$ or $^\circ R$).

EXAMPLE 5.1 Calculation of η_{max}

Problem Statement:
An inventor claims to have invented a unique work producing device and has provided the following measured quantities: The device, working in a cycle, operating between two reservoirs accepts $750kJ$ of heat from a source of temperature $127^\circ C$ and rejects $250kJ$ of heat to a sink at temperature of $27^\circ C$. Assess the inventor's claims.

Given:
Cyclic work producing device,
$Q_H = 750kJ$,
$Q_L = 250kJ$,
$T_H = 127^\circ C$,
$T_L = 27^\circ C$

Find:
$\eta_{claimed}$,
η_{max}

Assumptions:
None

Schematic:

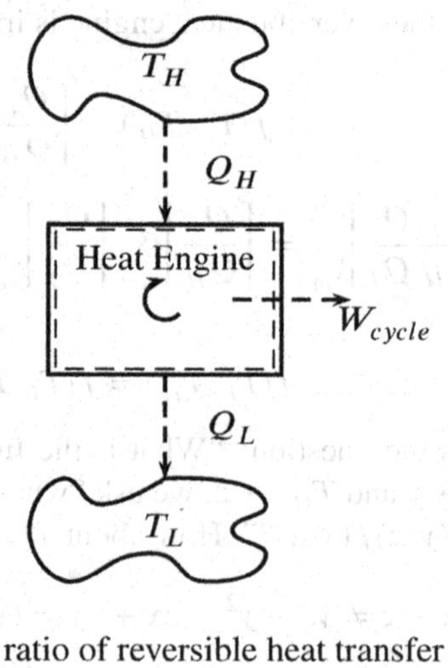

Analysis:
From (4.138), (5.8) and the expression for the ratio of reversible heat transfer (5.19) where the high and low thermal reservoir temperatures are *absolute* (i.e. in degrees K)

$$\eta_{max} = \eta_{rev} = \left[\frac{W_{cycle}}{Q_H}\right]_{rev} = 1 - \left[\frac{Q_L}{Q_H}\right]_{rev} = 1 - \left[\frac{T_L}{T_H}\right]$$

$$= 1 - \left[\frac{(273+27)^\circ K}{(273+127)^\circ K}\right] = 25\% \xleftarrow{\eta_{max}} \quad (5.20)$$

Now, from the definition of efficiency of a heat engine (4.138)

$$\eta_{claimed} = \left[\frac{W_{cycle}}{Q_H}\right]_{claimed} = 1 - \left[\frac{Q_L}{Q_H}\right]_{claimed} = 1 - \frac{250\cancel{kJ}}{750\cancel{kJ}} = 0.666 = 66.6\% \xleftarrow{\eta_{claimed}} \quad (5.21)$$

Thus, $\eta_{claimed} = 66.6\% > 25\% = \eta_{max}$ or $\eta_{claimed} > \eta_{max}$. The inventor's claim violates the Second-Law and is therefore impossible.

$\boxed{\eta_{claimed} > \eta_{max} \text{ (Inventor's claims violate Second-Law. Not possible!)}}$

□

5.5 THE CARNOT CYCLE: CARNOT HEAT-ENGINE

At this point the question presents itself: "Can we formulate or conceive, in detail, of a reversible heat engine?" Note that "conceive" does not mean "build"! Well, such a device must be free of the irreversibilities discussed in §5.1 both inside and outside the system. Or to repeat, such a device must be free of

1. Frictional processes
2. Unrestrained expansion
3. Heat transfer across a finite temperature difference

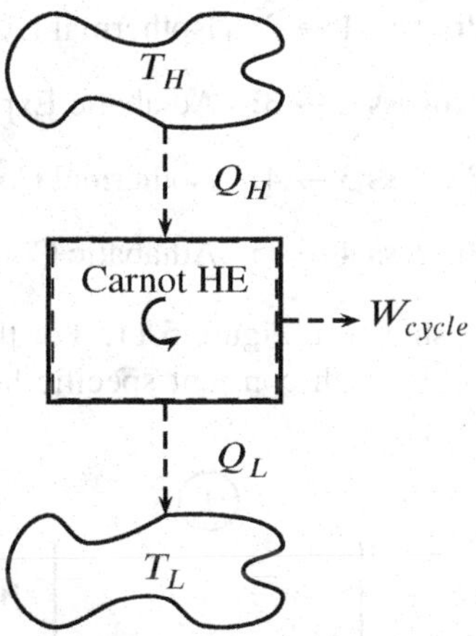

Figure 5.9 Pictorial illustration of a Carnot heat engine.

and hence, *internally and externally* reversible. From an engineering stand-point item #1 can be nearly achieved: First by careful machining of the component parts of the heat engine, and second by the effective use of lubricants. Item #2 can be nearly achieved by requiring that the device, say a piston-cylinder device, operates at a *sufficiently* slow piston speed relative to the working fluid speed of sound (the speed at which a pressure wave propagates through the fluid). Hence, the slower the heat engine operates the more likely condition #2 is approximately satisfied. This brings us to the final requirement #3: Heat transfer must not occur across a *finite* temperature difference! Unfortunately, the definition of heat transfer Q is the transfer of energy between systems due to a temperature difference (see §3.4). To reconcile the distinction between irreversible and reversible heat transfer we refer to Figure 5.10 a) which illustrates an energy transfer between a cold system A with temperature T_A and a hot system B with temperature $T_A + \Delta T$ in thermal contact. Thus, given that the temperature difference ΔT is *finite*, heat Q is transferred from system B to A irreversibly. However, if we let $\Delta T \to 0$ (i.e. the temperature difference becomes infinitesimal), the heat transfer becomes reversible in the limit as shown in Figure 5.10 b)! We shall return to this point in §7.1 to address the engineering implications of heat transfer across a vanishing temperature gradient, but let us now describe a work producing cycle which operates reversibly: The Carnot Cycle heat engine.

We start with the heat engine inter-acting with a high and low temperature source/sink as shown in Figure 5.9. This device operates reversibly, hence slowly and without friction. The device itself

a) Irreversible heat inter-action (ΔT is *finite*)

b) Reversible heat inter-action ($\Delta T \rightarrow 0$ is *infinitesimal*)

Figure 5.10 Illustration of irreversible and reversible heat inter-actions.

consists of a piston-cylinder device containing some working fluid (which may or may not be a gas). The distinct processes which define the Carnot cycle are the following:

Process $1 \rightarrow 2$**:** Isothermal Expansion

Process $2 \rightarrow 3$**:** Adiabatic Expansion

Process $3 \rightarrow 4$**:** Isothermal Compression

Process $4 \rightarrow 1$**:** Adiabatic Compression

as shown in Figure 5.11. For the purpose of illustration, let us assume the working fluid exhibits I.G.B. with constant specific heats. We can construct a $T - v$ and $P - v$ diagram by repeated

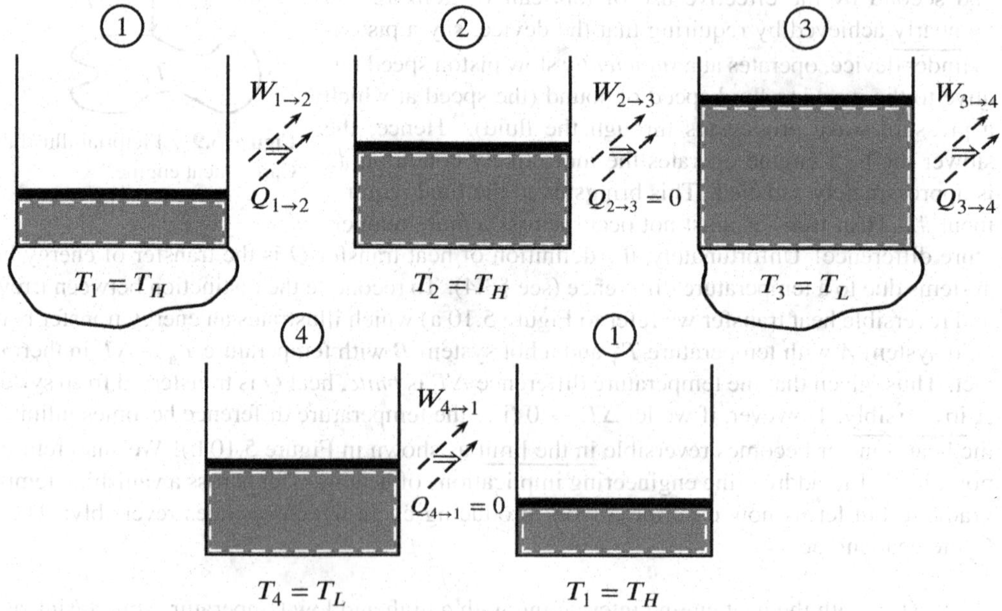

Figure 5.11 Process illustration of Carnot heat engine.

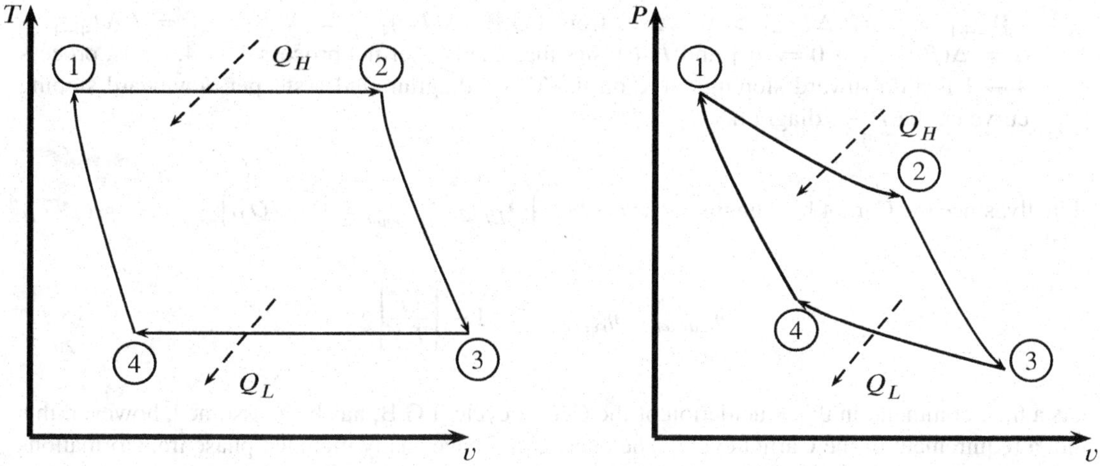

Figure 5.12 Carnot heat engine $T-v$ and $P-v$ plots for a gasseous working fluid.

application of the energy-balance and the ideal gas equation while neglecting changes in kinetic and potential energy:

Process $1\to2$ **Isothermal Expansion:** $Q_{1\to2} - \cancelto{W_{bound_{1\to2}}}{W_{1\to2}} = \Delta U_{1\to2} = mC_v\cancelto{0}{\Delta T_{1\to2}} \Rightarrow Q_{1\to2} = W_{bound_{1\to2}} = \int_{V_1}^{V_2} PdV$. Since $\Delta V_{1\to2} > 0 \Rightarrow Q_{1\to2} > 0$. Also, from I.G.B., $\Delta(Pv)_{1\to2} = \Delta(RT)_{1\to2} = R\cancelto{0}{\Delta T_{1\to2}} \Rightarrow \Delta(Pv)_{1\to2} = 0 \Rightarrow v\uparrow$ and $P\downarrow$. Thus, process $1\to2$ is a horizontal line on the $T-v$ diagram and a downward sloping curve on the $P-v$ diagram as shown in Figure 5.12.

Process $2\to3$ **Adiabatic Expansion:** $\cancelto{0}{Q_{2\to3}} - \cancelto{W_{bound_{2\to3}}}{W_{2\to3}} = \Delta U_{2\to3} = mC_v\Delta T_{2\to3} \Rightarrow -W_{2\to3} = -\int_{V_2}^{V_3} PdV = mC_v\Delta T_{2\to3}$. Since $\Delta V_{2\to3} > 0 \Rightarrow W_{2\to3} > 0$ and thus $mC_v\Delta T_{2\to3} = -W_{2\to3} < 0 \Rightarrow \Delta T_{2\to3} < 0$. Also, from I.G.B., $\Delta(Pv)_{2\to3} = \Delta(RT)_{2\to3} = R\Delta T_{2\to3} < 0 \Rightarrow \Delta(Pv)_{2\to3} < 0 \Rightarrow v\uparrow$ and $P\downarrow$ faster than the isothermal process $1\to2$. Thus, process $2\to3$ is a downward sloping curve on the $T-v$ diagram and a steep downward sloping curve on the $P-v$ diagram.

Process $3\to4$ **Isothermal Compression:** $Q_{3\to4} - \cancelto{W_{bound_{3\to4}}}{W_{3\to4}} = \Delta U_{3\to4} = mC_v\cancelto{0}{\Delta T_{3\to4}} \Rightarrow Q_{3\to4} = W_{bound_{3\to4}} = \int_{V_3}^{V_4} PdV$. Since $\Delta V_{1\to2} < 0 \Rightarrow Q_{3\to4} = \int_{V_3}^{V_4} PdV < 0$. Also, from I.G.B., $\Delta(Pv)_{3\to4} = \Delta(RT)_{3\to4} = R\cancelto{0}{\Delta T_{3\to4}} \Rightarrow \Delta(Pv)_{3\to4} = 0 \Rightarrow v\downarrow$ and $P\uparrow$. Thus, process $3\to4$ is a horizontal line on the $T-v$ diagram and, yet again, a downward sloping curve on the $P-v$ diagram as shown in Figure 5.12.

Process $4\to1$ **Adiabatic Compression:** $\cancelto{0}{Q_{4\to1}} - \cancelto{W_{bound_{4\to1}}}{W_{4\to1}} = \Delta U = mC_v\Delta T_{4\to1} \Rightarrow -W_{4\to1} = -\int_{V_1}^{V_2} PdV = mC_v\Delta T_{4\to1}$. Since $\Delta V_{4\to1} < 0 \Rightarrow W_{4\to1} < 0$ and thus $mC_v\Delta T_{4\to1} =$

$-W_{4\to1} > 0 \Rightarrow \Delta T_{4\to1} > 0$. Also, from I.G.B., $\Delta(Pv)_{4\to1} = \Delta(RT)_{4\to1} = R\Delta T_{4\to1} > 0 \Rightarrow \Delta(Pv)_{2\to3} > 0 \Rightarrow v \downarrow$ and $P \uparrow$ faster than the isothermal process $3 \to 4$. Thus, process $4 \to 1$ is a downward sloping curve on the $T - v$ diagram and a steeper downward sloping curve on the $P - v$ diagram.

Finally, since the Carnot heat engine is reversible, $[Q_L/Q_H]_{Carnot} = [Q_L/Q_H]_{reversible} = T_L/T_H$ and

$$\eta_{Carnot} = \eta_{reversible} = 1 - \left[\frac{T_L}{T_H}\right] \tag{5.22}$$

As a final comment, in this elucidation of the Carnot cycle, I.G.B. has been assumed, however, this not a requirement of the Carnot cycle. The cycle could have easily included phase transformations or non-ideal gas behavior: The resulting efficiency would be the same!

5.5.1 CARNOT REFRIGERATOR

Since the Carnot heat engine is reversible, (with efficiency equal to η_{max}), we can reverse the cycle, shown in Figure 5.12 producing the Carnot refrigerator/HP (with C.O.P. equal to $C.O.P._{max}$) as shown in Figure 5.13.

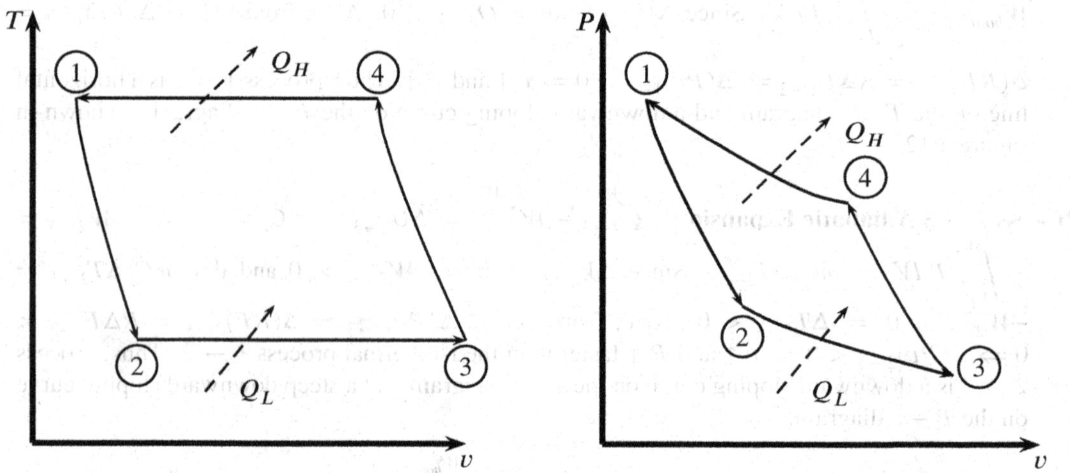

Figure 5.13 Carnot refrigeration cycle $T - v$ and $P - v$ plots for a gaseous working fluid.

EXAMPLE 5.2 Carnot Refrigerator: Calculation of $C.O.P._{ref}$ and W_{cycle}

Problem Statement:

A Carnot refrigerator operates between two reservoirs of temperature $300°K$ and $270°K$. Over a cycle, the refrigerator extracts $10kJ$ from the low temperature source. Determine the coefficient of performance and the cycle work.

Schematic:

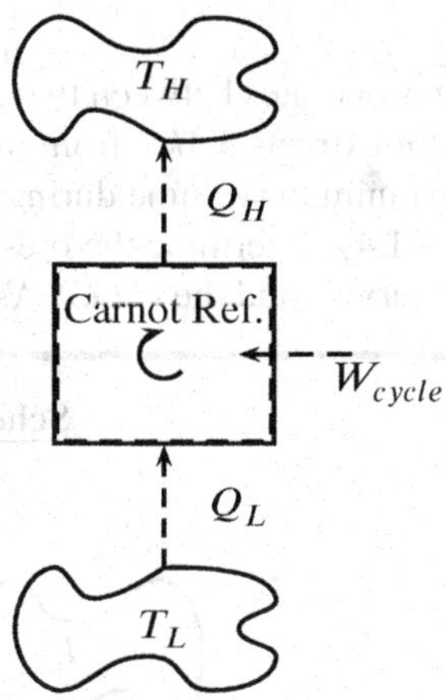

Given:
Carnot refrigerator,
$Q_L = 10kJ$,
$T_H = 300° K$,
$T_L = 270° K$

Find:
$C.O.P._{ref}$,
W_{cycle}

Assumptions:
None

Analysis:
Via the definition of C.O.P. for a refrigerator (4.139), noting that the Carnot cycle is reversible

$$C.O.P._{Carnot_{ref}} = C.O.P._{rev_{ref}} = \left[\frac{Q_L}{W_{cycle}}\right]_{rev} \tag{5.23}$$

Noting from the schematic that Q_L is into the system (positive), Q_H is out of the system (and contributes a negative heat transfer) while W_{cycle} is on the system (and contributes to a negative work transfer by the system), application of the energy-balance to the device over the cycle gives the cycle work

$$\overset{Q_L - Q_H}{\cancel{Q}} - \overset{-W_{cycle}}{\cancel{W}} = \overset{0}{\cancel{\Delta E}} = 0 \Rightarrow W_{cycle} = Q_H - Q_L \tag{5.24}$$

Thus, combining (5.23) and (5.24)

$$C.O.P._{Carnot_{ref}} = \left[\frac{Q_L}{W_{cycle}}\right]_{rev} = \left[\frac{Q_L}{Q_H - Q_L}\right]_{rev} = \left[\frac{1}{\frac{Q_H}{Q_L} - 1}\right]_{rev} = \frac{1}{\left[\frac{Q_H}{Q_L}\right]_{rev} - 1} \tag{5.25}$$

Combining (5.19) and (5.25)

$$C.O.P._{Carnot_{ref}} = \frac{1}{\left[\frac{Q_H}{Q_L}\right]_{rev} - 1} = \frac{1}{\frac{T_H}{T_L} - 1} = \frac{1}{\frac{300° \cancel{K}}{270° \cancel{K}} - 1} = 9 \overset{C.O.P._{Carnot_{ref}}}{\Longleftarrow} \tag{5.26}$$

Returning to (5.23) and solving for the cycle work we get

$$W_{cycle} = \frac{Q_L}{C.O.P._{Carnot_{ref}}} = \frac{10kJ}{9} = 1.111kJ \overset{W_{cycle}}{\Longleftarrow} \tag{5.27}$$

$C.O.P._{Carnot_{ref}} = 9$, $W_{cycle} = 1.11kJ$ (On the device)

□

EXAMPLE 5.3 Carnot refrigerator: Detailed cycle analysis

Problem Statement:
A Carnot refrigerator operates between two reservoirs of temperature $300°K$ and $250°K$. Over a cycle, the refrigerator extracts $3.4kJ$ from the low temperature reservoir. If the working fluid is $0.1kg$ of air with a minimum volume during the cycle of $0.01m^3$, assuming constant specific heats (with $k = C_P/C_v = 1.4$), determine the pressures at the various states, the heat rejection to the high temperature reservoir and the C.O.P. Assume ideal gas behavior

---

Given:
Carnot refrigerator,
air (I.G.B.),
$Q_L = 3.4kJ$,
$T_H = 300°K$,
$T_L = 250°K$,
$V_{min} = 0.01m^3$,
$m = 0.1kg$,
$k = \frac{C_P}{C_V} = 1.4$

Find:
P_1, P_2,
P_3, P_4,
Q_H,
$C.O.P.$

Assumptions:
$W = W_{bound}$,
$\Delta KE = \Delta PE = 0$

Schematic:

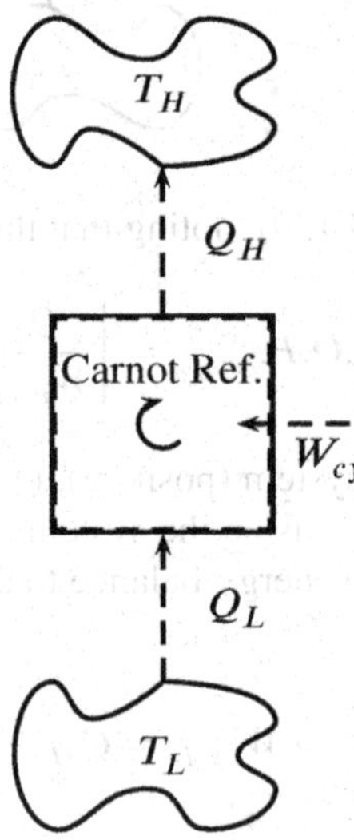

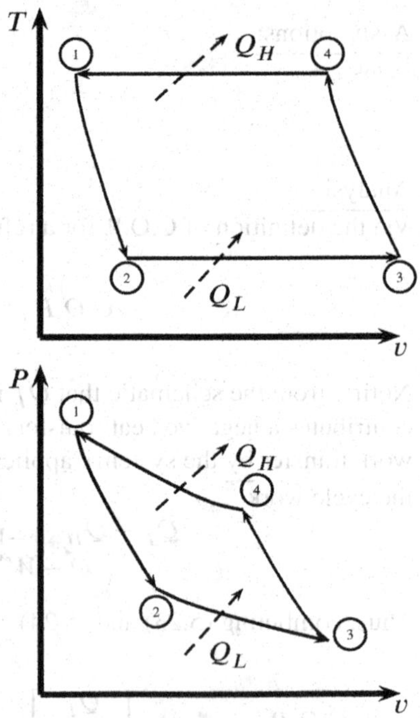

Analysis:
Via I.G.B. applied at state 1 (where $R = R_{air} = 286\frac{J}{kg-°K}$)

$$P_1 = \frac{mRT_1}{V_1} = \frac{0.1\cancel{kg}(0.286\frac{kJ}{\cancel{kg}-\cancel{°K}})300\cancel{°K}}{0.01m^3} = 858.0kPa \stackrel{P_1}{\Longleftarrow} \tag{5.28}$$

Process 1 → 2 (Adiabatic expansion):
Application of the energy-balance for a differential process while noting the property relationship for ideal gas behavior (4.40)

$$\overset{0}{\cancel{\delta Q}} - \overset{PdV}{\cancel{\delta W}} = dE = dU + \overset{0}{\cancel{dKE + dPE}} = dU = mC_v(T)dT \Rightarrow$$

$$-\overset{\frac{mRT}{V}}{\cancel{P}}dV = mC_v(T)dT \Rightarrow -\frac{\cancel{m}RT}{V}dV = \cancel{m}C_v(T)dT \Rightarrow -\frac{RT}{V}dV = C_v(T)dT$$

$$\Rightarrow -\frac{\overset{C_p-C_v}{\cancel{R}}}{C_v}\frac{dV}{V} = \frac{dT}{T} \Rightarrow -\frac{C_p - C_v}{C_v}\frac{dV}{V} = -(k-1)\frac{dV}{V} = \frac{dT}{T} \tag{5.29}$$

or

$$-(k-1)\frac{dV}{V} = \frac{dT}{T} \tag{5.30}$$

Integrating (5.30) from state 1 → 2 gives

$$-(k-1)ln\left[\frac{V_2}{V_1}\right] = ln\left[\frac{T_2}{T_1}\right] \tag{5.31}$$

Finally, via I.G.B

$$-(k-1)ln\left[\frac{V_2}{V_1}\right] = -(k-1)ln\left[\frac{\frac{\not{m}RT_2}{P_2}}{\frac{\not{m}RT_1}{P_1}}\right] = -(k-1)ln\left[\frac{T_2}{T_1}\frac{P_1}{P_2}\right] = ln\left[\frac{T_2}{T_1}\right]$$

$$\Rightarrow P_2 = P_1\left[\frac{T_2}{T_1}\right]^{\frac{k}{k-1}} = 858.0kPa\left[\frac{250°\not{K}}{300°\not{K}}\right]^{\frac{1.4}{1.4-1}} = 453.3kPa \overset{P_2}{\Longleftarrow} \tag{5.32}$$

Process 2 → 3 (Isothermal expansion):
Application of the energy-balance gives

$$\delta Q - \overset{PdV}{\not{\delta W}} = dU + \overset{0}{\not{dKE + dPE}} = dU = mC_v(T)\overset{0}{\not{dT}} = 0 \Rightarrow -\delta Q = PdV \tag{5.33}$$

Now, taking the total derivative of the ideal gas law

$$d(PV) = d(mRT) \Rightarrow VdP + PdV = mR\overset{0}{\not{dT}} \Rightarrow PdV = -VdP \tag{5.34}$$

Hence, combining (5.33) and (5.34) gives

$$\delta Q = PdV = -\overset{\frac{mRT}{P}}{\not{V}}dP = -\frac{mRT}{P}dP \tag{5.35}$$

Integrating (5.35) from state 2 → 3 where $Q_{2\to3} = Q_L$ and $T = T_2$ gives

$$Q_{2\to3} = -mRT_2 ln\left[\frac{P_3}{P_2}\right] = Q_L \Rightarrow ln\left[\frac{P_3}{P_2}\right] = \frac{-Q_L}{mRT} \tag{5.36}$$

Solving the above for P_3

$$P_3 = P_2 e^{\frac{-Q_L}{mRT_2}} = (453.3kPa)e^{\frac{-3.4\not{kJ}}{0.1\not{kg}\left(0.286\frac{\not{kJ}}{\not{kg}-°\not{K}}\right)250°\not{K}}} = 281.75kPa \overset{P_3}{\Longleftarrow} \tag{5.37}$$

Process 3 → 4 (Adiabatic compression):
Note, process 3 → 4, like process 1 → 2, is reversible and adiabatic. Thus, using our results from process 1 → 2, let state 1 ⇒ 3 and 2 ⇒ 4 (5.32) yields

$$P_4 = P_3\left[\frac{T_4}{T_3}\right]^{\frac{k}{k-1}} = 281.75kPa\left[\frac{300°\not{K}}{250°\not{K}}\right]^{\frac{1.4}{1.4-1}} = 533.3kPa \overset{P_4}{\Longleftarrow} \tag{5.38}$$

Process 4 → 1 (Isothermal compression):
Application of the energy-balance gives

$$\delta Q - \overset{PdV}{\not{\delta W}} = \overset{mC_v(T)\not{dT}^{0}}{\not{dU}} + \overset{0}{\not{dKE + dPE}} \Rightarrow \delta Q = PdV = \overset{\frac{mRT}{V}}{\not{P}}dV = mRT\frac{dV}{V} \tag{5.39}$$

Integrating from state 4 → 1 where $Q_{4\to1} = -Q_H$ gives via I.G.B.

$$Q_{4\to1} = mRT_1 ln\left[\frac{V_1}{V_4}\right] = mRT_1 ln\left[\frac{\frac{mRT_1}{P_1}}{\frac{mRT_4}{P_4}}\right] = mRT_1 ln\left[\frac{P_4}{P_1}\right]$$

$$= 0.1kg\left(0.286\frac{kJ}{kg-°K}\right)(300°K)ln\left[\frac{533.3kPa}{858.0kPa}\right] = -4.079kJ \tag{5.40}$$

or

$$Q_H = -Q_{4\to1} = -(-4.079kJ) = 4.079kJ \stackrel{Q_H}{\Longleftarrow} \tag{5.41}$$

Finally, from the definition of coefficient of performance

$$C.O.P._{Carnot_{ref}} = \left[\frac{Q_L}{W_{cycle}}\right] = \left[\frac{Q_L}{Q_H - Q_L}\right] = \left[\frac{1}{\frac{Q_H}{Q_L} - 1}\right]$$

$$= \left[\frac{1}{\frac{-Q_{4\to1}}{Q_{2\to3}} - 1}\right] = \left[\frac{1}{\frac{4.079kJ}{3.4kJ} - 1}\right] = 5.0 \stackrel{C.O.P._{Carnot_{ref}}}{\Longleftarrow} \tag{5.42}$$

$P_1 = 858.0kPa,\ P_2 = 453.3kPa,\ P_3 = 281.75kPa,\ P_4 = 533.3kPa,\ Q_{4\to1} = -4.079kJ,\ C.O.P._{Carnot_{ref}} = 5.0$

Comment: Comparing our result with the calculation of C.O.P. for a reversible cycle

$$C.O.P._{rev} = \left[\frac{Q_L}{W_{cycle}}\right]_{rev} = \frac{1}{\left[\frac{Q_H}{Q_L}\right]_{rev} - 1} = \frac{1}{\left[\frac{T_H}{T_L}\right] - 1} = \frac{1}{\left[\frac{300°K}{250°K}\right] - 1} = 5 \tag{5.43}$$

Our previous calculation of $C.O.P.$ based on the full cycle analysis agrees with this result.

□

5.6 THE CLAUSIUS INEQUALITY

We introduced the Second-Law at the beginning of this chapter as a pictorial representation describing what processes can, or cannot occur in nature. *We will now transform the pictorial statement of the Second-Law into a symbolic or mathematical equivalent* which will be used in the next chapter to establish the existence of an entirely new property: Entropy.

To begin, we present the composite device as shown in Figure 5.14 consisting of two heat engines: The first operates reversibly and accepts, at any given point in the cycle, a infinitesimal amount of heat δQ_R from a thermal reservoir of temperature T_R. At the same time this reversible heat engine rejects heat δQ at temperature T to a second heat engine which may or may not be reversible. Note, the second heat engine, according to Figure 5.14 does not apparently inter-act with a low temperature reservoir.

Now, by virtue of the fact that the first heat engine operates reversibly, a *differential* amount of heat is accepted and rejected according to (5.19)

$$\left[\frac{\delta Q_R}{\delta Q}\right]_{rev} = \frac{T_R}{T} \tag{5.44}$$

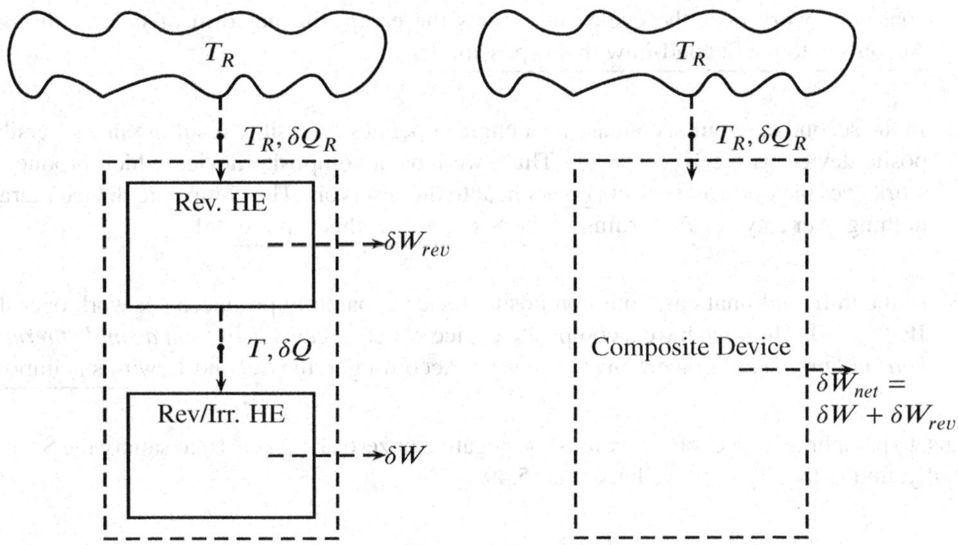

a) Reversible HE and heat engine b) Composite device

Figure 5.14 Reversible HE rejecting heat δQ at temperature T to a heat engine, and composite device

or

$$\delta Q_R = \delta Q \frac{T_R}{T} \tag{5.45}$$

Application of the energy-balance to the composite device ----, noting that $\delta W_{net} = \delta W_{rev} + \delta W$, yields

$$\delta Q_R - \delta W_{net} = dE \tag{5.46}$$

Integrating (5.46) over a cycle and noting that $\oint dE = 0$, we have

$$\oint \delta Q_R - \oint \delta W_{net} = \cancelto{0}{\oint dE} = 0 \Rightarrow \oint \delta Q_R = \oint \delta W_{net} \tag{5.47}$$

Finally, substituting (5.45) into (5.47) noting that the thermal reservoir temperature T_R is a constant gives

$$\oint \delta Q_R = \oint \frac{T_R}{T} \delta Q = \oint \delta W_{net} \Rightarrow T_R \oint \frac{\delta Q}{T} = \oint \delta W_{net} \tag{5.48}$$

Or since $\oint \delta W_{net} = W_{cycle}$

$$T_R \oint \frac{\delta Q}{T} = W_{cycle} \tag{5.49}$$

There are three possible values of W_{cycle}:

1. In the first case, our secondary heat engine operates irreversibly resulting in an irreversible composite device with $W_{cycle} < 0$. Thus, we have a composite device which

consumes work over the cycle and rejects the energy in the form of heat to the reservoir: According to the Second-Law this is possible![2]

2. In the second case our secondary heat engine operates reversibly resulting in a reversible composite device where $W_{cycle} = 0$. Thus, we have a composite device which produces no net work over the cycle and rejects no net heat to the reservoir. The composite device literally does nothing over a cycle! According to the Second-Law this is possible!

3. In the third and final case, our composite device apparently produces net work over the cycle $W_{cycle} > 0$. Thus, we have a composite device which *accepts heat from a single thermal reservoir and produces net work over the cycle*: According to the Second-Law this is impossible!

Thus, to paraphrase, the cycle work must be negative or zero ($W_{cycle} \leq 0$) to satisfy the Second-Law. Finally, noting that $T_R > 0$, we have, via (5.49)

$$T_R \oint \frac{\delta Q}{T} = W_{cycle} \leq 0 \Rightarrow \oint \frac{\delta Q}{T} \leq 0 \tag{5.50}$$

which yields the Clausius Inequality

$$\boxed{\oint \frac{\delta Q}{T} \leq 0, \text{ (Clausius Inequality)}} \tag{5.51}$$

where $\oint \frac{\delta Q}{T} = 0$ for a reversible process and $\oint \frac{\delta Q}{T} < 0$ for an irreversible process.

Note, the Clausius Inequality is mathematical equivalent to either the Clausius or Kelvin-Plank statements of the Second-Law.

Proof: We begin with a device as shown below which transfers heat from a low to a high temperature reservoir apparently without work (a violation of the Clausius statement of the Second-Law). For simplicity, this device operates in two stages:

Process 1 → 2: Heat Q_L is received from the low temperature reservoir with temperature T_L into the device.

Process 2 → 1: Heat Q_H is rejected from the device and flows into the high temperature reservoir with temperature T_H.

Thus, we have

$$\begin{aligned} \oint \frac{\delta Q}{T} &= \int_{1\to 2} \frac{\delta Q}{T} + \int_{2\to 1} \frac{\delta Q}{T} = \int_{1\to 2} \frac{\delta Q_L}{T_L} + \int_{2\to 1} \frac{-\delta Q_H}{T_H} \\ &= \frac{1}{T_L}\int_{1\to 2} \delta Q_L + \frac{1}{T_H}\int_{2\to 1} (-\delta Q_H) = \frac{Q_L}{T_L} - \frac{Q_H}{T_H} \end{aligned} \tag{5.52}$$

[2]Examples include: Electric work dissipation via a resistor. Transformation of mechanical work into heat via friction heating.

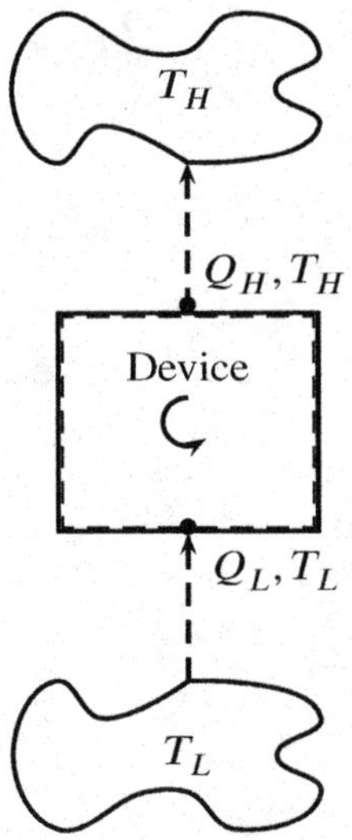

or

$$\oint \frac{\delta Q}{T} = \frac{Q_L}{T_L} - \frac{Q_H}{T_H} \tag{5.53}$$

Application of the energy-balance over the cycle gives

$$\overset{Q_L - Q_H}{\cancel{\oint \delta Q}} - \overset{0}{\cancel{\oint \delta W}} = \overset{0}{\cancel{\oint dE}} \Rightarrow Q_L - Q_H = 0 \tag{5.54}$$

or

$$Q = Q_L = Q_H \tag{5.55}$$

Substitution of (5.55) into (5.53) gives

$$\oint \frac{\delta Q}{T} = \frac{Q_L}{T_L} - \frac{Q_H}{T_H} = \frac{Q}{T_L} - \frac{Q}{T_H} = Q\left[\frac{1}{T_L} - \frac{1}{T_H}\right] \overset{?}{\leq} 0 \tag{5.56}$$

Since, Q is positive by definition, (5.56) implies $\left[\frac{1}{T_L} - \frac{1}{T_H}\right] \overset{?}{\leq} 0$, but since $T_L < T_H$ this implies $\left[\frac{1}{T_L} - \frac{1}{T_H}\right] > 0 \not\leq 0$. Thus, the device (a Clausius violator) violates the Clausius Inequality. ■

CHAPTER 6

ENTROPY AND THE ENTROPY-BALANCE

We ended the previous chapter by developing a mathematical restatement of the Second-Law as opposed to an "artsy-pooh" principle based on pictures: The result was the Clausius Inequality. We will next use this inequality as a motivator in hypothesizing and then establish the existence of a new property, entropy, and its corresponding conservation equation.

6.1 ENTROPY: EXISTENCE

We begin with a simple hypothetical question: "According to the Clausius Inequality, what can we say about an arbitrary reversible cyclic process $1 \to 2 \to 1$?" In §5.6 we noted that for a reversible cycle, the Clausius Inequality gives $\oint \frac{\delta Q}{T} = 0$. Figure 6.1 b) gives a control volume containing a closed system – – –– which, at any given point in the cycle, accepts heat δQ at an inter-action temperature T_{int}. Note, that if the cycle is *internally* reversible (no irreversibilities occur *within* the system), the temperature of the system is uniform and equal to the temperature at the system boundary (otherwise we would have heat transfer across a finite temperature difference which is not reversible). Thus, the inter-action temperature T_{int}, the temperature at the point on the system boundary where heat crosses the system control volume, must equal the system temperature T. Figure 6.1 a) gives two possible process paths from state $1 \to 2$ and the single return process $2 \to 1$ along a different, but still reversible path: We will refer to these paths as C, B and A respectively. Please note, *internally* reversible here means there are no effects due to friction, unrestrained expansion or heat transfer across a finite temperature difference *within* the system! Thus,

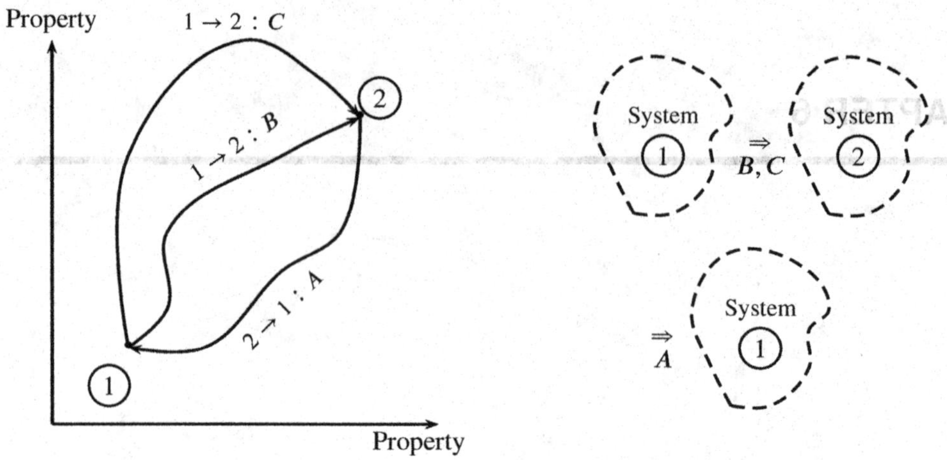

a) Reversible cyclic process 1 → 2 → 1

b) System evolution over cycle

Figure 6.1 Illustration of the cyclic reversible process paths 1 → 2 → 1.

the Clausius Inequality (5.51) for the cycle process along path $1 \to 2 : B$ and $2 \to 1 : A$ yields

$$\oint \left[\frac{\delta Q}{T}\right]_{rev} = \int_{1\to 2:B} \left[\frac{\delta Q}{T}\right]_{rev} + \int_{2\to 1:A} \left[\frac{\delta Q}{T}\right]_{rev} = 0 \tag{6.1}$$

or

$$\int_{1\to 2:B} \left[\frac{\delta Q}{T}\right]_{rev} = -\int_{2\to 1:A} \left[\frac{\delta Q}{T}\right]_{rev} \tag{6.2}$$

Now, this is an interesting result for let us vary the process $1 \to 2$ along path B by letting path $B \Rightarrow C$ as shown in Figure 6.1 a). The result is that no matter what reversible path is taken during the process $1 \to 2$ the result is always the same:

$$\overbrace{\int_{1\to 2:\text{Any Path}} \left[\frac{\delta Q}{T}\right]_{rev}}^{\text{Path independent}} = - \overbrace{\int_{2\to 1:A} \left[\frac{\delta Q}{T}\right]_{rev}}^{\text{Fixed Process}\Rightarrow\text{Fixed Value}} \tag{6.3}$$

Hence, the integral $\int_{1\to 2} \left[\frac{\delta Q}{T}\right]_{rev}$ is path independent: This *strongly suggests* the existence of a property we will henceforth refer to as entropy S defined initially via the expression

$$dS = \left[\frac{\delta Q}{T}\right]_{rev} \Rightarrow \Delta S_{1\to 2} = \int_{1\to 2} \left[\frac{\delta Q}{T}\right]_{rev} \tag{6.4}$$

Let us now relax our restriction on the reversibility of the cycle and let process $1 \to 2$ be irreversible. The Clausius Inequality gives (while relaxing the path subscripts A, B and C)

$$\oint \frac{\delta Q}{T} = \int_{1\to 2} \left[\frac{\delta Q}{T}\right]_{irr} + \underbrace{\int_{2\to 1} \left[\frac{\delta Q}{T}\right]_{rev}}_{\Delta S_{2\to 1}} \leq 0 \tag{6.5}$$

or

$$\Delta S_{1\to 2} = -\Delta S_{2\to 1} \geq \int_{1\to 2} \left[\frac{\delta Q}{T}\right]_{irr} \tag{6.6}$$

Well (6.6) looks promising except for the presence of the inequality. What do we do? Introduce a new quantity (not a property) call $S_{generated}$ or S_{gen} with the following characteristics

$$S_{gen_{1\to 2}} = \begin{cases} 0, \text{Process } 1 \to 2 \text{ is reversible} \\ > 0, \text{Process } 1 \to 2 \text{ is irreversible} \end{cases} \tag{6.7}$$

used in the closed system entropy-balance

$$\Delta S_{1\to 2} = \int_{1\to 2} \left[\frac{\delta Q}{T}\right] + S_{gen_{1\to 2}}, \text{ (Closed-system entropy-balance)} \tag{6.8}$$

Let's take a moment to look at how this balance behaves depending on whether cycle $1 \to 2 \to 1$ in Figure 6.1 is reversible or irreversible:

Process $1 \to 2 \to 1$ is reversible: In the case of a reversible process (6.7) gives $S_{gen_{1\to 2}}, S_{gen_{2\to 1}} = 0$ and the entropy-balance (6.8) for process $1 \to 2$ and $2 \to 1$ gives $\Delta S_{1\to 2} = \int_{1\to 2} \left[\frac{\delta Q}{T}\right]_{rev}$ and $\Delta S_{2\to 1} = \int_{2\to 1} \left[\frac{\delta Q}{T}\right]_{rev}$ respectively. This result allows us to recover the Clausius Inequality for reversible cyclic process $1 \to 2 \to 1$:

$$\oint \left[\frac{\delta Q}{T}\right]_{rev} = \int_{1\to 2} \left[\frac{\delta Q}{T}\right]_{rev} + \int_{2\to 1} \left[\frac{\delta Q}{T}\right]_{rev} = \Delta S_{1\to 2} + \cancelto{-\Delta S_{1\to 2}}{\Delta S_{2\to 1}} = 0 \tag{6.9}$$

Process $1 \to 2 \to 1$ is irreversible: In the case of an irreversible process (6.7) gives $S_{gen_{1\to 2}}, S_{gen_{2\to 1}} > 0$ and the entropy-balance (6.8) for process $1 \to 2$ and $2 \to 1$ gives $\Delta S_{1\to 2} = \int_{1\to 2} \left[\frac{\delta Q}{T}\right] + S_{gen_{1\to 2}}$ and $\Delta S_{2\to 1} = \int_{2\to 1} \left[\frac{\delta Q}{T}\right] + S_{gen_{2\to 1}}$ respectively. This result allows us to recover the Clausius Inequality for a irreversible cyclic process:

$$\oint \left[\frac{\delta Q}{T}\right]_{irr} = \int_{1\to 2} \left[\frac{\delta Q}{T}\right] + \int_{2\to 1} \left[\frac{\delta Q}{T}\right] = \left(\cancel{\Delta S_{1\to 2}} - S_{gen_{1\to 2}}\right)$$
$$+ \left(\cancel{\Delta S_{2\to 1}} - S_{gen_{2\to 1}}\right) = -\left(S_{gen_{1\to 2}} + S_{gen_{2\to 1}}\right) = -S_{gen_{1\to 2\to 1}} < 0 \tag{6.10}$$

6.2 ENTROPY-BALANCE: PHYSICAL INTERPRETATION

At this point *all* students are left in a state of profound puzzlement over the meaning of the entropy-balance (6.8). Like the energy-balance (the First -Law) for a closed system, the terms in the entropy-balance relate changes in a property, in this case entropy S, to the transfer of the property (entropy) in or out of the system via a heat inter-action. However, unlike energy E or mass m which can neither be created nor destroyed, the entropy S-balance contains a generation term S_{gen}: Entropy can be created according to (6.7). Thus, we interpret the S-balance (6.8) term-by-term as follows:

$$\overbrace{\Delta S_{1\to 2}}^{\text{Change in } S \text{ of closed system}} = \overbrace{\int_{1\to 2} \left[\frac{\delta Q}{T_{int}}\right]}^{S \text{ Transfer due to a heat inter-action}} + \overbrace{S_{gen_{1\to 2}}}^{S \text{ generated within closed system due to presence of irreversibilities}} \tag{6.11}$$

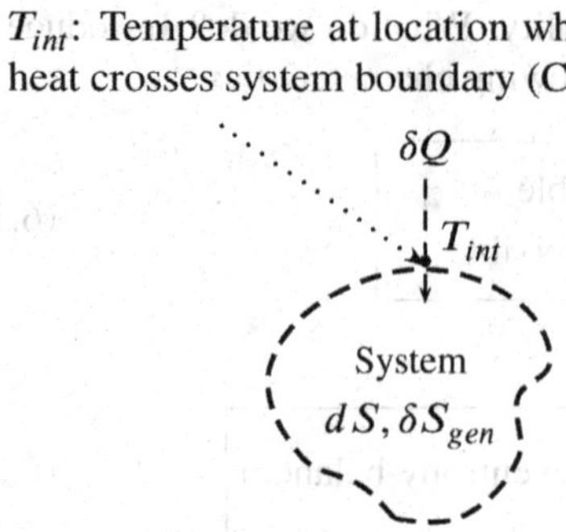

Figure 6.2 Differential process with infinitesimal entropy transfer $\delta Q/T_{int}$ due to infinitesimal heat transfer δQ.

At this juncture the student might appreciate some elaboration of the entropy transfer term due to a heat inter-action: $\int \left[\frac{\delta Q}{T_{int}}\right]$. Figure 6.2 illustrates a closed system receiving a transfer of an infinitesimal amount of heat δQ through a location at the system boundary at a temperature T_{int} resulting in a transfer into the system of an infinitesimal amount of entropy $\delta Q/T_{int}$: A heat inter-action transfers entropy as well as energy.

Finally, we should note that for reasons of general applicability, there may be multiple heat inter-actions occurring at various locations on the system boundary (control volume). Thus, we need to sum over all heat inter-actions yielding

$$\overbrace{\Delta S_{1\to2}}^{\text{Change in entropy } S \text{ of closed system}} = \overbrace{\sum \int_{1\to2} \left[\frac{\delta Q}{T_{int}}\right]}^{\text{Total entropy } S \text{ transfer due to all heat inter-actions}} + \overbrace{S_{gen_{1\to2}}}^{\text{Entropy } S \text{ generated within closed system due to irreversibilities}} \tag{6.12}$$

where the physical locations at which heat is transfer have corresponding temperatures T_{int}.

One last comment is warranted: The entropy-balance (6.12) does not include any dependency on a work inter-action! Heat transports energy and entropy, while work only transfers energy! Hence, as claimed at the end of §1.1, work and heat are not equivalent!

6.2.1 VERSIONS OF THE ENTROPY-BALANCE

Like work, heat, and the energy-balance, the entropy-balance for a closed system can take on a number of forms all of which are equivalent. Table 6.1 gives the various bases of the closed system entropy-balance:

Table 6.1 Various bases for the S-balance with dimensions and units

Basis	Symbol	Dimension	Unit (metric)	Unit (Imperial)
Overall process	$\Delta S = \sum \int \left[\frac{\delta Q}{T_{int}}\right] + S_{gen}$	Energy/Deg.	J/°K	BTU/°R
Specific overall	$\Delta s = \sum \int \left[\frac{\delta q}{T_{int}}\right] + s_{gen}$	$\frac{\text{Energy}}{\text{mass-Deg.}}$	$\frac{J}{kg-{}^\circ K}$	$\frac{BTU}{slug-{}^\circ R}$
Infinitesimal	$dS = \sum \left[\frac{\delta Q}{T_{int}}\right] + \delta S_{gen}$	Energy/Deg.	J/° K	BTU/°R
Rate	$\dot{S} = \sum \left[\frac{\dot{Q}}{T_{int}}\right] + \dot{S}_{gen}$	$\frac{\text{Energy}}{\text{time-Deg.}}$	$\frac{J}{s-{}^\circ K}$	$\frac{BTU}{s-{}^\circ R}$

6.2.2 SOURCES OF IRREVERSIBILITY

Recall that all "real or natural process" are irreversible, hence, $S_{gen} > 0$ for all natural process. Conversely, reversible processes result in no entropy generation, but do not spontaneously occur in nature: You might say, reversible processes are "for the black-board only".

Application of the closed system rate-based entropy-balance to the entire universe, an isolated system as shown in Figure 6.3, yields

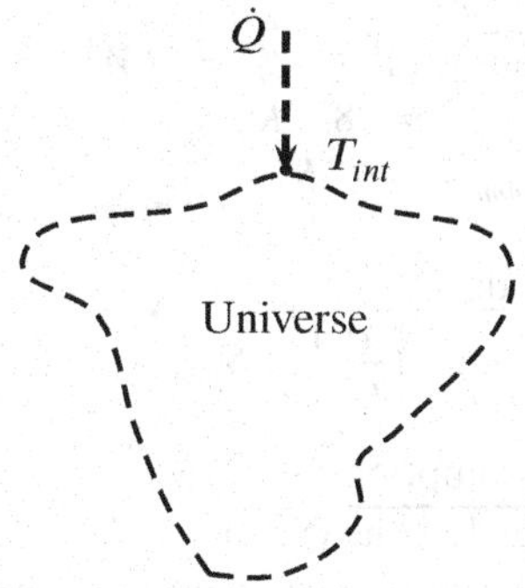

Figure 6.3 Control volume surrounding the entire Universe.

$$\dot{S}_{universe} = \frac{\overset{0,isolated}{\cancel{\dot{Q}_{universe}}}}{T_{int}} + \dot{S}_{universe_{gen}} = \dot{S}_{universe_{gen}} > 0 \tag{6.13}$$

Hence, the progression of events in the *universe*, *all* of which are irreversible, results in a continual generation of and an increase in entropy within the universe!

The next question now arises as to where the entropy is generated during a process. Given that the universe is composed of a system and the environment (see §5.1) there are several possibilities:

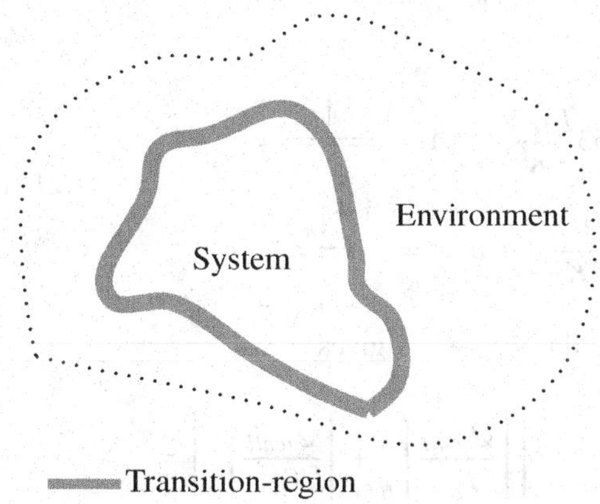

Figure 6.4 Illustration of transition region.

1. Irreversible processes occur within the system $\Rightarrow S_{system_{gen}} > 0$.
2. Irreversible processes occur within the environment $\Rightarrow S_{environment_{gen}} > 0$.
3. Irreversible processes occur within the thin boundary-region between the system and the environment.

The concept of the boundary or transition-region between systems, (in our case the system and the environment), is shown in Figure 6.4 and illustrated in the following example:

EXAMPLE 6.1 Flow of heat through a plane-wall: Entropy-balance

Problem Statement:
A building wall separates a room containing air at a temperature of 298° K from the outside air (the environment) at a temperature of 283° K. Due to the temperature difference between the inner and outer wall surface, the wall exhibits heat transfer at a rate of $1kW$. Determine the entropy transfer at the inner and outer wall surfaces due to the heat inter-action, and the rate of entropy generation within the wall.

Given:
Plane-wall, $\dot{Q}_{wall} = 1kW$,
$T_{outside} = 283°K$,
$T_{room} = 298°K$

Schematic:

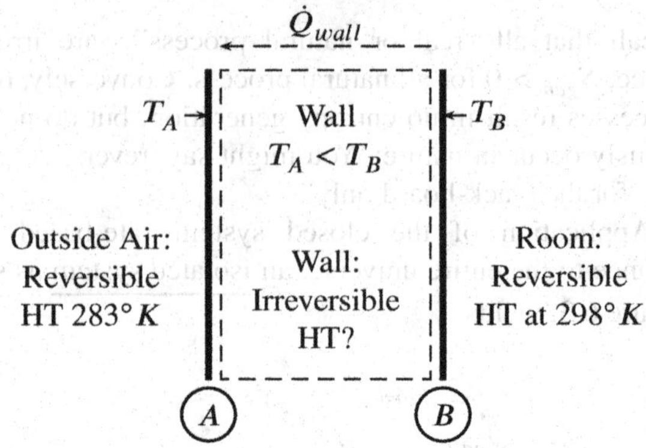

Find:
$\left[\frac{\dot{Q}}{T_{int}}\right]_A$, $\left[\frac{\dot{Q}}{T_{int}}\right]_B$, $\dot{S}_{gen}$

Assumptions:
Steady-state (S.S.),
HT from room air to inner wall surface is reversible,
HT from room wall outer surface to the outside air is reversible

Analysis:
From the entropy-balance (rate-based in Table 6.1) the term $\dot{Q}/T_{int}$ represents the entropy transfer due to a heat inter-action. If we assume that the heat transfer from the air to the inner wall surface is reversible, there must be no temperature difference between the inside air and the inner wall surface (else we would have heat transfer across a finite temperature difference ΔT). Hence, at the inner wall $T_{int} = T_{wall_B} = T_{room}$. Likewise, if we assume that the heat transfer at the outer wall surface to the outside air is reversible, then at the outer wall $T_{int} = T_{wall_A} = T_{outside}$. Thus, for our wall system the entropy transfer per-unit time is

$$\begin{cases} \left[\dfrac{\dot{Q}}{T_{int}}\right]_A = \dfrac{-\dot{Q}_{wall}}{T_{wall_A}} = \dfrac{-1kW}{283°K} = -0.00353\dfrac{kW}{°K}, \text{ (at } A) \xleftarrow{\left[\frac{\dot{Q}}{T_{int}}\right]_A} \\ \left[\dfrac{\dot{Q}}{T_{int}}\right]_B = \dfrac{\dot{Q}_{wall}}{T_{wall_B}} = \dfrac{1kW}{298°K} = 0.00335\dfrac{kW}{°K}, \text{ (at } B) \xleftarrow{\left[\frac{\dot{Q}}{T_{int}}\right]_B} \end{cases} \tag{6.14}$$

Application of the rate-based entropy-balance to the wall CV – – – – gives

$$\overset{0,\,S.S.}{\cancel{\dot{S}}} = \sum \frac{\dot{Q}}{T_{int}} + \dot{S}_{gen} \Rightarrow \dot{S}_{gen} = -\sum \frac{\dot{Q}}{T_{int}} = -\left[\left[\frac{\dot{Q}_{wall}}{T_{int}}\right]_B + \left[\frac{\dot{Q}_{wall}}{T_{int}}\right]_A\right] =$$

$$-\left[\frac{\dot{Q}_{wall}}{T_{wall_B}} + \frac{-\dot{Q}_{wall}}{T_{wall_A}}\right] = -0.00335\frac{kW}{°K} + 0.00353\frac{kW}{°K} = 0.00018\frac{kW}{°K} \xleftarrow{\dot{S}_{gen}} \tag{6.15}$$

$$\boxed{\left[\frac{\dot{Q}}{T_{int}}\right]_A = -0.00353\frac{kW}{°K}, \left[\frac{\dot{Q}}{T_{int}}\right]_B = 0.00335\frac{kW}{°K}, \dot{S}_{gen} = 0.00018\frac{kW}{°K}}$$

Comment: Heat transfer through the wall (which exhibits a finite temperature difference) is irreversible, resulting in entropy generation within the wall at a rate of $0.00018\frac{kW}{°K}$.

□

EXAMPLE 6.2 Isothermal heat rejection to environment: Entropy-balance

Problem Statement:
A piston-cylinder device containing air is compressed isothermally at a temperature of $25°C$ receiving boundary work at a rate of $12kW$. The piston-cylinder device is allowed to inter-act with the environment which is at a temperature $10°C$. Determine the rate of entropy transfer from the air and to the environment.

Given:
Air,
isothermal compression,
$T_{air} = 25°C = 298°K$,
$\dot{W}_{bound} = -12kW$,
$T_{env} = 10°C = 283°K$

Find:
$\left[\frac{\dot{Q}}{T_{int}}\right]_{air}$,
$\left[\frac{\dot{Q}}{T_{int}}\right]_{env}$

Assumptions:
Reversible heat transfer in air and environmental system

Schematic:

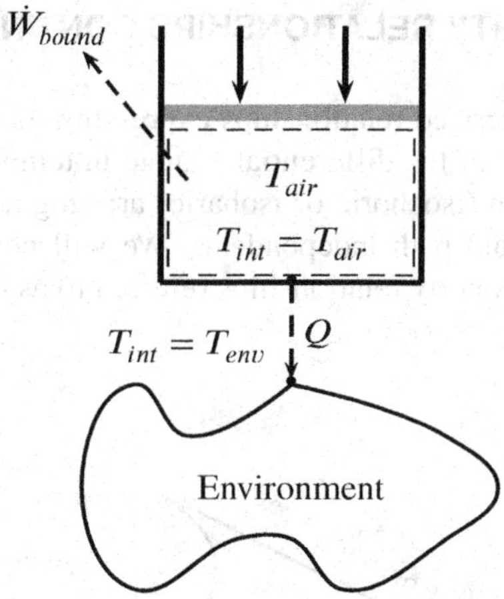

Analysis:
From Table 6.1 the entropy-balance (rate-based) applied to the air-system, we note that the entropy transfer to the air is given by $\frac{\dot{Q}}{T_{int}}$. Now, the heat transfer within the air system is assumed to be internally reversible, hence $T_{int} = T_{air}$ and $\frac{\dot{Q}}{T_{int}} = \frac{\dot{Q}_{air}}{T_{air}}$. Application of the rate-based energy-balance to the air system gives

$$\overset{\dot{Q}_{air}}{\not{\dot{Q}}} - \dot{W}_{bound} = \dot{E} = \overset{0,\, isothermal}{\not{\dot{U}(T)}} \Rightarrow \dot{Q}_{air} = \dot{W} = -12kW \tag{6.16}$$

Hence, the entropy transfer into the air-system is

$$\left[\frac{\dot{Q}}{T_{int}}\right]_{air} = \frac{\dot{Q}_{air}}{T_{air}} = \frac{-12kW}{298°K} = -0.040\frac{kW}{°K} \xLeftarrow{\left[\frac{\dot{Q}}{T_{int}}\right]_{air}} \tag{6.17}$$

(i.e. entropy is transferred from the air).

From the entropy-balance (rate-based) applied to the environmental-system we note that the entropy transfer to the environment is given by $\frac{\dot{Q}_{env}}{T_{int}}$. The heat transfer within the environment is also assumed to be internally reversible, hence $T_{int} = T_{env}$ and

$$\frac{\dot{Q}_{env}}{T_{int}} = \frac{\dot{Q}_{env}}{T_{env}} = -\frac{\dot{Q}_{air}}{T_{env}} = -\frac{(-12kW)}{283°K} = \frac{12kW}{283°K} = 0.0424\frac{kW}{K} \xLeftarrow{\left[\frac{\dot{Q}}{T_{int}}\right]_{env}} \tag{6.18}$$

$$\boxed{\left[\frac{\dot{Q}}{T_{int}}\right]_{air} = -0.040\frac{kW}{^{o}K}, \left[\frac{\dot{Q}}{T_{int}}\right]_{env} = 0.0424\frac{kW}{K}}$$

Comment: Since the rate of entropy transfer from the air is less than the rate of transfer to the environment, where is the excess entropy being generated? The answer is within the transition-region between the air-system and the environmental-system which exhibits heat transfer across a finite temperature difference: An irreversible process.

□

6.3 PROPERTY RELATIONSHIPS CONTINUED: THE Tds-EQUATIONS

In §4.3 we derived relationships expressing the differential change in internal energy and enthalpy as a function of the differential change in temperature. We did this by choosing a convenient path of integration (isochoric or isobaric) arriving at a set of differential equations which turned out to be general and path independent. We will now repeat this procedure for entropy S to arrive at additional property relationships referred to as the Tds-equations.

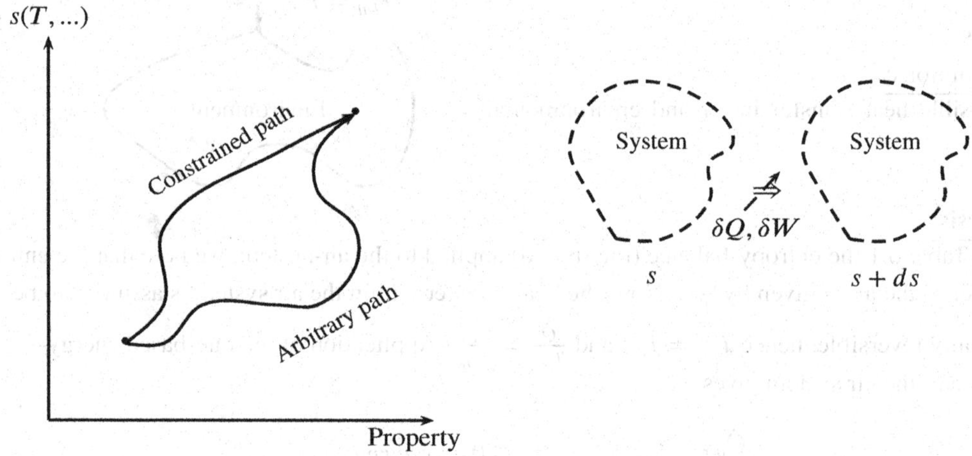

Figure 6.5 Pictorial illustration of constrained differential process.

Let us begin by evaluating the infinitesimal change in entropy between two states along a constrained path. The constraints on the path are as follows:

1. Process is internally reversible (no irreversibilities occur within the system). Thus $\delta S_{gen} = 0$ and $T_{int} = T$.

2. Over the process $\delta W = \delta W_{bound}$

3. No changes in potential and kinetic energy over the process

This constrained path is shown in Figure 6.5. The energy and entropy-balances give

$$\delta Q - \overset{\delta W_{bound} = PdV}{\delta W} = \overset{dU}{dE} \Rightarrow \delta Q - PdV = dU \Rightarrow \delta q - Pdv = du \qquad (6.19)$$

and

$$dS = \frac{\delta Q}{T}_{int} + \delta S_{gen} \Rightarrow dS = \frac{\delta Q}{T} \Rightarrow ds = \frac{\delta q}{T} \tag{6.20}$$

Solving (6.20) for δq and substituting into (6.19) gives

$$Tds - Pdv = du \Rightarrow Tds = du + Pdv \tag{6.21}$$

Thus (6.21) relates changes in specific entropy s as a function of changes in specific volume v and internal energy u. Note, we can recast (6.21) in terms of changes in enthalpy h by taking the total derivative of enthalpy

$$h = u + Pv \Rightarrow dh = d(u + Pv) = du + d(Pv) = du + Pdv + vdP \tag{6.22}$$

or

$$du + Pdv = dh - vdP \tag{6.23}$$

Substituting of (6.23) into (6.21) gives

$$Tds = du + Pdv = dh - vdP \Rightarrow Tds = dh - vdP \tag{6.24}$$

Now, it was shown by the Clausius inequality that specific entropy s, like h, u, v and p is a property or point-function, thus (6.21) and (6.24) are path independent and valid for any process! That is, although the appearance of Pdv in equation (6.24) would imply that we need the path, i.e. P(v), we would discover that when solving for Δs for any and every path, we would in fact obtain the same value. Hence, we arrive a the Tds equations valid for all processes:

$$Tds = \begin{cases} du + Pdv \\ dh - vdP \end{cases} \tag{6.25}$$

We will now proceed to determine the reduced forms of (6.25) for various substances (e.g. incompressibles, substances exhibiting I.G.B., processes with and without phase changes, etc.).

6.3.1 PROCESSES WITHOUT PHASE CHANGE

If the system is compressible and the process does not exhibit a phase change, we can utilize specific heats via (4.34) $du = C_v(T, P)dT$ and (4.38) $dh = C_P(T, v)dT$. Thus, (6.25) reduces to

$$Tds = \begin{cases} C_v(T, P)dT + Pdv \\ C_P(T, v)dT - vdP \end{cases} \tag{6.26}$$

6.3.2 INCOMPRESSIBLE SYSTEMS

For an incompressible system where v is a constant we have $dv = 0$ for all processes. Thus, combining (4.34) with (6.25), recalling that for incompressible system $C_P = C_v$, gives

$$Tds = du + Pdv \Rightarrow ds = \frac{du}{T} = \frac{C_v(T)dT}{T} = \frac{C_P(T)dT}{T} \tag{6.27}$$

or

$$ds = \frac{C_P(T)dT}{T}, \text{ (Valid for incompressible systems)} \tag{6.28}$$

To evaluate the overall change in entropy of an incompressible system we simply integrate (6.28) yielding

$$\Delta s_{1\to 2} = \int_{1\to 2} ds = \int_{T_1}^{T_2} \frac{C_P(T)dT}{T} \tag{6.29}$$

If temperature variations are sufficiently small, a constant or average specific heat $C_{P_{av.}}$ may be assumed and (6.29) can be integrated giving

$$\Delta s_{1\to 2} = \int_{T_1}^{T_2} \frac{C_P(T)dT}{T} = C_{P_{av.}} \int_{T_1}^{T_2} \frac{dT}{T} = C_{P_{av.}} ln\left[\frac{T_2}{T_1}\right] \tag{6.30}$$

6.3.3 IDEAL GAS BEHAVIOR

Beginning with (6.26) and solving for ds we have

$$ds = \begin{cases} \dfrac{C_v(T,P)}{T}dT + \dfrac{P}{T}dv \\ \\ \dfrac{C_P(T,v)}{T}dT - \dfrac{v}{T}dP \end{cases} \tag{6.31}$$

In the case of I.G.B. we have specific heats as a function of temperature alone ($C_v(T)$ and $C_P(T)$) and via the ideal gas law $P/T = R/v$ and $v/T = R/P$. Thus (6.31) becomes

$$ds = \begin{cases} \dfrac{C_v(T)}{T}dT + \dfrac{R}{v}dv, \text{ (Valid for I.G.B.)} \\ \\ \dfrac{C_P(T)}{T}dT - \dfrac{R}{P}dP, \text{ (Valid for I.G.B.)} \end{cases} \tag{6.32}$$

Integrating (6.32) gives

$$\Delta s_{1\to 2} = \int_{1\to 2} ds = \begin{cases} \displaystyle\int_{T_1}^{T_2} \frac{C_v(T)}{T}dT + \int_{v_1}^{v_2} \frac{R}{v}dv = \int_{T_1}^{T_2} \frac{C_v(T)}{T}dT + Rln\left[\frac{v_2}{v_1}\right] \\ \\ \displaystyle\int_{T_1}^{T_2} \frac{C_P(T)}{T}dT - \int_{P_1}^{P_2} \frac{R}{P}dP = \int_{T_1}^{T_2} \frac{C_P(T)}{T}dT - Rln\left[\frac{P_2}{P_1}\right] \end{cases} \tag{6.33}$$

At this point we may choose to assume constant specific heats if the temperature varies only slightly over the process (as discussed in §4.3.1 where I.G.B. specific heats as a function of temperature for various gases are provided in Table B.10). If not, polynomial expressions for specific heats as a function of temperature are given in Table B.9. Alternatively, we note that $\int_{T_1}^{T_2} \frac{C_P(T)}{T}dT$ and $\int_{T_1}^{T_2} \frac{C_v(T)}{T}dT$ are both a function of temperature alone and can be evaluated. Thus, we define

$$s_0(T_2) - s_0(T_1) = \int_{T_1}^{T_2} \frac{C_P(T)}{T}dT \tag{6.34}$$

where the tabulated quantity $s_0(T)$ is a function of temperature alone and provided in I.G.B. property Table B.11. Substituting (6.34) into the second expression in (6.33) gives

$$\Delta s_{1\to 2} = \int_{T_1}^{T_2} \frac{C_P(T)}{T} dT - Rln\left[\frac{P_2}{P_1}\right] = s_0(T_2) - s_0(T_1) - Rln\left[\frac{P_2}{P_1}\right] \quad (6.35)$$

which is valid for I.G.B. for variable specific heats.

One final note, it is often useful to recast the property relationships on a per-unit mole (as opposed to a per-unit mass) basis. In that case, u, h, s_o and s are replaced by $\overline{u}$, $\overline{h}$, $\overline{s}_o$ and $\overline{s}$, and if applicable, specific heats and the gas constant C and R are replaced by $\overline{C}$ and R_U.

EXAMPLE 6.3 Irreversibility of unrestrained expansion

Problem Statement:

Two rigid insulated containers of equal volume are connected via a valve: One container contains $5kmol$ of a gas while the other contains a vacuum. The valve is opened and the gas is allowed to expand to fill the empty container. Find the change in temperature and entropy of the gas. Also find the entropy generated due to the expansion of the gas. Assume I.G.B.

Given:

Connected rigid insulated containers of equal volume,

Initial state:

$N_{gas} = 5kmol$ (container one),

vacuum (container two)

Find:

$\Delta T_{1\to 2}$,

$\Delta S_{1\to 2}$,

$S_{gen_{1\to 2}}$

Assumptions:

I.G.B., $Q_{1\to 2} = 0$ (insulated),

$W_{1\to 2} = W_{bound_{1\to 2}} = 0$ (isochoric),

$\Delta KE_{1\to 2} = \Delta PE_{1\to 2} = 0$

Schematic:

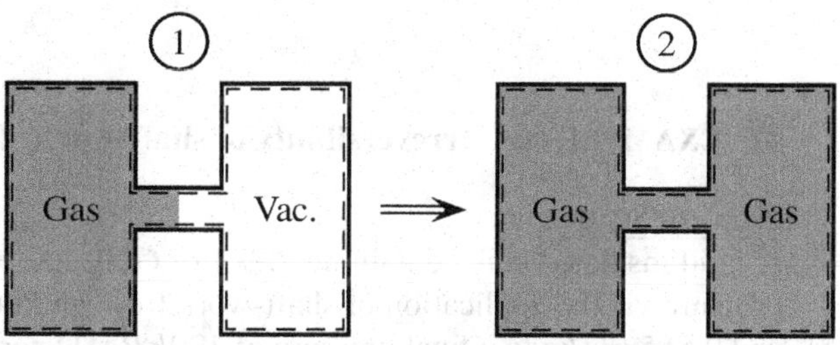

$V_{gas_1} = V_{vacuum}$ $\qquad$ $V_{gas_2} = 2V_{gas_1}$

Analysis:

Application of the closed system energy-balance to gas-vacuum system – – –– gives (where $\overline{u}$ is on a per-kmol basis)

$$\cancelto{0}{Q}_{1\to 2} - \cancelto{0}{W}_{1\to 2} = \Delta E_{1\to 2} = \Delta U_{1\to 2} = N_{gas}(\overline{u}_2 - \overline{u}_1) \Rightarrow 0 = \overline{u}_2 - \overline{u}_1 \Rightarrow \overline{u}_2 = \overline{u}_1 \quad (6.36)$$

or since $u(T)$ for I.G.B.

$$\overline{u}(T_2) = \overline{u}(T_1) \Rightarrow \Delta T_{1\to 2} = 0 \xLeftarrow{\Delta T_{1\to 2}} \quad (6.37)$$

Thus, the process 1 → 2 is isothermal. Turning to entropy, for I.G.B. (6.33) is valid and can be rewritten in terms of moles (where $\overline{s}$ and $\overline{C}_v$ are on a per-unit mole basis)

$$\Delta S_{1\to2} = N_{gas}\Delta\overline{s}_{1\to2} = N_{gas}\left[\int_{T_1}^{T_2}\frac{\overline{C}_v(T)}{T}dT + R_u ln\left(\frac{V_2}{V_1}\right)\right] \quad (0,\ T_2 = T_1;\ V_2 = 2V_1)$$

$$= N_{gas}R_U ln\left(\frac{2V_1}{V_1}\right) = 5kmol\left(8.3\frac{kJ}{kmol-^\circ K}\right)ln(2) = 28.7\frac{kJ}{^\circ K} \xLeftarrow{\Delta S_{1\to2}} \quad (6.38)$$

Application of overall S-balance Table 6.1 (p.108) to gas-vacuum system — — — gives

$$\Delta S_{1\to2} = \int\frac{\delta Q}{T_{int}}^{\nearrow 0,adia.} + S_{gen_{1\to2}} = S_{gen_{1\to2}} \quad (6.39)$$

or

$$S_{gen_{1\to2}} = \Delta S_{1\to2} = 28.7\frac{kJ}{^\circ K} \xLeftarrow{S_{gen_{1\to2}}} \quad (6.40)$$

$$\boxed{\Delta T_{1\to2} = 0,\ \Delta S_{1\to2} = 28.7\frac{kJ}{^\circ K},\ S_{gen_{1\to2}} = 28.7\frac{kJ}{^\circ K}}$$

Comment: Since $S_{gen_{1\to2}}$ is positive, unrestrained expansion of a gas is irreversible (as expected). Finally, the choice of system, the inner surface of the rigid (connected) containers, renders the process isochoric and serves to exclude the possibility of boundary work.

□

EXAMPLE 6.4 Irreversibility of shaft-work: Entropy-balance

Problem Statement:
A rigid insulated tank containing $2.7kg$ of CO_2 gas exhibits an increase in temperature and pressure via the application of shaft-work from an initial pressure and specfic volume of $100kPa$ and $0.555m^3/kg$ to a final pressure of $120kPa$. Determine the shaft work, change in entropy and entropy generated inside the tank due to the application of shaft-work. Assume constant specific heats evaluated at $325^\circ K$ where I.G.B. is valid.

Given:
Rigid and insulated tank with shaft-work,
CO_2,
$m = 2.7kg$,
$P_1 = 100kPa$,
$v_1 = 0.555\frac{m^3}{kg}$,
$C_P, C_V|_{T=325^\circ K}$,
$P_2 = 120kPa$

Schematic:

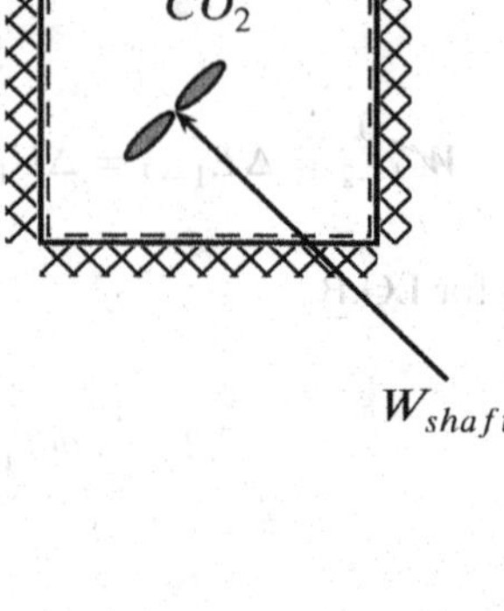

Find:
$W_{shaft_{1\to2}}$,
$\Delta S_{1\to2}$,
$S_{gen_{1\to2}}$

Assumptions:
I.G.B.,
$Q_{1\to2} = 0$ (insulated),
$W_{electric_{1\to2}} = W_{bound_{1\to2}} = 0$,
isochoric process (rigid container),
$\Delta KE_{1\to2} = \Delta PE_{1\to2} = 0$,
$C_v\big|_{325^\circ K} = 0.68 \frac{kJ}{kg-^\circ K}$ via I.G.B. Table B.10

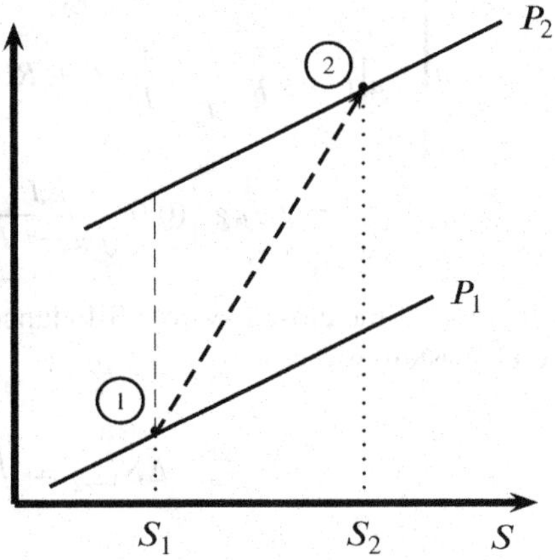

Analysis:
Application of the energy-balance to CO_2 system – – – – gives

$$\overset{0}{\cancel{Q}}_{1\to2} - \overset{W_{shaft_{1\to2}}}{\cancel{W}}_{1\to2} = \Delta E_{1\to2} = \Delta U_{1\to2} \Rightarrow$$

$$W_{shaft_{1\to2}} = -\Delta U_{1\to2} = -m(\Delta u)_{1\to2} = -\ mC_v\big|_{T=325^\circ K}\Delta T_{1\to2} \tag{6.41}$$

or

$$W_{shaft_{1\to2}} = -\ mC_v\big|_{T=325^\circ K}(T_2 - T_1) \tag{6.42}$$

From I.G.B. where $R_{CO2} = \frac{R_U}{MW_{CO2}} = \frac{8.314\frac{kJ}{\cancel{kmol}-^\circ K}}{44.01\frac{kg}{\cancel{kmol}}} = 0.188\frac{kJ}{kg-^\circ K}$

$$T_1 = \frac{P_1 v_1}{R_{CO_2}} = \frac{100\cancel{kPa}\left(0.555\frac{\cancel{m^3}}{\cancel{kg}}\right)}{0.188\frac{\cancel{kJ}}{\cancel{kg}-^\circ K}} = 295^\circ K \tag{6.43}$$

Now, process $1 \to 2$ is isochoric, thus via I.G.B.

$$V_1 = \frac{P_1}{\cancel{R_{CO_2}}T_1} = \frac{P_2}{\cancel{R_{CO_2}}T_2} = V_2 \Rightarrow T_2 = \left(\frac{P_2}{P_1}\right)T_1 = \left(\frac{120\cancel{kPa}}{100\cancel{kPa}}\right)295^\circ K = 354^\circ K \tag{6.44}$$

Hence, returning to the energy-balance (6.42)

$$W_{shaft_{1\to2}} = -\ mC_v\big|_{T=325^\circ K}(T_2 - T_1)$$

$$= -2.7\cancel{kg}\left(0.68\frac{kJ}{\cancel{kg}-^\circ \cancel{K}}\right)(354^\circ\cancel{K} - 295^\circ\cancel{K}) = -108kJ \overset{W_{shaft_{1\to2}}}{\Longleftarrow} \tag{6.45}$$

and we find that the work is negative (i.e. work is done on the system). Turning to entropy, for I.G.B. (6.33) is valid and

$$\Delta S_{1\to2} = m\Delta s_{1\to2} = m\left[\int_{T_1}^{T_2}\frac{\overset{C_v|_{325^\circ K}}{\cancel{C_v(T)}}}{T}dT + Rln\left(\frac{\overset{v_1}{\cancel{v_2}}}{v_1}\right)\right] =$$

$$m\left[C_v|_{T=325^\circ K}\int_{T_1}^{T_2}\frac{1}{T}dT + Rln\left(\frac{v_1}{v_1}\right)^{\nearrow 0}\right] = m\left[C_v|_{T=325^\circ K}\, ln\left(\frac{T_2}{T_1}\right)\right]$$

$$= 2.7kg\left[(0.68\frac{kJ}{kg-^\circ K})ln\left(\frac{354^\circ K}{295^\circ K}\right)\right] = 0.335\frac{kJ}{^\circ K} \xLeftarrow{\Delta S_{1\to 2}} \tag{6.46}$$

Application of the closed system S-balance for the overall process $1 \to 2$ from Table 6.1 (p.108) to the CO_2 system gives

$$\Delta S_{1\to 2} = \int \frac{\delta Q}{T}^{\nearrow 0,\, adia.} + S_{gen_{1\to 2}} = S_{gen_{1\to 2}} \tag{6.47}$$

or

$$S_{gen_{1\to 2}} = \Delta S_{1\to 2} = 0.335\frac{kJ}{^\circ K} \xLeftarrow{\Delta S_{gen_{1\to 2}}} \tag{6.48}$$

$$\boxed{W_{shaft_{1\to 2}} = -108kJ,\ \Delta S_{1\to 2} = 0.335\frac{kJ}{^\circ K},\ S_{gen_{1\to 2}} = 0.335\frac{kJ}{^\circ K}}$$

Comment: Since $S_{gen_{1\to 2}}$ is positive, processes involving shaft-work on a closed system are irreversible. Note that this is in contrast to boundary work which, provided the process is slow and frictionless, is generally reversible.

□

6.4 ISENTROPIC PROCESSES

In addition to isobaric, isochoric and isothermal processes, processes with constant entropy are referred to as isentropic processes.

Definition 6.1 ***Isentropic Processes:*** *A process defined by constant entropy of the system.*

EXAMPLE 6.5 Reversible isentropic compression of H_2O system

Problem Statement:

A piston-cylinder device containing $1kg$ of H_2O is compressed isentropically and reversibly from an initial temperature of $100°C$ and a quality of $x = 0.76$ till the final temperature is $400°C$. Determine the final pressure, heat transfer, work and the change in internal energy over the process.

Given:

H_2O,
$m = 1kg$,
isentropic,
reversible compression,
$T_1 = 100°C, T_2 = 400°C$,
$x_1 = 0.76$

Schematic:

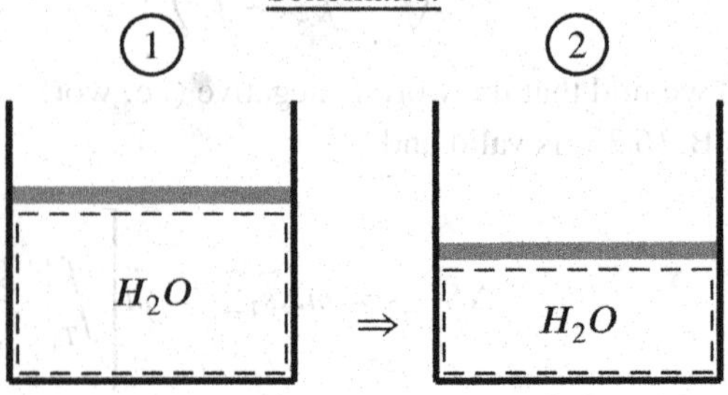

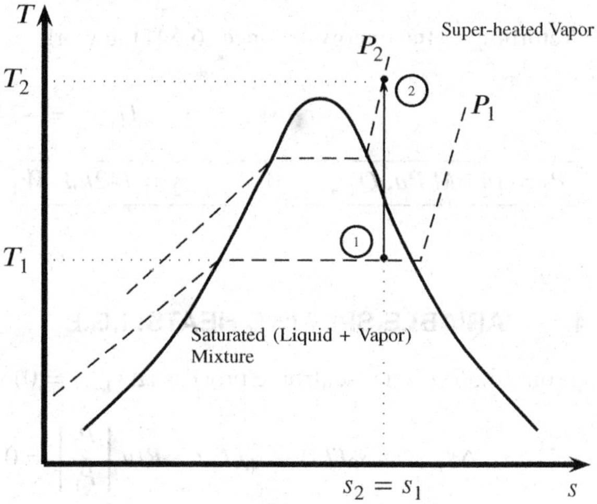

Find:
P_2,
$\Delta U_{1\to2}$,
$Q_{1\to2}$,
$W_{1\to2}$

Assumptions:
$W = W_{bound}$,
$\Delta KE_{1\to2} = \Delta PE_{1\to2} = 0$,
H_2O Saturation and steam Tables B.2 and B.4

Analysis:
Process 1 → 2 (isentropic compression): Applying the expression for quality (2.1) to specific entropy (a property of the system thus letting $v \to s$) gives, using saturation Tables B.2

$$s_2 = s_1 = s|_{\substack{T=100^\circ C,\\ x=0.76}} = s_{liq}^{sat}\Big|_{T=100^\circ C} + x_1\,(s_{vap}^{sat} - s_{liq}^{sat})\Big|_{T=100^\circ C}$$

$$= 1.307\frac{kJ}{kg-^\circ K} + 0.76(7.35 - 1.307)\frac{kJ}{kg-^\circ K} = 5.89\frac{kJ}{kg-^\circ K} \tag{6.49}$$

Now, $T_2 > T^{crit}$ (via Table B.1) so state 2 is super-heated vapor. Interpolating from super-heated vapor Tables B.4 we get

$$P_2 = P|_{\substack{T=400^\circ C,\\ s=5.89\frac{kJ}{kg-K}}} = 14.6MPa \xLeftarrow{P_2} \tag{6.50}$$

Applying the differential entropy-balance from Table 6.1 (p.108) for process 1 → 2 gives

$$\overset{0,isen.}{\cancel{dS}} = \frac{\delta Q}{T_{int}} + \overset{0,rev.}{\cancel{\delta S_{gen}}} \Rightarrow \frac{\delta Q}{T_{int}} = 0 \Rightarrow \delta Q = 0 \tag{6.51}$$

Integrating from state 1 → 2 gives

$$Q_{1\to2} = \int_{1\to2} \overset{0}{\cancel{\delta Q}} = 0 \xLeftarrow{Q_{1\to2}} \tag{6.52}$$

Thus, process 1 → 2 is adiabatic. Hence, from the closed system energy-balance

$$\overset{0}{\cancel{Q}}_{1\to2} - W_{1\to2} = \Delta E_{1\to2} = \Delta U_{1\to2} \tag{6.53}$$

or

$$W_{1\to2} = -\Delta U_{1\to2} \tag{6.54}$$

From saturation and super-heated vapor Tables B.2 and B.4 (6.54) we have

$$\Delta U_{1\to2} = m(\Delta u)_{1\to2} = m\left(u|_{\substack{T=400^\circ C,\\ s=5.83\frac{kJ}{kg-K}}} - u|_{\substack{T=100^\circ C,\\ x=0.76}} \right) =$$

$$1\cancel{kg}\left(2747\frac{kJ}{\cancel{kg}} - 2005\frac{kJ}{\cancel{kg}}\right) = 742kJ \xLeftarrow{\Delta U_{1\to2}} \tag{6.55}$$

Returning to the energy-balance (6.54) the work is found to be

$$W_{1\to 2} = -\Delta U_{1\to 2} = -742kJ \xLongleftarrow{W_{1\to 2}} \tag{6.56}$$

$$\boxed{P_2 = 14.6MPa,\ Q_{1\to 2} = 0,\ \Delta U_{1\to 2} = 742kJ,\ W_{1\to 2} = -742kJ}$$

□

6.4.1 VARIABLE SPECIFIC HEATS: I.G.B.

From (6.35) for an isentropic process ($\Delta s_{1\to 2} = 0$) we have

$$\Delta s_{1\to 2} = s_0(T_2) - s_0(T_1) - Rln\left[\frac{P_2}{P_1}\right] = 0 \Rightarrow s_0(T_2) = s_0(T_1) + Rln\left[\frac{P_2}{P_1}\right] \tag{6.57}$$

Hence, for an isentropic process $T_2(T_1, P_2, P_1)$. Now, to determine, for example T_2 as a function of final pressure and initial temperature and pressure, one could look-up $s_0(T_1)$ in ideal gas Table B.11, then calculate $s_0(T_2)$ and interpolate T_2. Alternatively, Table B.11 also provides reduced pressure P_r and specific volume v_r applicable for calculating isentropic process properties via the following relationships:

$$\boxed{\frac{P_2}{P_1} = \frac{P_{r_2}}{P_{r_1}},\ \frac{v_2}{v_1} = \frac{v_{r_2}}{v_{r_1}},\ \text{(Using Table B.11 for isentropic processes: I.G.B.)}} \tag{6.58}$$

For example, if the process $1 \to 2$ is isentropic and the temperature at state 2 is to be found using (6.58), proceed as follows:

i In Table B.11 locate row corresponding to $T = T_1$ and read $P_{r_1} = P_r$: Interpolation may be required.

ii Use (6.58) to calculate P_{r_2}: $P_{r_2} = P_{r_1}\frac{P_2}{P_1}$.

iii Find the row corresponding to $P_r = P_{r_2}$ and read $T_2 = T$: Interpolation may be required.

The above procedure also applies to reduced volume v_r.

6.4.2 CONSTANT SPECIFIC HEATS: I.G.B.

For sufficiently small variations in temperature, constant specific heats evaluated at some averaged value may be assumed. From (6.32) with $C_v(T) = C_v$ we have, noting that $C_P = C_v + R$ and $k = C_P/C_v$

$$ds = \frac{C_v}{T}dT + \frac{R}{v}dv \Rightarrow \overset{0}{\Delta s} = \int_{T_1}^{T_2}\frac{C_v}{T}dT + \int_{v_1}^{v_2}\frac{R}{v}dv = 0$$

$$\Rightarrow C_v ln\left[\frac{T_2}{T_1}\right] = -Rln\left[\frac{v_2}{v_1}\right] \Rightarrow \frac{C_v}{R}ln\left[\frac{T_2}{T_1}\right] = \frac{C_v}{C_P - C_v}ln\left[\frac{T_2}{T_1}\right] =$$

$$\frac{1}{k-1}ln\left[\frac{T_2}{T_1}\right] = -ln\left[\frac{v_2}{v_1}\right] \Rightarrow ln\left[\frac{T_2}{T_1}\right] = -(k-1)ln\left[\frac{v_2}{v_1}\right] \tag{6.59}$$

From the definition of the natural log

$$ln\left[\frac{T_2}{T_1}\right] = -(k-1)ln\left[\frac{v_2}{v_1}\right] \Rightarrow ln\left[\frac{T_2}{T_1}\right] = ln\left[\frac{v_2}{v_1}\right]^{-k+1} \Rightarrow \frac{T_2}{T_1} = \left[\frac{v_2}{v_1}\right]^{1-k} \tag{6.60}$$

Thus, for an isentropic process assuming constant specific heats

$$\frac{T_2}{T_1} = \left[\frac{v_2}{v_1}\right]^{1-k}, \text{ (Isentropic process: Constant } C_{P,v} \text{ for I.G.B.)} \tag{6.61}$$

Also, combining the ideal gas law $v = RT/P$ with (6.61) yields

$$\frac{T_2}{T_1} = \left[\frac{v_2}{v_1}\right]^{1-k} = \left[\frac{\frac{RT_2}{P_2}}{\frac{RT_1}{P_1}}\right]^{1-k} = \left[\frac{T_2}{T_1}\frac{P_1}{P_2}\right]^{1-k} = \left[\frac{T_2}{T_1}\right]^{1-k}\left[\frac{P_1}{P_2}\right]^{1-k} \Rightarrow$$

$$\frac{T_2}{T_1} = \left[\frac{T_2}{T_1}\right]^{1-k}\left[\frac{P_1}{P_2}\right]^{1-k} \Rightarrow \left[\frac{P_2}{P_1}\right]^{1-k} = \left[\frac{T_2}{T_1}\right]^{-k} \Rightarrow \left[\frac{P_2}{P_1}\right]^{\frac{k-1}{k}} = \left[\frac{T_2}{T_1}\right] \tag{6.62}$$

Thus, for an isentropic process assuming constant specific heats, we also have

$$\frac{T_2}{T_1} = \left[\frac{P_2}{P_1}\right]^{\frac{k-1}{k}}, \text{ (Isentropic process: Constant } C_{P,v} \text{ for I.G.B.)} \tag{6.63}$$

EXAMPLE 6.6 Carnot Heat Engine: I.G.B.

Problem Statement:
A Carnot heat engine operates between two reservoirs of temperature $1000°K$ and $500°K$. Over a cycle, the heat engine extracts $100kJ$ of heat per kg of working fluid from the high temperature reservoir. If the working fluid is air (exhibiting I.G.B.) with a minimum specific volume during the cycle of $0.1m^3/kg$, determine the pressure and specific volume after the heat addition process, pressure before the heat rejection process and the change in specific entropy over the heat addition process. Perform calculations for:
a) constant specific heats with $k = 1.4$
b) variable specific heats using I.G.B air property Tables B.11
Include a $P - v$ and $T - s$ diagrams.

Given:
Carnot Heat Engine,
air (I.G.B.),
$q_H = 100\frac{kJ}{kg}$,
$T_H = 1000°K$,
$T_L = 500°K$,
$v_{min} = v_1 = 0.1\frac{m^3}{kg}$,
$k = \frac{C_P}{C_V} = 1.4$ (part (a)),
I.G.B. Tables B.11 (part (b))

Schematic:

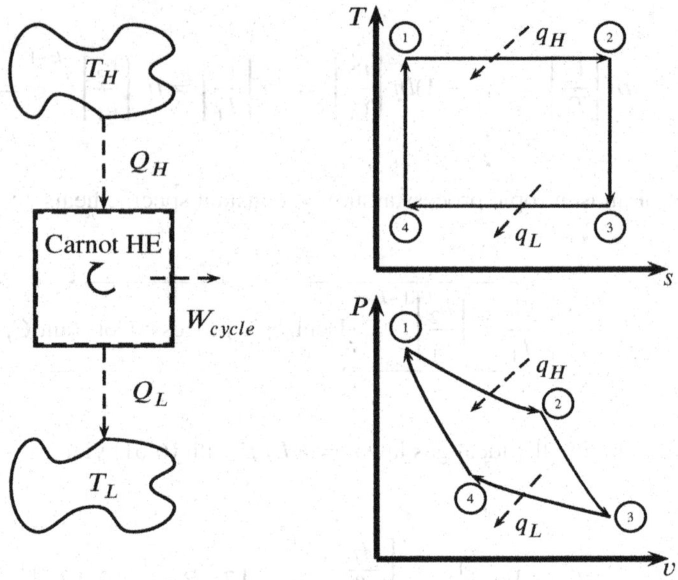

Find:
P_2,
P_3,
v_2,
$\Delta s_{1\to2}$

Assumptions:
$w = w_{bound}$,
$\Delta ke = \Delta pe = 0$

Analysis:
For the air-system within the piston-cylinder device we have the following process(es) analysis:

a) Constant specific heats with $k = 1.4$
Process $1 \to 2$ *(Isothermal expansion)*:
Via the entropy-balance for an overall process Table 6.1 (p.108), noting that $T_1 = T_2 = T_H$

$$\Delta s_{1\to2} = \int_{1\to2} \frac{\delta q}{\cancelto{T_H}{T_{int}}} + \cancelto{0,\,rev.}{s_{gen}} = \int_{1\to2} \frac{\delta q}{T_H} = \frac{1}{T_H}\cancelto{q_H}{\int_{1\to2} \delta q} = \frac{q_H}{T_H} = \frac{100\frac{kJ}{kg}}{1000°K}$$

$$= 0.1\frac{kJ}{kg-°K} \xLeftarrow{\Delta s_{1\to2}} \tag{6.64}$$

Application of the energy-balance for a differential process gives

$$\delta q - \cancelto{Pdv}{\delta w} = de = du + \cancelto{0}{dke + dpe} = du = C_v(T)\cancelto{0}{dT} \tag{6.65}$$

or

$$Pdv = \delta q \tag{6.66}$$

Integrating (6.66) over the process $1 \to 2$ (noting I.G.B. is valid) gives

$$\int_{v_1}^{v_2} \cancelto{\frac{RT}{v}}{P}dv = \int_{v_1}^{v_2} \frac{RT}{v}dv = RT_H\int_{v_1}^{v_2} \frac{1}{v}dv = RT_H ln\left(\frac{v_2}{v_1}\right) = \int_{1\to2} \delta q = q_H \tag{6.67}$$

or solving for v_2

$$v_2 = v_1 e^{\frac{q_H}{RT_H}} = 0.1\frac{m^3}{kg} e^{\left(\frac{100\frac{kJ}{kg}}{0.287\frac{kJ}{kg-°K}1000°K}\right)} = 0.141\frac{m^3}{kg} \xLeftarrow{v_2} \tag{6.68}$$

Finally, via the ideal gas law applied to state 2

$$P_2 v_2 = RT_2 \Rightarrow P_2 = \frac{R\overset{T_H}{\cancel{T_2}}}{v_2} = \frac{0.287\frac{kJ}{\cancel{kg}-°\cancel{K}}1000°\cancel{K}}{0.141\frac{m^3}{\cancel{kg}}} = 2030kPa \overset{P_2}{\Longleftarrow} \tag{6.69}$$

Process 2 → 3 *(Adiabatic expansion)*:

Via the entropy-balance for an overall process Table 6.1 (p.108)

$$\Delta s_{2\to3} = \int_{2\to3} \frac{\overset{0,adia.}{\cancel{\delta q}}}{T_{int}} + \overset{0,rev.}{\cancel{s}_{gen_{2\to3}}} = 0 \Rightarrow \Delta s_{2\to3} = 0 \tag{6.70}$$

Hence, process 2 → 3 is isentropic and from (6.63) noting that $T_2 = T_H$ and $T_3 = T_L$

$$\frac{T_3}{T_2} = \left[\frac{P_3}{P_2}\right]^{\frac{k-1}{k}} \Rightarrow P_3 = P_2\left[\frac{\overset{T_L}{\cancel{T_3}}}{\underset{T_H}{\cancel{T_2}}}\right]^{\frac{k}{k-1}} = P_2\left[\frac{T_L}{T_H}\right]^{\frac{k}{k-1}}$$

$$= 2030kPa\left[\frac{500°\cancel{K}}{1000°\cancel{K}}\right]^{\frac{1.4}{1.4-1}} = 179.4kPa \overset{P_3}{\Longleftarrow} \tag{6.71}$$

Process 4 → 1 *(Adiabatic compression)*:

Like process 2 → 3, $\Delta s_{4\to1} = 0$ and process 4 → 1 is isentropic. Thus, in $T - s$ diagram, process 1 → 2 and 3 → 4 are isothermal, and 2 → 3 and 4 → 1 are lines of constant entropy.

$$\boxed{\Delta s_{1\to2} = 0.1\frac{kJ}{kg-°K},\ v_2 = 0.141\frac{m^3}{kg},\ P_2 = 2030kPa,\ P_3 = 179.4kPa}$$

b) Variable specific heats using I.G.B air property Tables B.11

Process 1 → 2 *(Isothermal expansion)*:

Solution procedure for $\Delta s_{1\to2}$, v_2 and P_2 are identical to that of constant specific heats.

Process 2 → 3 *(Adiabatic expansion)*:

Via the entropy-balance for an overall process Table 6.1 (p.108)

$$\Delta s_{2\to3} = \int \frac{\overset{0,adia.}{\cancel{\delta q}}}{T_{int}} + \overset{0,rev.}{\cancel{s}_{gen_{2\to3}}} = 0 \Rightarrow \Delta s_{2\to3} = 0 \tag{6.72}$$

Hence, process 2 → 3 is isentropic and from (6.58)

$$\frac{P_3}{P_2} = \frac{P_{r_3}}{P_{r_2}} \Rightarrow P_3 = P_2\frac{P_{r_3}}{P_{r_2}} \tag{6.73}$$

From Table B.11 $P_{r_3} = P_r|_{T=500°K} = 8.411$ and $P_{r_2} = P_r|_{T=1000°K} = 114.0$. Hence,

$$P_3 = P_2\frac{P_{r_3}}{P_{r_2}} = 2030kPa\frac{8.411}{114.0} = 149.77kPa \overset{P_3}{\Longleftarrow} \tag{6.74}$$

$$\boxed{\Delta s_{1\to2} = 0.1\frac{kJ}{kg-°K},\ v_2 = 0.141\frac{m^3}{kg},\ P_2 = 2030kPa,\ P_3 = 149.77kPa}$$

Table 6.2 Summary of property relationships

Basis/Relationship	Internal energy u, enthalpy h, entropy s property relationships (valid everywhere)	Introducing specific heats and I.G.B. (not valid inside saturation dome!)	For constant specific heats and I.G.B. (not valid inside saturation dome!)
Differential process	$TdS = dU + PdV = dH - VdP \Rightarrow$ $dS = \frac{dU}{T} + \frac{P}{T}dV = \frac{dH}{T} - \frac{V}{T}dP$	$dU = mC_v(T)dT$, $dH = mC_p(T)dT$, $dS = m\frac{C_v(T)}{T}dT + mR\frac{dV}{V} = m\frac{C_p(T)}{T}dT - mR\frac{dP}{P}$	$dU = mC_v dT$, $dH = mC_p dT$, $dS = m\frac{C_v}{T}dT + mR\frac{dV}{V} = m\frac{C_p}{T}dT - mR\frac{dP}{P}$
Differential process (per-unit mass)	$Tds = du + Pdv = dh - vdP \Rightarrow$ $ds = \frac{du}{T} + \frac{P}{T}dv = \frac{dh}{T} - \frac{v}{T}dP$	$du = C_v(T)dT$, $dh = C_p(T)dT$, $ds = \frac{C_v(T)}{T}dT + R\frac{dV}{V} = \frac{C_p(T)}{T}dT - R\frac{dP}{P}$	$du = C_v dT$, $dh = C_p dT$, $ds = \frac{C_v}{T}dT + R\frac{V}{T}dV = \frac{C_p}{T}dT - R\frac{P}{T}dP$
Overall process (obtained via integration of differential process equations)	$\Delta S_{i\to f} = \int_{i\to f}\frac{dU}{T} + \int_{i\to f}\frac{P}{T}dV$ $= \int_{i\to f}\frac{dH}{T} - \int_{i\to f}\frac{V}{T}dP$	$\Delta U_{i\to f} = m\int_{T_i}^{T_f} C_v(T)dT$, $\Delta H_{i\to f} = m\int_{T_i}^{T_f} C_p(T)dT$, $\Delta S_{i\to f} = m\int_{T_i}^{T_f}\frac{C_v(T)}{T}dT + mRln\frac{V_f}{V_i} = m\int_{T_i}^{T_f}\frac{C_p(T)}{T}dT - mRln\frac{P_f}{P_i}$	$\Delta U_{i\to f} = mC_v\Delta T_{i\to f}$, $\Delta H_{i\to f} = mC_p\Delta T_{i\to f}$, $\Delta S_{i\to f} = mC_v ln\frac{T_f}{T_i} + mRln\frac{V_f}{V_i} = mC_p ln\frac{T_f}{T_i} - mRln\frac{P_f}{P_i}$
Overall process per-unit mass (obtained via integration of differential process equations)	$\Delta s_{i\to f} = \int_{i\to f}\frac{du}{T} + \int_{i\to f}\frac{P}{T}dv$ $= \int_{i\to f}\frac{dh}{T} - \int_{i\to f}\frac{v}{T}dP$	$\Delta u_{i\to f} = \int_{T_i}^{T_f} C_v(T)dT$, $\Delta h_{i\to f} = \int_{T_i}^{T_f} C_p(T)dT$, $\Delta s_{i\to f} = \int_{T_i}^{T_f}\frac{C_v(T)}{T}dT + Rln\frac{V_f}{V_i} = \int_{T_i}^{T_f}\frac{C_p(T)}{T}dT - mRln\frac{P_f}{P_i}$	$\Delta u_{i\to f} = C_v\Delta T_{i\to f}$, $\Delta h_{i\to f} = C_p\Delta T_{i\to f}$, $\Delta s_{i\to f} = C_v ln\frac{T_f}{T_i} + Rln\frac{V_f}{V_i} = C_p ln\frac{T_f}{T_i} - Rln\frac{P_f}{P_i}$
Isentropic process (overall)	——	$0 = s_0(T_f) - s_0(T_i) - mRln\frac{P_f}{P_i} \Rightarrow \frac{P_f}{P_i} = \frac{P_{r_f}}{P_{r_i}}, \frac{v_f}{v_i} = \frac{v_{r_f}}{v_{r_i}}$	$\frac{T_f}{T_i} = \left[\frac{v_f}{v_i}\right]^{1-k}, \frac{T_f}{T_i} = \left[\frac{P_f}{P_i}\right]^{\frac{k-1}{k}}, k = \frac{C_p}{C_v}$

6.5 REVIEW OF PROPERTY RELATIONSHIPS

Perhaps it is time to collect our thoughts and condense the various property relationships in the form of a table: Table 6.2 gives differential and integral property relationships for processes outside the saturation dome.

6.6 ENTROPY-BALANCE: OPEN SYSTEMS

Table 6.3 Closed system conservation equations (including entropy-balance)

Ext. prop.	Conservation equation	β	b
Mass	$0 = \frac{d}{dt} m_{\substack{closed\\system}} = \frac{d}{dt}\beta_{\substack{closed\\system}}$	$\beta = m$	$b = \frac{\beta}{m} = \frac{m}{m} = 1$
Momentum $(i = x, y, z)$	$\sum F_{i:\substack{closed\\system}} = \frac{d}{dt}(mV)_{i:\substack{closed\\system}} = \frac{d}{dt}\beta_{\substack{closed\\system}}$	$\beta = (mV)_i$	$b = \frac{\beta}{m} = \frac{(mV)_i}{m} = V_i$
Energy	$(\dot{Q} - \dot{W})_{\substack{closed\\system}} = \dot{E}_{\substack{closed\\system}} = \frac{d}{dt}\beta_{\substack{closed\\system}}$	$\beta = E$	$b = \frac{\beta}{m} = \frac{E}{m} = e$
Entropy	$\sum (\frac{\dot{Q}}{T_{int}})_{\substack{closed\\system}} + \dot{S}_{gen} = \dot{S}_{\substack{closed\\system}} = \frac{d}{dt}\beta_{\substack{closed\\system}}$	$\beta = S$	$b = \frac{\beta}{m} = \frac{S}{m} = s$

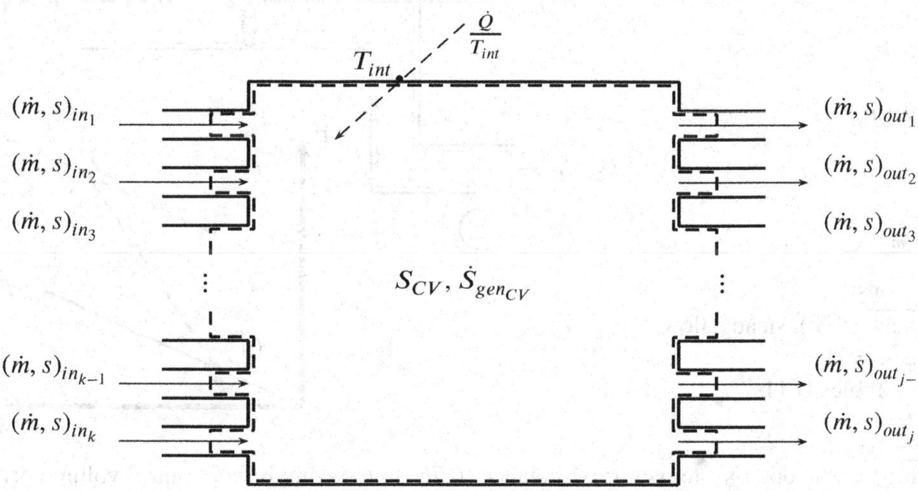

Figure 6.6 Illustration of open system control volume – – –– with a multiple inlets and outlets. Entropy inter-actions shown.

In §4.4.3 we applied the Reynolds Transport Theorem (RTT) to the extensive property total energy E to arrive at an open system energy conservation equation. We will now extend this procedure to the extensive property entropy S. To this end, Table 6.3 repeats the closed system conservation laws including the entropy-balance. Application of the RTT (4.69) for $\beta = S$ and $b = s$ to the rate-based conservation law for entropy S given in Table 6.3 yields

$$\sum (\frac{\dot{Q}}{T_{int}})_{\substack{closed\\system}} + \dot{S}_{gen} = \frac{dS_{CV}}{dt} + \sum (\dot{m}s)_{outlets} - \sum (\dot{m}s)_{inlets} \tag{6.75}$$

Moving the generation term to the right-hand side of the, now open system entropy-balance (6.75) and noting that the control volume coincides with the location of the closed system (in the limit) gives

$$\underbrace{\sum\left(\frac{\dot{Q}}{T_{int}}\right)}_{\substack{\text{Net rate of } S\text{-transfer due to}\\ \text{heat inter-actions}}} = \underbrace{\frac{dS_{CV}}{dt}}_{\substack{\text{Rate of change in}\\ S \text{ in CV}}} + \underbrace{\sum(\dot{m}s)_{outlets} - \sum(\dot{m}s)_{inlets}}_{\substack{\text{Net outflow of } S \text{ due}\\ \text{to mass inter-actions}}} - \underbrace{\dot{S}_{gen}}_{\substack{\text{Rate of } S \text{ generated in CV due to}\\ \text{irriversibilities}}} \tag{6.76}$$

EXAMPLE 6.7 Steady-state adiabatic air compressor: Entropy-balance

Problem Statement:

An adiabatic air compressor accepts air at a pressure and temperature of $95kPa$ and $27°C$ and produces compressed air at $600kPa$ and $277°C$. Assuming I.G.B., determine the rate of entropy generated per unit mass-flow within the air compressor.

Given:

Adiabatic Compressor,
air,
I.G.B.,
$T_1 = 27°C$,
$P_1 = 95kPa$,
$T_2 = 277°C$,
$P_2 = 600kPa$

Find:

$s_{gen} = \frac{\dot{S}_{gen}}{\dot{m}}$

Assumptions:

Steady-state (S.S.), steady-flow,
$\Delta pe = \Delta ke = 0$,
I.G.B. air Tables B.11

Schematic:

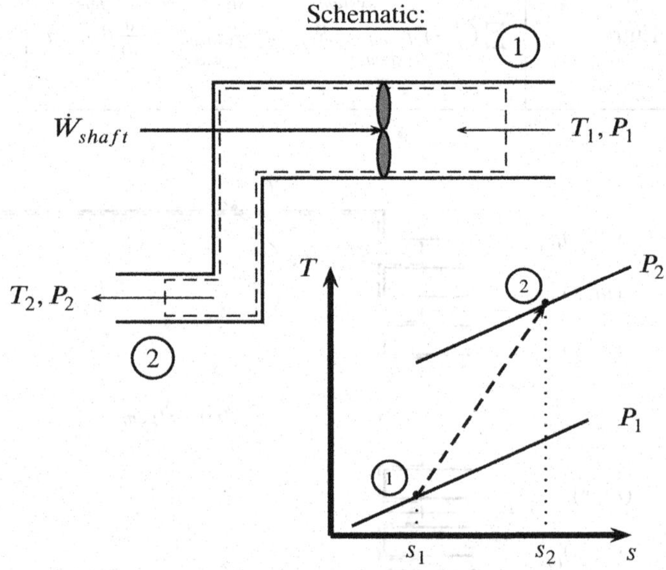

Analysis:

Application of the open-system entropy-balance (6.76) to the compressor control volume gives

$$\sum\left(\frac{\cancelto{0,adiabatic}{\dot{Q}}}{T_{int}}\right)_{system} = \cancelto{0,S.S.}{\frac{dS_{CV}}{dt}} + \sum\cancelto{\dot{m}s_2}{(\dot{m}s)}_{outlets} - \sum\cancelto{\dot{m}s_1}{(\dot{m}s)}_{inlets} - \dot{S}_{gen} \tag{6.77}$$

or

$$\dot{S}_{gen} = (\dot{m}s)_2 - (\dot{m}s)_1 \tag{6.78}$$

Applying the open-system mass-balance to the compressor control volume

$$\cancelto{0,S.S.}{\frac{dm_{CV}}{dt}} = \sum\cancelto{\dot{m}_1}{\dot{m}}_{inlets} - \sum\cancelto{\dot{m}_2}{\dot{m}}_{outlets} \Rightarrow \dot{m} = \dot{m}_1 = \dot{m}_2 \tag{6.79}$$

Combining this result with the entropy-balance (6.78), noting the integrated property relationship for I.G.B. for variable specific heats (6.35) gives

$$s_{gen} = \frac{\dot{S}_{gen}}{\dot{m}} = s_2 - s_1 = \Delta\, s|_{P_1,T_1}^{P_2,T_2} = s_o(T_2) - s_o(T_1) - Rln\left(\frac{P_2}{P_1}\right) \tag{6.80}$$

From I.G.B. Tables B.11 $s_o|_{T_2=277°C} = 2.31\frac{kJ}{kg-°K}$ and $s_o|_{T_1=27°C} = 1.7\frac{kJ}{kg-°K}$. Thus, from (6.80)

$$s_{gen} = s_o(T_2) - s_o(T_1) - Rln\left(\frac{P_2}{P_1}\right) =$$

$$2.31\frac{kJ}{kg-°K} - 1.7\frac{kJ}{kg-°K} - 0.287\frac{kJ}{kg-°K}ln\left(\frac{600\cancel{kPa}}{95\cancel{kPa}}\right) = 0.081\frac{kJ}{kg-°K} \overset{s_{gen}}{\Longleftarrow} \tag{6.81}$$

$$\boxed{s_{gen} = \frac{\dot{S}_{gen}}{\dot{m}} = 0.081\frac{kJ}{kg-°K}}$$

Comment: Thus, due to irreversible processes present inside the compressor (e.g. friction, unrestrained expansion, etc.) $s_{gen} > 0$.

□

6.7 PHYSICAL INTERPRETATION OF CONSERVATION LAWS

After having developed your skills in thermodynamics perhaps we should rest for a moment and look at the "big picture". In the previous chapters we have introduced and developed conservation equations and property relationships side-by-side: How are these related? Figure 6.7 illustrates closed systems experiencing energy and entropy inter-actions. For example, the closed system energy-balance (1st-Law) describes a *cause and effect* relationship between heat Q, work W (the causes) and a change in total energy ΔE (the effect): In words "the transfer of energy via work or heat results in a change in total energy of the system." This is in contrast to the property-relationships which merely describe the inter-dependence of system properties (e.g. $du = C_v dT$). The same goes for entropy.

6.8 ISENTROPIC EFFICIENCIES FOR WORK PRODUCING/CONSUMING DEVICES

Given that reversible devices are the most efficient, we are now in the position to define two efficiencies for open system devices: one for work producing devices (turbines) and another for work consuming devices (pumps and compressors).

Definition 6.2 ***Isentropic Efficiency*** η^{isen}**:**[1] *For an adiabatic work producing device, the maximum power produced $\dot{W}_{max}$ corresponds to the reversible and thus isentropic work producing device $\dot{W}^{isen}$. Hence,*

$$\boxed{\eta^{isen} = \frac{\dot{W}_{produced}}{\dot{W}_{max}} = \frac{\dot{W}_{produced}}{\dot{W}^{isen}}} \tag{6.82}$$

[1] Also referred to as adiabatic efficiency

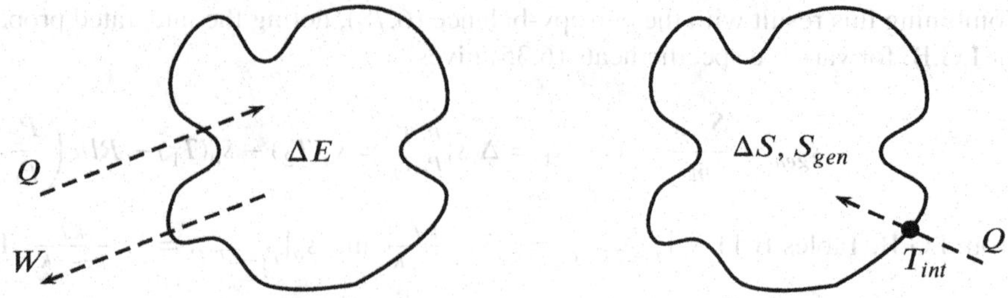

	a)	b)
	Cause! Effect!	Effect! Cause!
Conservation Equation	$\overbrace{Q - W}^{\text{Cause!}} = \overbrace{\Delta E = \Delta U + \ldots}^{\text{Effect!}}$	$\overbrace{\Delta S}^{\text{Effect!}} = \overbrace{\sum \int \frac{\delta Q}{T_{int}} + S_{gen}}^{\text{Cause!}}$
Property Relationship(s)	$\Delta U = m\Delta u = m \int C_v dT$	$\Delta S = m\Delta s$ $= m\left[s_{0_f} - s_{0_i} - Rln[\frac{P_f}{P_i}]\right]$ ...
	a) Property Energy E	b) Property Entropy S

Figure 6.7 Energy E and entropy S causal & property relationships.

For an adiabatic work consuming device, the minimum power consumed $\dot{W}_{min}$ corresponds to the reversible and thus isentropic work consuming device $\dot{W}^{isen}$. Hence,

$$\eta^{isen} = \frac{\dot{W}_{min}}{\dot{W}_{consumed}} = \frac{\dot{W}^{isen}}{\dot{W}_{consumed}} \tag{6.83}$$

To justify these definitions Figure 6.8 a) shows an adiabatic compressor/pump with a single inlet/single outlet (a work *consuming* device). Application of the mass and entropy-balance gives $\dot{m} = \dot{m}_{in} = \dot{m}_{out}$ and $s_{out} = s_{in} + \frac{\dot{S}_{gen}}{\dot{m}}$. If the compressor/pump is reversible (and hence consumes the minimum power) $\dot{S}_{gen} = 0$ and the device operates isentropically. Thus, for a single inlet/singe outlet device, the isentropic work consuming device will be the most efficient and required the least power input. Figure 6.8 b) shows the $T - s$ diagram for the real vs. the reversible (i.e. isentropic) device operating between the same pressure limits with end-states $1, 2$ and $1, 2_s$ respectively. As we can see from the $T - s$ diagram Figure 6.8 b), the real compressor consumes more power and thus adds more energy to the flow, resulting in a higher outlet flow temperature T_2 than the reversible (i.e. isentropic) device T_{2_s}. Similarly, Figure 6.8 c) shows an adiabatic turbine with a single inlet/single outlet (a work *producing* device). As in the case of a compressor or pump, application of the mass and entropy-balance gives $\dot{m} = \dot{m}_{in} = \dot{m}_{out}$ and $s_{out} = s_{in} + \dot{S}_{gen}/\dot{m}$. If the turbine is reversible (and hence produces the maximum power) $\dot{S}_{gen} = 0$ and the device operates isentropically. Thus, for a single inlet/singe outlet, work producing device, the isentropic device will be the most efficient and produce the most power. Figure 6.8 d) shows the $T - s$ diagram for the real vs. the reversible (i.e. isentropic) device operating between the same pressure limits with end-states $1, 2$ and $1, 2_s$ respectively. Again, as we can see, from the $T - s$ diagram, the real turbine produces less power, extracts less energy from the flow resulting in a higher outlet flow temperature T_2 than the reversible (i.e. isentropic) turbine T_{2_s}.

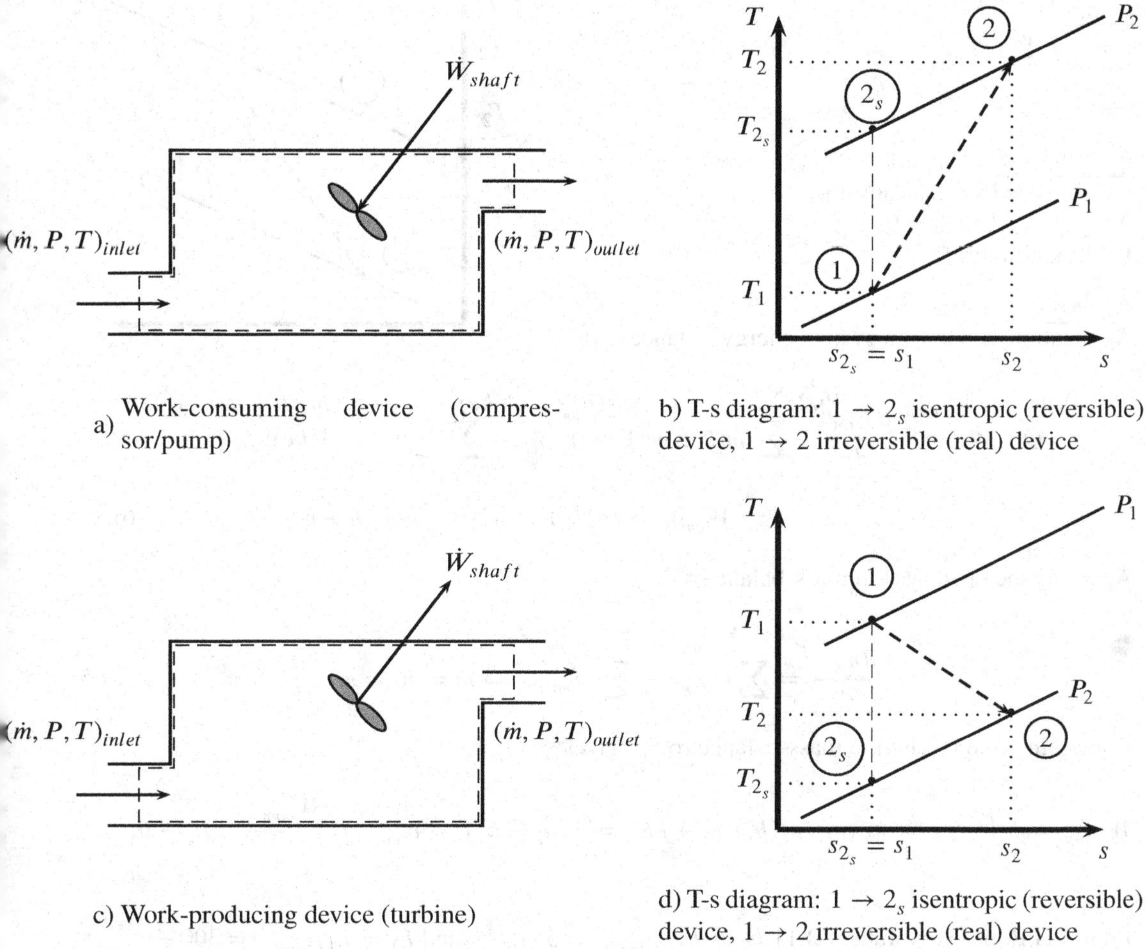

a) Work-consuming device (compressor/pump)

b) T-s diagram: $1 \rightarrow 2_s$ isentropic (reversible) device, $1 \rightarrow 2$ irreversible (real) device

c) Work-producing device (turbine)

d) T-s diagram: $1 \rightarrow 2_s$ isentropic (reversible) device, $1 \rightarrow 2$ irreversible (real) device

Figure 6.8 Illustration of an adiabatic compressor/pump and turbine operation with $T - s$ diagrams.

EXAMPLE 6.8 Steady-state adiabatic air compressor: Isentropic efficiency

Problem Statement:

An adiabatic air compressor accepts air at a pressure and temperature of $95kPa$ and $27°C$ while producing compressed air at $600kPa$ and $277°C$. Assuming I.G.B., determine the specific work (power-per unit mass-flow) as well as the isentropic compressor efficiency.

Given:

Adiabatic air compressor,
I.G.B.,
$T_1 = 27°C$,
$P_1 = 95kPa$,
$T_2 = 277°C$,
$P_2 = 600kPa$

Schematic:

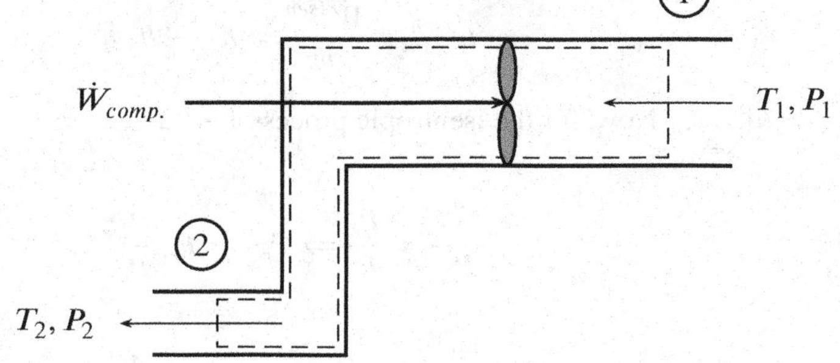

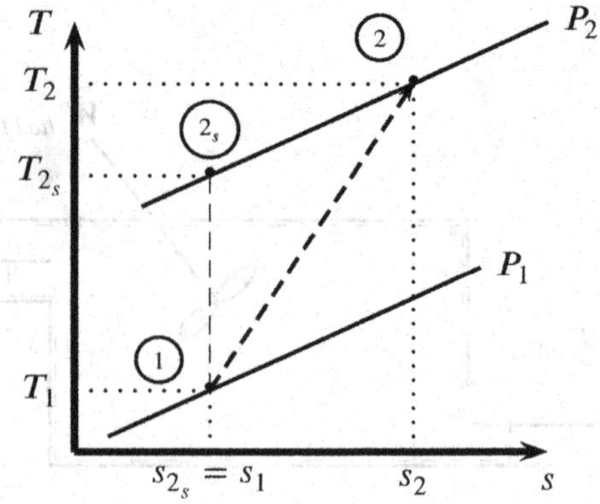

Find:

$w_{comp} = \frac{\dot{W}_{comp}}{\dot{m}}$,
η_{comp}^{isen}

Assumptions:

Steady-state (S.S.), steady-flow,
$\Delta pe_{1\to 2} = \Delta ke_{1\to 2} = 0$,
I.G.B. air Tables B.11

Analysis:

Application of the open system energy-balance gives

$$\overset{0,\,adia.}{\dot{Q}} - \overset{-\dot{W}_{comp.}}{\dot{W}} = \overset{0,\,S.S.}{\frac{dE_{CV}}{dt}} + \sum \overset{\dot{m}_2(h+pe+ke)_2}{(\dot{m}(h+pe+ke))_{outlets}} - \sum \overset{\dot{m}_1(h+pe+ke)_1}{(\dot{m}(h+pe+ke))_{inlets}}$$

$$\Rightarrow \dot{W}_{comp.} = \dot{m}_2(h+pe+ke)_2 - \dot{m}_1(h+pe+ke)_1 \tag{6.84}$$

Applying the open system mass-balance

$$\overset{0,\,S.S.}{\frac{dm_{CV}}{dt}} = \sum \overset{\dot{m}_1}{\dot{m}_{inlets}} - \sum \overset{\dot{m}_2}{\dot{m}_{outlets}} \Rightarrow \dot{m} = \dot{m}_1 = \dot{m}_2 \tag{6.85}$$

Combining (6.84) with the mass-balance (6.85) gives

$$\dot{W}_{comp.} = \overset{\dot{m}}{\dot{m}_2}(h + pe + ke)_2 - \overset{\dot{m}}{\dot{m}_1}(h + pe + ke)_1 = \dot{m}(h_2 - h_1) \Rightarrow w_{comp} = \frac{\dot{W}_{comp.}}{\dot{m}} = h_2 - h_1 \tag{6.86}$$

From I.G.B. property Tables B.11 $h_2 = h|_{T=277°C} = 554.7\frac{kJ}{kg}$ and $h_1 = h|_{T=27°C} = 300\frac{kJ}{kg}$. Hence, via (6.86)

$$w_{comp} = h_2 - h_1 = 554.7\frac{kJ}{kg} - 300\frac{kJ}{kg} = 254.7\frac{kJ}{kg} \text{ (done \underline{on} the device)} \xLeftarrow{w_{comp}} \tag{6.87}$$

Now, we can rewrite the definition of isentropic efficiency (6.83) in terms of specific work $w = \dot{W}/\dot{m}$ yielding

$$\eta_{comp}^{isen} = \frac{\dot{W}^{isen}}{\dot{W}_{comp}} = \frac{\dot{m}w^{isen}}{\dot{m}w_{comp}} = \frac{w^{isen}}{w_{comp}} \tag{6.88}$$

From (6.84), application of the open system energy-balance to the <u>isentropic</u> device gives

$$w^{isen} = \frac{\dot{W}^{isen}}{\dot{m}} = h_{2s} - h_1 \tag{6.89}$$

where $h_{2s} = h|_{\substack{P=P_2 \\ s=s_1}}$. Now, for the isentropic process $1 \to 2_s$

$$\frac{P_{r_{2s}}}{P_{r_1}} = \frac{P_2}{P_1} \Rightarrow P_{r_{2s}} = P_{r_1}\frac{P_2}{P_1} \tag{6.90}$$

From I.G.B. property Tables B.11 $P_{r_1} = P_r|_{T=27°C} = 1.386$. Hence from (6.90)

$$P_{r_{2s}} = P_{r_1}\frac{P_2}{P_1} = 1.386\frac{600kPa}{95kPa} = 8.75 \tag{6.91}$$

Interpolation from Table B.11 $h_{2s} = h|_{P_r=8.75} = 508.6\frac{kJ}{kg}$. Returning to the energy-balance applied to the isentropic device (6.89)

$$w^{isen} = \frac{\dot{W}^{isen}}{\dot{m}} = h_{2s} - h_1 = 508.6\frac{kJ}{kg} - 300\frac{kJ}{kg} = 208.6\frac{kJ}{kg} \tag{6.92}$$

Hence, from the expression for isentropic efficiency (6.88)

$$\eta_{comp}^{isen} = \frac{w^{isen}}{w_{comp}} = \frac{208.6\frac{kJ}{kg}}{255.7\frac{kJ}{kg}} = 0.81 = 81\% \Longleftarrow \eta_{comp}^{isen} \tag{6.93}$$

$w_{comp} = 254.7\frac{kJ}{kg}, \eta_{comp}^{isen} = 81\%$

Comment: Note that the real compressor exhaust temperature ($277°C$) is higher than the reversible compressor ($T_{2_s} = T|_{P_r=8.75} = 232°C$) operating between the same pressure limits. Hence, the additional specific work input required of the real compressor results in a hotter exhaust temperature.

□

EXAMPLE 6.9 Steady-state adiabatic steam turbine: Isentropic efficiency

Problem Statement:
An adiabatic steam turbine with inlet steam at $8MPa$ and $500°C$ and an outlet pressure of $30kPa$ has an isentropic efficiency of 90% and a steam mass-flow rate of $3\frac{kg}{s}$. Determine the exhaust temperature and power produced.

Schematic:

Given:
Adiabatic turbine,
H_2O,
$T_1 = 500°C$,
$P_1 = 8MPa$,
$\eta_{turbine}^{isen} = 90\%$,
$P_2 = 30kPa$,
$\dot{m} = 3\frac{kg}{s}$

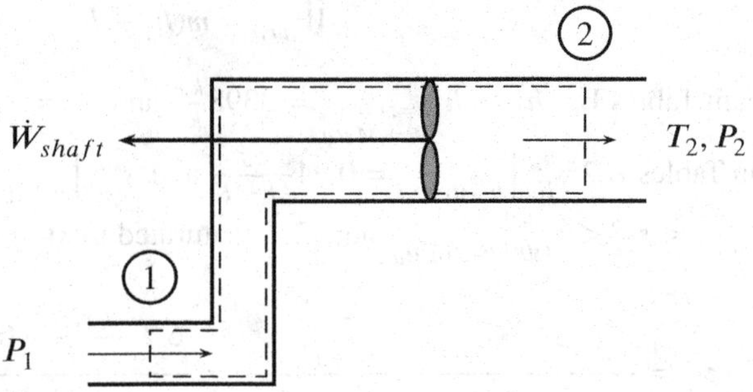

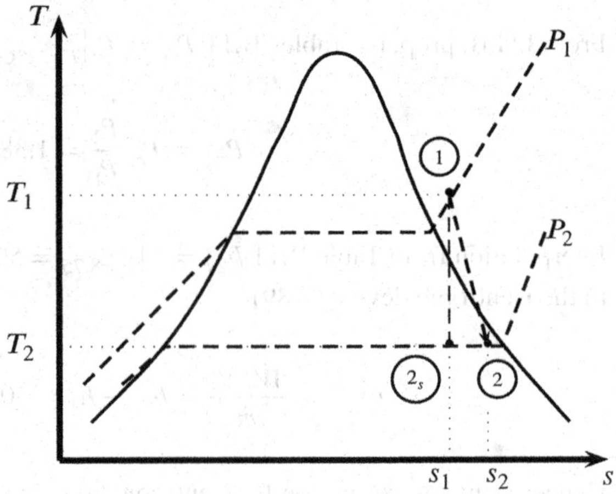

Find:
$\dot{W}_{turb}$,
T_2

Assumptions:
Steady-state (S.S.),
steady-flow,
$\Delta pe_{1\to 2} = \Delta ke_{1\to 2} = 0$,
saturation and steam Tables B.2, B.3 and B.4

Analysis:
From the definition of isentropic efficiency for a work producing device (6.82)

$$\eta_{turb}^{isen} = \frac{\dot{W}_{turb}}{\dot{W}^{isen}} \Rightarrow \dot{W}_{turb} = \eta_{turb}^{isen}\dot{W}^{isen} \tag{6.94}$$

Applying the open system mass-balance for the real or isentropic turbine gives

$$\frac{dm_{CV}}{dt}^{0,S.S.} = \sum \dot{m}_{inlets}^{\dot{m}_1} - \sum \dot{m}_{outlets}^{\dot{m}_{2_s,2}} \tag{6.95}$$

or

$$\dot{m} = \dot{m}_1 = \dot{m}_{2_s,2} \tag{6.96}$$

Application of the open system energy-balance to the isentropic turbine gives

$$\dot{Q}^{0,adia.} - \dot{W}^{\dot{W}^{isen}} = \frac{dE_{CV}}{dt}^{0,S.S.} + \sum(\dot{m}(h+pe+ke))_{outlets}^{\dot{m}(h+pe+ke)_{2_s}} - \sum(\dot{m}(h+pe+ke))_{inlets}^{\dot{m}(h+pe+ke)_1}$$

$$= -\dot{m}\left[(h_{2_s} - h_1) + (ke_{2_s} - ke_1)^{0} + (pe_{2_s} - pe_1)^{0}\right] = -\dot{m}(h_{2_s} - h_1) \tag{6.97}$$

or

$$\dot{W}^{isen} = \dot{m}(h_1 - h_{2_s}) \tag{6.98}$$

Also, via the energy-balance, for the actual (real) turbine

$$\dot{W}_{turb} = \dot{m}(h_1 - h_2) \tag{6.99}$$

From steam Tables B.4 $h_1 = h|_{T=500^\circ C,\ P=8MPa} = 3398\frac{kJ}{kg}$ and $s_{2_s} = s_1 = s|_{T=500^\circ C,\ P=8MPa} = 6.7\frac{kJ}{kg-{}^\circ K}$. From saturation Tables B.2 $s_{liq}^{sat}\Big|_{P=30kPa} = 0.94\frac{kJ}{kg-{}^\circ K}$ and $s_{vap}^{sat}\Big|_{P=30kPa} = 7.77\frac{kJ}{kg-{}^\circ K}$. Since, $s_{liq}^{sat}\Big|_{P=30kPa} < s_{2_s} < s_{vap}^{sat}\Big|_{P=30kPa}$ state 2_s is saturated mixture. Thus,

$$x_{2_s} = \frac{s_{2_s} - s_{liq}^{sat}\Big|_{P=30kPa}}{s_{vap}^{sat}\Big|_{P=30kPa} - s_{liq}^{sat}\Big|_{P=30kPa}} = \frac{6.7\frac{kJ}{kg-{}^\circ K} - 0.94\frac{kJ}{kg-{}^\circ K}}{7.77\frac{kJ}{kg-{}^\circ K} - 0.94\frac{kJ}{kg-{}^\circ K}} = 84.2\% \tag{6.100}$$

From saturation Tables B.3 $h_{liq}^{sat}\Big|_{P=30kPa} = 289\frac{kJ}{kg}$ and $h_{fg}\Big|_{P=30kPa} = 2334\frac{kJ}{kg}$. Hence

$$h_{2_s} = h_{liq}^{sat}\Big|_{P=30kPa} + x_{2_s} h_{fg}\Big|_{P=30kPa} = 289\frac{kJ}{kg} + 0.845\left(2334\frac{kJ}{kg}\right) = 2255.9\frac{kJ}{kg} \quad (6.101)$$

Thus, from the energy-balance applied to the isentropic turbine (6.98)

$$\dot{W}^{isen} = \dot{m}(h_1 - h_{2_s}) = 3\frac{kg}{s}\left(3398\frac{kJ}{kg} - 2255.9\frac{kJ}{kg}\right) = 3426kW \quad (6.102)$$

and via the isentropic efficiency (6.94)

$$\dot{W}_{turb} = \eta_{turb}^{isen}\dot{W}^{isen} = 0.9(3426kW) = 3084kW \xLeftarrow{\dot{W}_{turb}} \quad (6.103)$$

Returning to the energy-balance for the real turbine (6.99)

$$\dot{W}_{turb} = \dot{m}(h_1 - h_2) \Rightarrow h_2 = -\frac{\dot{W}_{turb}}{\dot{m}} + h_1 = -\frac{3084kW}{3\frac{kg}{s}} + 3398\frac{kJ}{kg} = 2370\frac{kJ}{kg} \quad (6.104)$$

Finally, $h_{liq}^{sat}\Big|_{P=30kPa} = 292\frac{kJ}{kg} < h_2 < h_{vapor}^{sat}\Big|_{P=30kPa} = 2626\frac{kJ}{kg}$ and state 2 is saturated mixture. Thus, from saturation Tables B.3

$$T_2 = T^{sat}\big|_{P=30kPa} = 69.1°C \xLeftarrow{T_2} \quad (6.105)$$

$$\boxed{\dot{W}_{turb} = 3084kW, T_2 = 69.1°C}$$

Comment: Calculation of the exit quality for the turbine gives

$$x_2 = \frac{h_2 - h_{liq}^{sat}\Big|_{30kPa}}{h_{fg}\Big|_{30kPa}} = \frac{2370\frac{kJ}{kg} - 289\frac{kJ}{kg}}{2336\frac{kJ}{kg}} = 89\% \quad (6.106)$$

Thus, the real turbine, which extracts less energy from the steam, has an exit state closer to saturated vapor than the isentropic (i.e. reversible) device with quality $x_{2_s} = 84.5\%$.

□

6.9 ISENTROPIC EFFICIENCY FOR ENERGY CONVERSION DEVICES

Extension of the previous definitions for isentropic efficiency to work-less devices is relatively straight-forward. For example, the purpose of a nozzle is to produce a kinetic energy rise via a reduction in flow pressure and internal energy. Furthermore, one expects the reversible adiabatic (isentropic) nozzle to produce the maximum kinetic energy rise. Likewise, for the diffuse, one expects the reversible adiabatic (isentropic) diffuser to produce the greatest kinetic energy reduction. With this in mind, we define the following isentropic efficiencies for nozzles and diffusers:[2]

$$\boxed{\eta_{\substack{nozzle,\\diffuser}}^{isen} = \frac{\|\Delta ke\|}{\|\Delta ke^{isen}\|}} \quad (6.107)$$

[2]Although this definition is general, the common definition of isentropic efficiency for a diffuser is the ratio of the actual to isentropic pressure rise $\eta_{diffuser}^{isen} = \frac{\Delta P}{\Delta P^{isen}}$.

For example, in the case of a nozzle, neglecting inlet kinetic energy for the isentropic process, (6.107) reduces to $\eta_{nozzle} = \frac{\|\Delta ke\|}{ke_{2_s}}$. Likewise, for the diffuser, neglecting outlet kinetic energy for the isentropic process, (6.107) reduces to $\eta_{diffuser} = \frac{\|\Delta ke\|}{ke_{1_s}}$. Finally, as in case of turbines and compressors, it is implied that the outlet pressure of the isentropic and real nozzle and diffuser are the same.

EXAMPLE 6.10 Steady-state adiabatic nozzle: Isentropic efficiency

Problem Statement:
An adiabatic air nozzle installed on a jet engine is used to accelerate the exhaust flow (modeled as ideal gas air) has an inlet pressure and temperature of $200kPa$ and $950°K$, and outlet pressure of $80kPa$. If the nozzle isentropic efficiency is 92%, determine the outlet flow velocity.

Schematic:

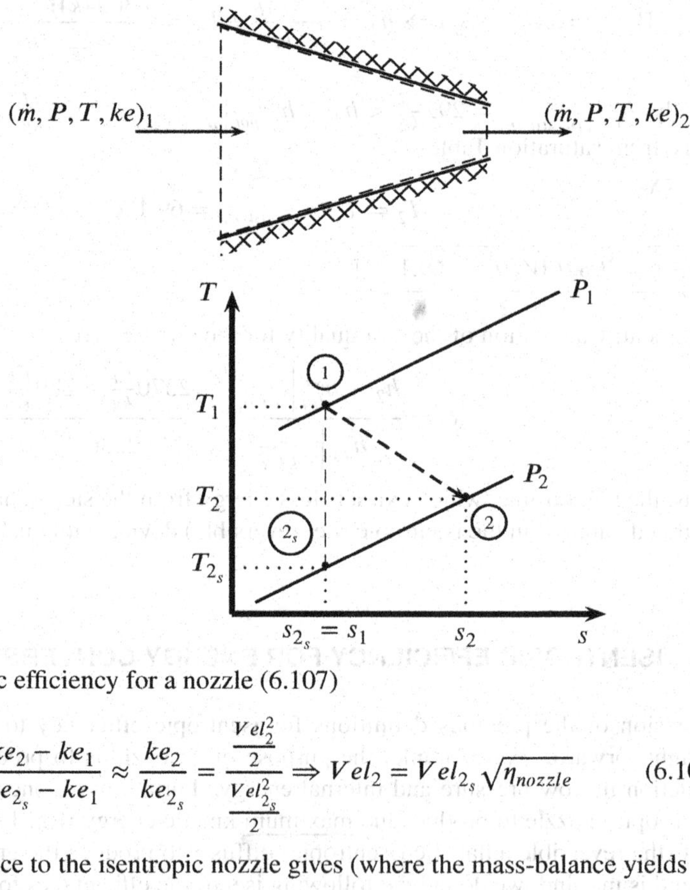

Given:
Adiabatic nozzle,
air,
I.G.B.,
$T_1 = 950°K$,
$P_1 = 250kPa$,
$P_2 = 80kPa$,
$\eta = 0.92$

Find:
Vel_2

Assumptions:
Steady-state (S.S.), steady-flow,
$\Delta pe = 0$,
$V_1 \ll V_2 \Rightarrow ke_1 \ll ke_2$,
I.G.B. air Tables B.11

Analysis:
From the definition of isentropic efficiency for a nozzle (6.107)

$$\eta_{nozzle}^{isen} = \frac{\|\Delta ke\|}{\|\Delta ke^{isen}\|} = \frac{ke_2 - ke_1}{ke_{2_s} - ke_1} \approx \frac{ke_2}{ke_{2_s}} = \frac{\frac{Vel_2^2}{2}}{\frac{Vel_{2_s}^2}{2}} \Rightarrow Vel_2 = Vel_{2_s}\sqrt{\eta_{nozzle}} \qquad (6.108)$$

Application of the energy-balance to the isentropic nozzle gives (where the mass-balance yields $\dot{m} = \dot{m}_1 = \dot{m}_2 = \dot{m}_{2_s}$)

$$\overset{0}{\cancel{\dot{Q}}} - \overset{0}{\cancel{\dot{W}}} = \overset{0,\,S.S.}{\cancel{\frac{dE_{CV}}{dt}}} + \sum \overset{\dot{m}_2(h+ke+pe)_{2_s}}{\cancel{(\dot{m}(h+ke+pe))_{outlets}}} - \sum \overset{\dot{m}_1(h+ke+pe)_1}{\cancel{(\dot{m}(h+ke+pe))_{inlets}}} \qquad (6.109)$$

or

$$0 = \dot{m}_2(h+ke)_2 - \dot{m}_1(h+ke)_1 \Rightarrow Vel_{2_s} = \sqrt{2(h_1 - h_{2_s})} \qquad (6.110)$$

From I.G.B. Tables B.11 $h_1 = h|_{T_2=950^\circ K} = 989.24\frac{kJ}{kg}$. Noting that $h_{2_s} = h|_{P_{2_s}=P_2, s_{2_s}=s_1}$ we calculate the reduced pressure $P_{r_{2_s}}$

$$\frac{P_{r_{2_s}}}{P_{r_1}} = \frac{P_2}{P_1} \Rightarrow P_{r_{2_s}} = P_{r_1}\frac{P_2}{P_1} \tag{6.111}$$

From I.G.B. Tables B.11 $P_{r_1} = P_r|_{T_1=950^\circ K} = 93.14$ which yields from (6.111)

$$P_{r_{2_s}} = P_{r_1}\frac{P_2}{P_1} = 93.14\frac{80kPa}{200kPa} = 37.26 \Rightarrow h_{2_s} = h|_{P_{r_{2_s}}} = 766.74\frac{kJ}{kg} \tag{6.112}$$

Returning to our result from application of the energy-balance to the isentropic nozzle (6.110)

$$Vel_{2_s} = \sqrt{2(h_1 - h_{2_s})} = \sqrt{2(989.24\frac{kJ}{kg} - 766.74\frac{kJ}{kg})\frac{1000J}{1kJ}} = 667.1\frac{m}{s} \tag{6.113}$$

Finally, via our result from the definition of isentropic efficiency (6.108)

$$Vel_2 = Vel_{2_s}\sqrt{\eta_{nozzle}} = 667.1\frac{m}{s}\sqrt{0.92} = 639.8\frac{m}{s} \xLeftarrow{Vel_{2_s}} \tag{6.114}$$

$$\boxed{Vel_2 = 639.8\frac{m}{s}}$$

Comment: The real and isentropic nozzle outlet temperatures are $T_2 = T|_{h_2=784.6\frac{kJ}{kg}} = 765.9^\circ K$ and $T_{2_s} = T|_{h_{2_s}=766.74} = 749^\circ K$: $T_2 > T_{2_s}$. Hence, due to irreversible processes present inside the nozzle (e.g. friction, unrestrained expansion, etc.), conversion of internal energy and pressure work to kinetic energy is lower in the case of the real nozzle, resulting in a hotter exhaust temperature.

□

6.10 AVAILABILITY/EXERGY

Thus far, we have encountered conservation laws for properties: Conservation of mass, energy and entropy. We now will take a slight detour to illuminate the physical significance of, up to this point, nebulous quantities such as the generation of entropy $\dot{S}_{gen}$ (which we have discussed in detail) and the importance of the environment as a reservoir (which we have not discussed at all). To this end, we will investigate, separately, the inter-action of an open and a closed-system with the environment.

6.10.1 THE ENVIRONMENTAL SYSTEM AND MAXIMUM WORK

Figure 6.9 a) shows a closed-system device which receives heat $\dot{Q}_{source}$ from a source at temperature T_{source} and rejects heat $\dot{Q}_{env}$ to the environment at a temperature T_{env}. Of course the maximum possible rate of work produced $\dot{W}_{max}$ is given by the maximum or reversible power

$$\dot{W}_{max} = \dot{Q}_{source}\eta_{max} = \dot{Q}_{source}\eta_{reversible} = \dot{Q}_{source}\left[1 - \frac{T_{env}}{T_{source}}\right] \tag{6.115}$$

What about for an open-system? Well, Figure 6.9 b) shows an adiabatic steady-state device interacting with an inlet fluid source and the environment (where the outlet-flow is rejected). Application

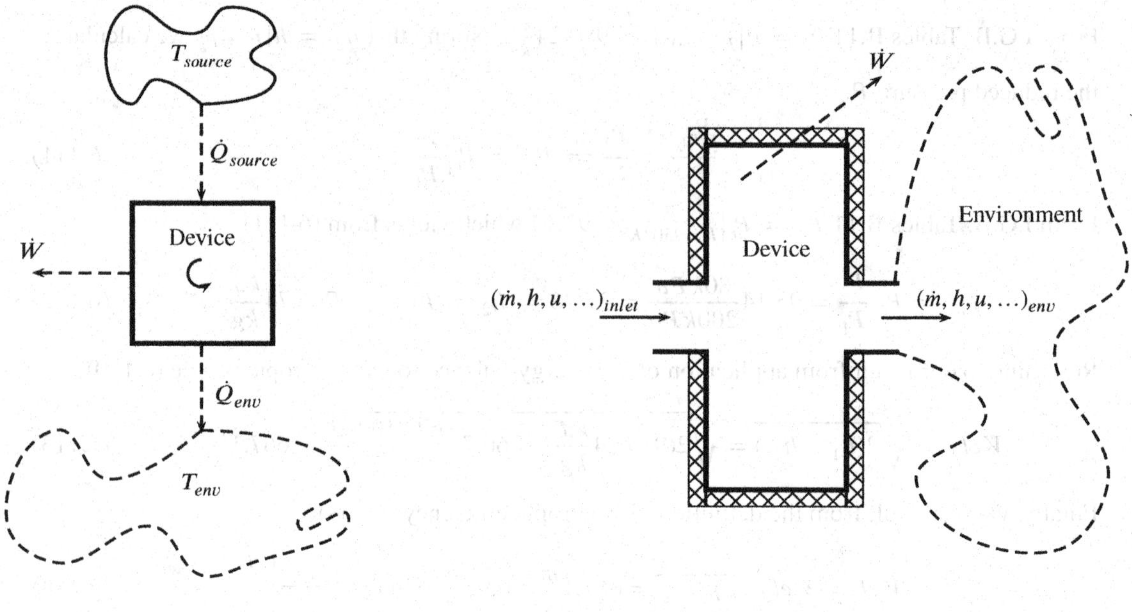

a) Device with source and environmental heat inter-actions

b) Open-system device inter-acting with source and the environment

Figure 6.9 Open and closed-system inter-action with the environment

of the energy and entropy-balances gives

$$\dot{W} = -\dot{m}(h + ke + pe)_{outlet} + \dot{m}(h + ke + pe)_{inlet}, 0 = s_{outlet} - s_{inlet} - \frac{\dot{S}_{gen}}{\dot{m}} \tag{6.116}$$

The most efficient device will, of course, be *internally reversible* (i.e. $\dot{S}_{gen} = 0$) *and eternally reversible*. What is the implication of external reversibility? Well, firstly the outlet mass-flow from the device *rejected to the environment must be at the environmental temperature* T_{env}, *otherwise we would have heat-transfer over a finite temperature difference within the environmental system* as the warm outflow (ultimately) transfers heat to the cold environment: An externally irreversible process! Secondly, the outflow *must also be at the environmental pressure, otherwise there would be* un-restrained expansion occurring within the environmental system: An externally irreversible process! Therefore, the maximum possible power produced by this device corresponds to internally reversible operation ($\dot{S}_{gen} = 0$) where the outlet flow is at the environmental state: T_{env}, P_{env} as well as pe_{env} and ke_{env}. Thus, the energy and entropy-balances (6.116) yield

$$\dot{W}_{max} = -\dot{m}(h_{env} + ke_{env} + pe_{env}) + \dot{m}(h + ke + pe)_{inlet}, 0 = s_{env} - s_{inlet} \tag{6.117}$$

The point here is that the environment, directly or indirectly, plays an important role given that *all devices eventually inter-act with the environment.* Hence, the environmental state plays a role in determining the work or power available from a given energy source. An implication of (6.117) is that if the inlet flow is <u>also</u> at the environmental state, maximum work or power is zero! Hence, as far as our device is concerned, an inlet mass-flow with energy (and all other properties) at the environment state, is ultimately "dead energy".

6.10.2 THE AVAILABLE WORK-BALANCE

The previous section illustrates the importance of the environment in determining the *maximum* work or power available from a given energy source. To proceed in a more rigorous and general way, we will now *combine* the energy and entropy-balance to arrive at an available work or exergy-balance: Multiplying the open-system entropy-balance (6.76) by the environmental temperature T_{env} and subtracting the result from the open-system energy-balance (4.85) gives

$$\left[\sum \dot{Q} - \dot{W}\right] - \sum\left[\dot{Q}(\frac{T_{env}}{T_{int}})\right] = \left[\frac{dE_{CV}}{dt}\right] - \left[T_{env}\frac{dS_{CV}}{dt}\right] + \left[\sum(\dot{m}(h+ke+pe))_{outlets} - \sum(\dot{m}(h+ke+pe))_{inlets}\right] - \left[\sum(\dot{m}T_{env}s)_{outlets} - \sum(\dot{m}T_{env}s)_{inlets}\right] + \left[T_{env}\dot{S}_{gen}\right] \quad (6.118)$$

Or, grouping terms according to inter-action (heat transfer, mass transfer, etc.), noting that the environmental temperature is a constant (hence, $T_{env}\frac{dS_{CV}}{dt} = \frac{d(T_{env}S)_{CV}}{dt}$) and solving for power $\dot{W}$ gives the "combined rate-balance"

$$\dot{W} = \sum\left[\dot{Q}(1-\frac{T_{env}}{T_{int}})\right] - \frac{d(E-T_{env}S)_{CV}}{dt} + \sum(\dot{m}(h+ke+pe-T_{env}s))_{inlets} - \sum(\dot{m}(h+ke+pe-T_{env}s))_{outlets} - T_{env}\dot{S}_{gen} \quad (6.119)$$

Let us now apply (6.119) to the special case of a device which is

1. Internally reversible
2. At the environmental state (i.e. $T = T_{env}$, $P = P_{env}$, ... or $e = e_{env}$ and $s = s_{env}$)
3. All heat inter-actions occur at the environmental temperature ($T_{int} = T_{env}$)
4. Inlet and outlet flows are at the environmental state (i.e. $h|_{T_{env}, P_{env}}$, $s|_{T_{env}, P_{env}}$, $ke|_{T_{env}, P_{env}}$, $pe|_{T_{env}, P_{env}}$, ...)

yielding

$$\dot{W} = \sum\dot{Q}\left[1-\frac{T_{env}}{T_{env}}\right]^{\nearrow 0} - \frac{d(me_{env}-T_{env}ms_{env})_{CV}}{dt}^{\nearrow 0} + \sum(\dot{m}(h+ke+pe-T_{env}s)_{env})_{inlets} - \sum(\dot{m}(h+ke+pe-T_{env}s)_{env})_{outlets} - T_{env}\dot{S}_{gen}^{\nearrow 0,\, int.rev.} \Rightarrow \dot{W} = 0 \quad (6.120)$$

Thus, an internally reversible device at the environmental state, inter-acting only with the environment produces no work! Or,

$$-\frac{d(me_{env}-T_{env}ms_{env})_{CV}}{dt} + \sum(\dot{m}(h+ke+pe-T_{env}s)_{env})_{inlets} - \sum(\dot{m}(h+ke+pe-T_{env}s)_{env})_{outlets} = 0 \quad (6.121)$$

Let's subtract the previous (6.121) (which is zero) from the combined rate-balance (6.119) giving

$$\dot{W} = \sum \left[\dot{Q}(1 - \frac{T_{env}}{T_{int}})\right] - \frac{d\left[(E - E_{env}) - T_{env}(S - S_{env})\right]_{CV}}{dt}$$
$$+ \sum \left[\dot{m}((h - h_{env}) + (ke - ke_{env}) + (pe - pe_{env}) - T_{env}(s - s_{env}))\right]_{inlets}$$
$$- \sum \left[\dot{m}((h - h_{env}) + (ke - ke_{env}) + (pe - pe_{env}) - T_{env}(s - s_{env}))\right]_{outlets} - T_{env}\dot{S}_{gen} \quad (6.122)$$

Staring at (6.122) we appear to have what looks to be a conservation equation: From the CV time-rate of change term, conservation of $[(E - E_{env}) - T_{env}(S - S_{env})]$! The next question to ask is "Is $[(E-E_{env})-T_{env}(S-S_{env})]$ a property of the system"? The quick answer is "no", given that the term contains the environmental temperature T_{env}, a property of the environment not the system, as are the energy and entropy evaluated at the environmental state ($E_{env} = m_{CV}e_{env}$ and $S_{env} = m_{CV}s_{env}$)! Next, note that the heat source term $\left[\dot{Q}(1 - \frac{T_{env}}{T_{int}})\right]$ is identical to (6.115) and thus represents the maximum power available for conversion to work from a given heat inter-action. Also, referring to (6.117) we see that the mass-flow terms $\dot{m}((h-h_{env})+(ke-ke_{env})+(pe-pe_{env})-(T_{env}s-T_{env}s_{env}))$ represent the flow of energy available for conversion to work. Finally, $T_{env}\dot{S}_{gen}$ represents the rate of destruction of energy convertible to work due to entropy generation within the system. Hence we define the following:

Definition 6.3 ***Availability/Exergy:*** *The work potential or maximum possible energy available for conversion to useful work (or power) is referred to as "Availability"or "Exergy".*

Again, Exergy or Availability is *not* a property of the system due to the dependence on system *and* environmental properties. The Exergy or Availability-balance (6.122), is restated below along with the physical significance of each term:

$$\dot{W} = \overbrace{\sum \left[\dot{Q}(1 - \frac{T_{env}}{T_{int}})\right]}^{\text{Total Availability transfer due to heat inter-actions}} - \overbrace{\frac{d\left[(E - E_{env}) - T_{env}(S - S_{env})\right]_{CV}}{dt}}^{\text{Time-rate of change in Availability within the system } CV} +$$
$$\overbrace{\sum \left[\dot{m}((h - h_{env}) + (ke - ke_{env}) + (pe - pe_{env}) - T_{env}(s - s_{env}))\right]_{inlets}}^{\text{Total inflow of Availability due to mass inter-actions}} -$$
$$\overbrace{\sum \left[\dot{m}((h - h_{env}) + (ke - ke_{env}) + (pe - pe_{env}) - T_{env}(s - s_{env}))\right]_{outlets}}^{\text{Total outflow of Availability due to mass inter-actions}} - \overbrace{T_{env}\dot{S}_{gen}}^{\text{Destruction of Availability } \dot{W}_{lost} \text{ due to internal irreversibilities}} \quad (6.123)$$

The last term in (6.123) $\dot{W}_{lost}$ representing the rate of destruction of available work is central to the concept of Availability and will be calculated in the following examples.

EXAMPLE 6.11 Availability/Exergy destruction from an electric wall heater

Problem Statement:

A building is heated via a base-board heater which consumes electricity at a rate of $2kW$. The building interior is kept at 27°. The inner-wall of the building is maintained at 23° while the outer wall is kept at $12°C$. The environment (outside the building) is at $7°C$. Find the rate of Availability destruction in the room, in the wall, within the environment and the total Availability destruction.

Given:

$\dot{W}_{electric} = -2kW$,

$T_{interior} = 27°C, T_{\substack{inner\\wall}} = 23°C$,

$T_{\substack{outer\\wall}} = 12°C, T_{env} = 7°K$

Find:

$\dot{W}_{\substack{lost\\interior}}, \dot{W}_{\substack{lost\\wall}}$,

$\dot{W}_{\substack{lost\\environment}}, \dot{W}_{\substack{lost\\total}}$

Assumptions:

Steady-state (S.S.),

Potential and kinetic energy of the environment, room and wall are negligible.

Schematic:

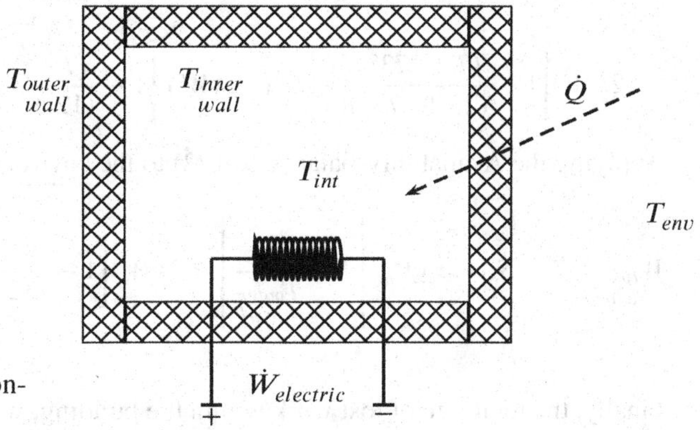

Analysis:

Solving the Availability-balance (6.122) for the rate of Availability destruction $T_{env}\dot{S}_{gen}$ or $\dot{W}_{lost}$ and applying to the building interior system gives

$$\overset{\dot{W}_{electric}}{\cancel{\dot{W}}} = \sum\left[\dot{Q}(1 - \frac{T_{env}}{T_{int}})\right] - \overbrace{\frac{d\left[(E - E_{env}) - T_{env}(S - S_{env})\right]_{CV}}{dt}}^{0,\ S.S.}$$

$$+\overbrace{\sum\left[\dot{m}((h - h_{env}) + (ke - ke_{env}) + (pe - pe_{env}) - T_{env}(s - s_{env}))\right]_{inlets}}^{0}$$

$$-\overbrace{\sum\left[\dot{m}((h - h_{env}) + (ke - ke_{env}) + (pe - pe_{env}) - T_{env}(s - s_{env}))\right]_{outlets}}^{0} - T_{env}\dot{S}_{gen}$$

$$\Rightarrow \dot{W}_{\substack{lost\\interior}} = T_{env}\dot{S}_{\substack{gen\\interior}} = -\dot{W}_{electric} + \sum\left[\dot{Q}(1 - \frac{T_{env}}{T_{int}})\right] \qquad (6.124)$$

Noting that we have a single heat inter-action (rejection from the room at the inner-wall surface, $\dot{Q} = \dot{W}_{electric}$ via the energy-balance where the inter-action temperature is $T_{\substack{inner\\wall}}$) we get

$$\dot{W}_{\substack{lost\\interior}} = T_{env}\dot{S}_{\substack{gen\\interior}} = -\dot{W}_{electric} + \dot{Q}\left[1 - \frac{T_{env}}{T_{\substack{inner\\wall}}}\right] =$$

$$-(-2kW) + (-2kW)\left[1 - \frac{(7 + 273°K)}{(23 + 273°K)}\right] = 1.892kW \Longleftarrow \dot{W}_{\substack{lost \\ interior}} \quad (6.125)$$

Applying the Availability-balance (6.122) to the building wall and again solving for the rate of Availability destruction $T_{env}\dot{S}_{gen}$ gives

$$\dot{W}_{\substack{lost \\ wall}} = T_{env}\dot{S}_{\substack{gen \\ wall}} = \dot{Q}_{\substack{inner \\ wall}}\left[1 - \frac{T_{env}}{T_{\substack{inner \\ wall}}}\right] + \dot{Q}_{\substack{outer \\ wall}}\left[1 - \frac{T_{env}}{T_{\substack{outer \\ wall}}}\right] =$$

$$2kW\left[1 - \frac{(7 + 273°K)}{(23 + 273°K)}\right] + (-2kW)\left[1 - \frac{(7 + 273°K)}{(12 + 273°K)}\right] = 0.07298kW \Longleftarrow \dot{W}_{\substack{lost \\ wall}} \quad (6.126)$$

Applying the Availability-balance (6.122) to the environmental system gives

$$\dot{W}_{\substack{lost \\ env}} = T_{env}\dot{S}_{\substack{gen \\ env}} = \dot{Q}_{env}\left[1 - \frac{T_{env}}{T_{\substack{outer \\ wall}}}\right] = 2kW\left[1 - \frac{(7 + 273°K)}{(12 + 273°K)}\right] = 0.03507kW \Longleftarrow \dot{W}_{\substack{lost \\ env}} \quad (6.127)$$

Finally, the total rate of lost work within the building, wall and environmental system is

$$\dot{W}_{\substack{lost \\ total}} = \dot{W}_{\substack{lost \\ interior}} + \dot{W}_{\substack{lost \\ wall}} + \dot{W}_{\substack{lost \\ env}} = (1.892 + 0.07298 + 0.03507)kW = 2kW \Longleftarrow \dot{W}_{\substack{lost \\ total}} \quad (6.128)$$

$$\boxed{\dot{W}_{\substack{lost \\ interior}} = 1.892kW,\ \dot{W}_{\substack{lost \\ wall}} = 0.07298kW,\ \dot{W}_{\substack{lost \\ env}} = 0.03507kW,\ \dot{W}_{\substack{lost \\ total}} = 2kW}$$

Comment: Note that $-\dot{W}_{electric} = \dot{W}_{\substack{lost \\ total}} = 2kW$, hence *all* the heater power is lost heating the building interior due to entropy generation.

□

EXAMPLE 6.12 Availability Analysis: Gas-Turbine

Problem Statement:
A power producing device (gas-turbine) composed of a compressor, heat-exchanger/combustor and turbine inter-acts with a heat source and the environment at temperatures $1500°K$ and $300°K$ respectively. Specifically, air is compressed in an adiabatic compressor from the environmental state (at $1atm$) to a pressure and temperature of $50bar$ and $1000°K$. The pressurized air then passes through a heat- exchanger/combustor where heat is added at a rate of $5MW$, heating the air to $1400°K$. The heat-exchanger exhibits a pressure drop (due to friction) of $20kPa$ and has a heat inter-action temperature equal to the combustor outlet temperature. The pressurized, heated air then expands through an adiabatic turbine to a temperature and pressure of $600°K$ and $1atm$ respectively. The outlet exhaust is rejected to the environment. You are to first calculate the air mass-flow rate, isentropic efficiencies for the compressor and turbine, as well as the compressor, turbine and the net power, and the overall device efficiency. Next, calculate the maximum possible

device efficiency (assuming inter-actions with the source and the environment) and the overall power (or availability) lost due to irreversibilities. Finally, determine the availability lost within the compressor, heat-exchanger and turbine, as well as the availability lost in the transition region *between* the heat-source and the heat-exchanger/combustor, and *between* the turbine exit and the environment. You may assume the air behaves as an Ideal Gas.

Given:

Gas-turbine power-cycle (air (I.G.B.)),
Adiabatic compressor and turbine,
$T_{source} = 1500°K$,
$T_{env} = 300°K$,
$P_{env} = 1atm$,
$\dot{Q}_{source} = 5MW$,
$T_2 = 1000°K$,
$T_3 = 1400°K$,
$T_4 = 600°K$,
$P_2 = 50bar$,
$P_4 = P_{env}$,
$\Delta P_{2\to 3} = -20kPa$,
$T_{int.HE} = T_3 = 1400°K$

Schematic:

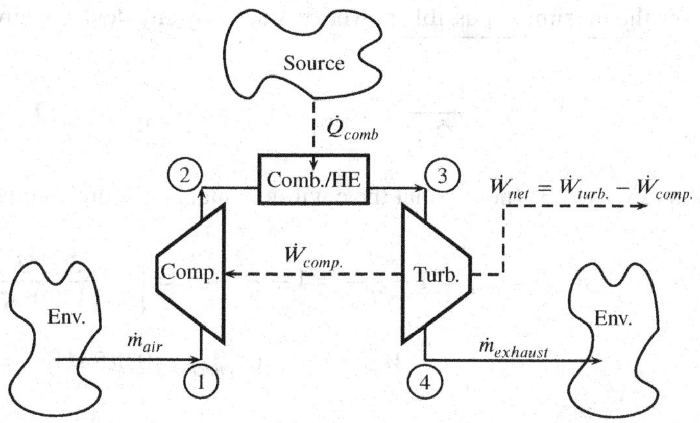

Find:

a) $\dot{m}$, $\eta^{isen}_{turb.}$, $\eta^{isen}_{comp.}$
b) $\dot{W}_{turb.}$, $\dot{W}_{comp.}$, $\dot{W}_{net}$, η
c) η_{max}, $\dot{W}_{max}$
d) $\dot{W}_{lost\ overall}$
e) $\dot{W}_{lost\ comp.}$, $\dot{W}_{lost\ turb}$, $\dot{W}_{lost\ HE}$, $\dot{W}_{lost\ source\to HE}$, $\dot{W}_{lost\ 4\to env}$

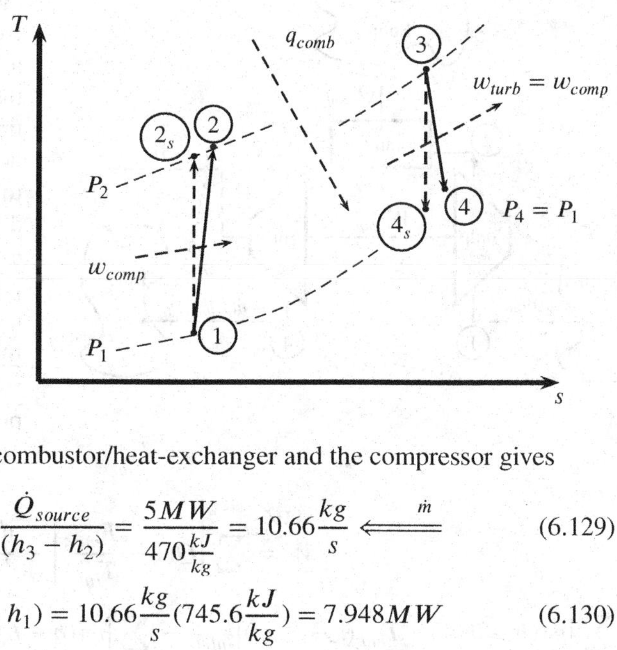

Assumptions:

$T_1 = T_{env}$, $P_1 = P_{env}$,
Steady-state, steady-flow

Analysis:

Application of the energy-balance to the combustor/heat-exchanger and the compressor gives

$$\dot{Q}_{source} = \dot{m}(h_3 - h_2) \Rightarrow \dot{m} = \frac{\dot{Q}_{source}}{(h_3 - h_2)} = \frac{5MW}{470\frac{kJ}{kg}} = 10.66\frac{kg}{s} \overset{\dot{m}}{\Longleftarrow} \tag{6.129}$$

$$\dot{W}_{comp.} = \dot{m}(h_2 - h_1) = 10.66\frac{kg}{s}(745.6\frac{kJ}{kg}) = 7.948MW \tag{6.130}$$

Applying the energy-balance to the isentropic compression process $1 \to 2_s$ gives the isentropic compressor power $\dot{W}^{isen}_{comp.} = \dot{m}(h_{2_s} - h_1)$ yielding a compressor isentropic efficiency

$$\eta^{isen}_{comp} = \frac{\dot{W}^{isen}_{comp.}}{\dot{W}_{comp.}} = \frac{\dot{m}(h_{2_s} - h_1)}{\dot{W}_{comp.}} = \frac{10.66\frac{kg}{s}(608.6\frac{kJ}{kg})}{\dot{W}_{comp.}} = \frac{6.485\cancel{MW}}{7.948\cancel{MW}} = 0.8159 \overset{\eta^{isen}_{comp}}{\Longleftarrow} \tag{6.131}$$

Similarly, the the turbine power and isentropic efficiency are found to be $9.678MW$ and 0.9082 respectively. The net power and efficiency are

$$\dot{W}_{net} = \dot{W}_{turb} - \dot{W}_{comp} = 9.678MW - 7.948MW = 1.730MW \overset{\dot{W}_{net}}{\Longleftarrow} \tag{6.132}$$

$$\eta = \frac{\dot{W}_{net}}{\dot{Q}_{source}} = \frac{1.730MW}{5MW} = 0.346 \overset{\eta}{\Longleftarrow} \tag{6.133}$$

Now the maximum possible power produced by any device is given by

$$\frac{\dot{W}_{max}}{\dot{Q}_{in}} = \eta_{max} = \eta_{rev} \Rightarrow \dot{W}_{max} = \eta_{rev}\dot{Q}_{in} = \eta_{rev}\dot{Q}_{source} \tag{6.134}$$

where via (5.13) (noting that the environmental and source temperature correspond to T_L and T_H)

$$\eta_{max} = \eta_{rev} = 1 - \frac{T_L}{T_H} = 1 - \frac{T_{env}}{T_{source}} = \left[1 - \frac{300°K}{1500°K}\right] = 0.8 \overset{\eta_{max}}{\Longleftarrow} \tag{6.135}$$

$$\dot{W}_{max} = \eta_{rev}\dot{Q}_{source} = 0.8(5MW) = 4MW \overset{\dot{W}_{max}}{\Longleftarrow} \tag{6.136}$$

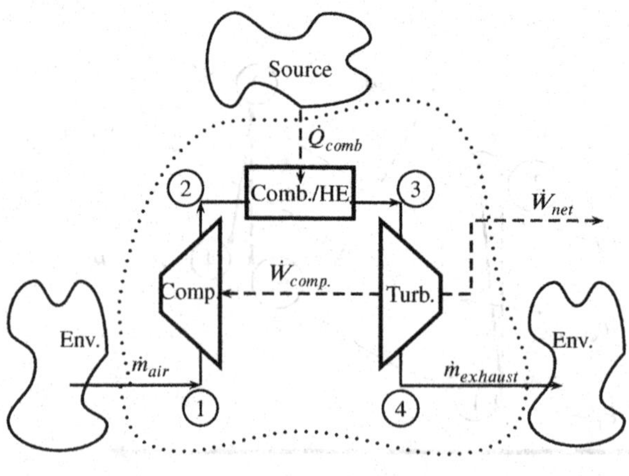

To find the total rate of lost power $\dot{W}_{lost}$ (availability destroyed due to irreversibility) from the operation of the device, we choose a control-volume (dotted-line shown in the inset) which includes the device, the transition regions between the source and heat-exchanger (entropy is generated due to heat-transfer across a finite temperature difference), and between the hot turbine exit-flow and the cool environment. Applying the Availability-balance (6.123) noting that we are in steady-state with inlet and outlet mass-flows at the environment state P_{env} and T_{env}, neglecting changes in potential and kinetic energy gives

$$\dot{W}_{net} = \sum\left[\dot{Q}(1 - \frac{T_{env}}{T_{int}})\right] - \overbrace{\frac{d\left[(E - E_{env}) - T_{env}(S - S_{env})\right]_{CV}}{dt}}^{0,\,S.S.} +$$

$$\sum\left[\dot{m}((h - h_{env}) - T_{env}(s - s_{env}))\right]_{inlets} - \sum\left[\dot{m}((h - h_{env}) - T_{env}(s - s_{env}))\right]_{outlets} - T_{env}\dot{S}_{gen} \tag{6.137}$$

or solving for the overall rate of power (rate of availability) lost due to irreversibility $T_{env}\dot{S}_{gen}$

$$\dot{W}_{\substack{lost\\overall}} = T_{env}\dot{S}_{gen} = -\dot{W}_{net} + \sum\left[\dot{Q}(1 - \frac{T_{env}}{T_{int}})\right] =$$

$$-\dot{W}_{net} + \dot{Q}_{source}\left(1 - \frac{T_{env}}{T_{source}}\right) - \dot{Q}_{env}\overbrace{\left(1 - \frac{T_{env}}{T_{env}}\right)}^{0} = -\dot{W}_{net} + \dot{Q}_{source}\left(1 - \frac{T_{env}}{T_{source}}\right)$$

$$= -1.730MW + 5MW\left(1 - \frac{300°K}{1500°K}\right) = 2.27MW \overset{\dot{W}_{\substack{lost\\overall}}}{\Longleftarrow} \quad (6.138)$$

To isolate the source(s) of Availability destruction, we now turn our attention to each individual component within the gas turbine including the transition regions between the source and heat-exchanger, and the turbine exit and the environmental reservoir. Beginning with the compressor we apply the Availability-balance (neglecting kinetic and potential energy terms)

$$\overset{-\dot{W}_{comp.}}{\cancel{\dot{W}}} = \sum\left[\overset{0,\,adia.}{\cancel{\dot{Q}}}\left(1 - \frac{T_{env}}{T_{int}}\right)\right] - \overbrace{\frac{d\left[(E - E_{env}) - T_{env}(S - S_{env})\right]_{CV}}{dt}}^{0,\,S.S.} +$$

$$\left[\dot{m}((h - \cancel{h_{env}}) - T_{env}(s - \cancel{s_{env}}))\right]_1 - \left[\dot{m}((h - \cancel{h_{env}}) - T_{env}(s - \cancel{s_{env}}))\right]_2 - T_{env}\dot{S}_{gen}$$

$$\Rightarrow \dot{W}_{\substack{lost\\comp.}} = T_{env}\dot{S}_{gen} = \dot{W}_{comp.} + \dot{m}[h_1 - h_2 - T_{env}(s_1 - s_2)] =$$

$$7948kW + 10.66\frac{kg}{s}\left[-745.6\frac{kJ}{kg} - 300°K\left(-0.146\frac{kJ}{kg - °K}\right)\right] = 467.7kW \overset{\dot{W}_{\substack{lost\\comp.}}}{\Longleftarrow} \quad (6.139)$$

Next, application of the Availability-balance to the combustor/HE (*excluding the transition region between the combustor and heat-source*) gives (neglecting kinetic and potential energy terms)

$$\overset{0}{\cancel{\dot{W}}} = \sum\left[\dot{Q}\left(1 - \frac{T_{env}}{T_{int}}\right)\right] - \overbrace{\frac{d\left[(E - E_{env}) - T_{env}(S - S_{env})\right]_{CV}}{dt}}^{0,\,S.S.} +$$

$$\left[\dot{m}((h - \cancel{h_{env}}) - T_{env}(s - \cancel{s_{env}}))\right]_2 - \left[\dot{m}((h - \cancel{h_{env}}) - T_{env}(s - \cancel{s_{env}}))\right]_3 - T_{env}\dot{S}_{gen}$$

$$\Rightarrow \dot{W}_{\substack{lost\\HE}} = T_{env}\dot{S}_{gen} = \dot{Q}_{source}\left[1 - \frac{T_{env}}{T_{source}}\right] + \dot{m}\left[h_2 - h_3 - T_{env}(s_2 - s_3)\right] =$$

$$5MW\left[1 - \frac{300°K}{1500°K}\right] + 10.66\frac{kg}{s}\left[-530\frac{kJ}{kg} - 300°K\left(-0.395\frac{kJ}{kg - °K}\right)\right] = 191.3kW \overset{\dot{W}_{\substack{lost\\HE}}}{\Longleftarrow} \quad (6.140)$$

Likewise for the turbine (neglecting kinetic and potential energy terms)

$$\overset{\dot{W}_{turb}}{\cancel{\dot{W}}} = \sum\left[\overset{0,\,adia.}{\cancel{\dot{Q}}}\left(1 - \frac{T_{env}}{T_{int}}\right)\right] - \overbrace{\frac{d\left[(E - E_{env}) - T_{env}(S - S_{env})\right]_{CV}}{dt}}^{0,\,S.S.} +$$

$$\left[\dot{m}((h - \cancel{h_{env}}) - T_{env}(s - \cancel{s_{env}}))\right]_3 - \left[\dot{m}((h - \cancel{h_{env}}) - T_{env}(s - \cancel{s_{env}}))\right]_4 - T_{env}\dot{S}_{gen}$$

$$\Rightarrow \dot{W}_{\substack{lost \\ turb}} = T_{env}\dot{S}_{gen} = -\dot{W}_{turb} + \dot{m}[h_3 - h_4 - T_{env}(s_3 - s_4)] =$$

$$-(9678kW) + 10.66\frac{kg}{s}\left[908.7\frac{kJ}{kg} - 300^\circ K\left(-0.165\frac{kJ}{kg-^\circ K}\right)\right] = 528.7kW \xleftarrow{\dot{W}_{\substack{lost \\ turb}}} \quad (6.141)$$

Next, we turn our attention to the lost power associated with the operation, but outside of the device itself: The regions in-between the combustor/HE and the source, and the turbine exit and the environment. Choosing a control-volume surrounding the region with boundary adjacent to the source and the combustor/HE, we apply the Availability-balance yielding (where the heat-exchanger inter-action temperature is T_3)

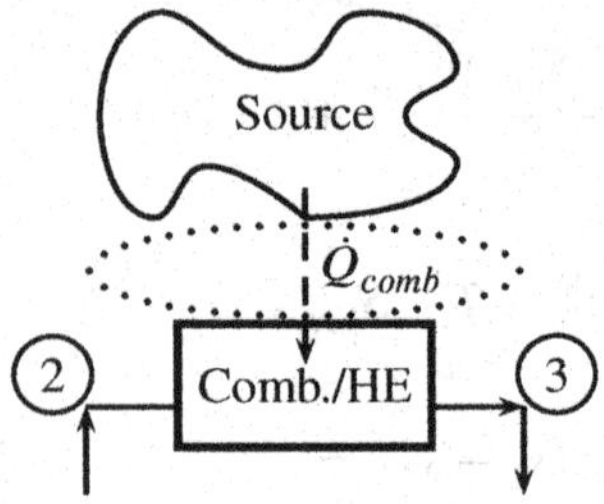

$$\dot{W}_{\substack{lost \\ source \to HE}} = T_{env}\dot{S}_{\substack{gen. \\ source \to HE}} =$$

$$\dot{Q}_{source}\left[1 - \frac{T_{env}}{T_{source}}\right] - \dot{Q}_{source}\left[1 - \frac{T_{env}}{T_{Int.HE} \to T_3}\right] =$$

$$5MW\left[1 - \frac{300^\circ K}{1500^\circ K}\right] - 5MW\left[1 - \frac{300^\circ K}{1400^\circ K}\right] = 71.43kW \xleftarrow{\dot{W}_{\substack{lost \\ source \to HE}}}$$

(6.142)

Next, applying the Availability-balance to the region between the turbine exit and the environment (where the exhaust energy is ultimately dissipated) gives (assuming that the mass-crossing the system boundary into the environment is at the environment state T_{env} and P_{env})

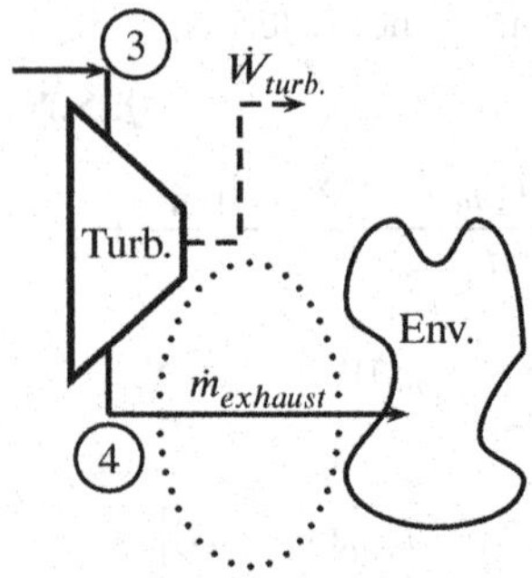

$$\dot{W}_{\substack{lost \\ 4 \to env.}} = T_{env}\dot{S}_{\substack{gen. \\ 4 \to env.}} =$$

$$\dot{Q}_{env}\left[1 - \frac{T_{evn}}{T_{env}}\right]^{\nearrow 0} + \dot{m}\left[h_4 - h_{env}^{\nearrow h|_{T_{env}}} + T_{env}(s_4 - s_{env}^{\nearrow s|_{T_{env},P_{env}}})\right] =$$

$$10.66\frac{kg}{s}\left[306.9\frac{kJ}{kg} + 300^\circ K\left(0.706\frac{kJ}{kg-^\circ K}\right)\right] = 1011kW \xleftarrow{\dot{W}_{\substack{lost \\ 4 \to env.}}}$$

(6.143)

$\dot{m} = 10.66\frac{kg}{s}$, $\eta^{isen}_{turb.} = 0.9082$, $\eta^{isen}_{comp.} = 0.8159$, $\dot{W}_{turb.} = 9678kW$, $\dot{W}_{comp.} = -7948kW$, $\dot{W}_{net} = 1730kW$, $\eta = 0.346$, $\eta_{max} = 0.8$, $\dot{W}_{max} = 4MW$, $\dot{W}_{\substack{lost \\ overall}} = 2270kW$, $\dot{W}_{\substack{lost \\ comp.}} = 467.7kW$, $\dot{W}_{\substack{lost \\ turb}} = 528.7kW$, $\dot{W}_{\substack{lost \\ HE}} = 191.3kW$, $\dot{W}_{\substack{lost \\ source \to HE}} = 71.43kW$, $\dot{W}_{\substack{lost \\ 4 \to env.}} = 1011kW$

Comment: Note that summing <u>all</u> component and region losses gives the total lost power $2270kW$; the same as found for the calculation of $\dot{W}_{\substack{lost \\ overall}}$. In addition, the turbine is responsible for the largest fraction of the lost-power, hence, improvements in turbine isentropic efficiency would significantly improve overall gas-turbine efficiency.

□

CHAPTER 7

CYCLES: GAS AND VAPOR POWER DEVICES

7.1 CYCLE ANALYSIS: MOTIVATION

Recall that the Carnot cycle (an internally and externally reversible cycle), exhibits the following characteristics:

- Frictionless operation
- No un-restrained expansions of working fluid
- No heat-transfer across a finite temperature difference

Needless to say, from what we now know of entropy, these constraints guarantee that S_{gen} is zero at all locations and at all times anywhere within Carnot device or in the transition region between the device control volume and the inter-acting reservoirs.

As described in §5.5 the first two constraints can be approached in nature by using lubricants (to minimize friction) and sufficiently slow operation of the device (to minimize un-restrained expansion). However, the last constraint presents a problem. Recall that for the Carnot cycle we are restricted to reversible heat-transfer processes as discussed in §5.5: This required a vanishingly small temperature difference between the working fluid and the heat source or sink: $\Delta T = T_{system} - T_{reservior} \rightarrow 0$. The distinction is shown in Figure 7.1. Now, the primary problem with the Carnot cycle, from an engineering stand-point, is that the rate of heat-transfer $\dot{Q}$ to or from the system is proportional to the temperature difference ΔT: $\dot{Q} \propto \Delta T$ and a high temperature difference between the system and reservor implies high rates of heat transfer. Thus, in the limit at $\Delta T \rightarrow 0$

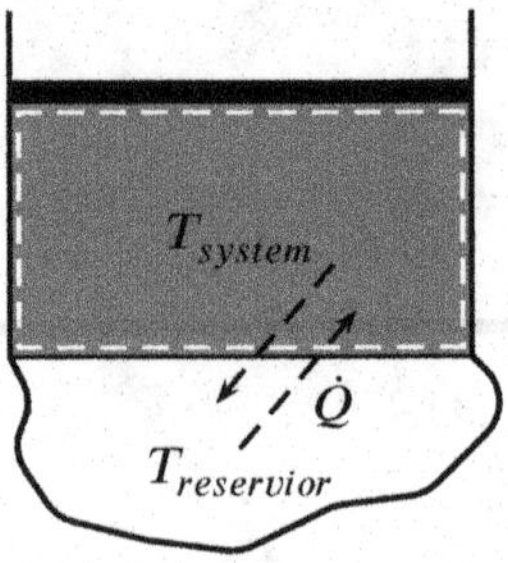

$$\Delta T = T_{system} - T_{reservior} \begin{cases} \nrightarrow 0, & \text{irreversible heat transfer} \\ \rightarrow 0, & \text{reversible heat transfer} \end{cases}$$

Figure 7.1 Reversible and irreversible heat transfer $\dot{Q}$ to/from piston cyclinder device

(reversible heat transfer), $\dot{Q} \rightarrow 0$. Any attempt to construct a device which approximates the Carnot cycle results in the maximum possible efficiency but an (infinitely) slow rate of heat transfer. Thus, the rate of work done $\dot{W}$ (or power produced) tends to zero!

What is the engineering solution? Well, we proceed to relax the constraint on reversible heat-transfer (to increase cycle power), but retain friction-less slow operation.

7.2 AIR-STANDARD CYCLE ANALYSIS: CLOSED SYSTEMS

Because we have relaxed the reversibility of the heat transfer process, the number of cycles available for analysis are infinite. Thus, we will introduce a standardized set of constraints which will lay the groundwork for our analysis of any number of possible closed system cycles. These constraints, the air-standard cycle assumptions, are given below:

Closed system air-standard cycle assumptions:

1. Closed system
2. System (working fluid) composed of air
3. Internally reversible processes : $S_{gen_{system}} = 0$
4. Combustion/Exhaust processes $\Rightarrow$ Heat addition/rejection processes
5. $\Delta PE = \Delta KE = 0$ for all processes
6. $W_{electric} = W_{shaft} = 0$ for all processes

Note that condition #3 requires that no irreversibilities exist within the system: All processes are internally reversible. Adding this constraint mearly requires that there be no friction, un-restrained expansion or heat transfer across a finite temperature difference within the system. Irreversibilities outside the system are perfectly legal and the cycle need not be externally reversible. This means, for example, a finite temperature jump between the system and heat source/sink may be present: This is in contrast to the Carnot cycle which is fully (internally and externally) reversible!

A schematic representating a piston-cylinder device complete with valve-train, crank-shaft and connecting-rod are shown in Figure 7.2. The piston imparts a motion (and power) to a rotating crank-shaft via a connecting-rod. This cycle consists of four strokes executed over two crankshaft rotations: Compression, power, exhaust and an intake-stroke. Hence, Figure 7.2 illustrates a four-stroke cycle. Also, min and max cylinder volumes are indicated as are the piston positions bottom-

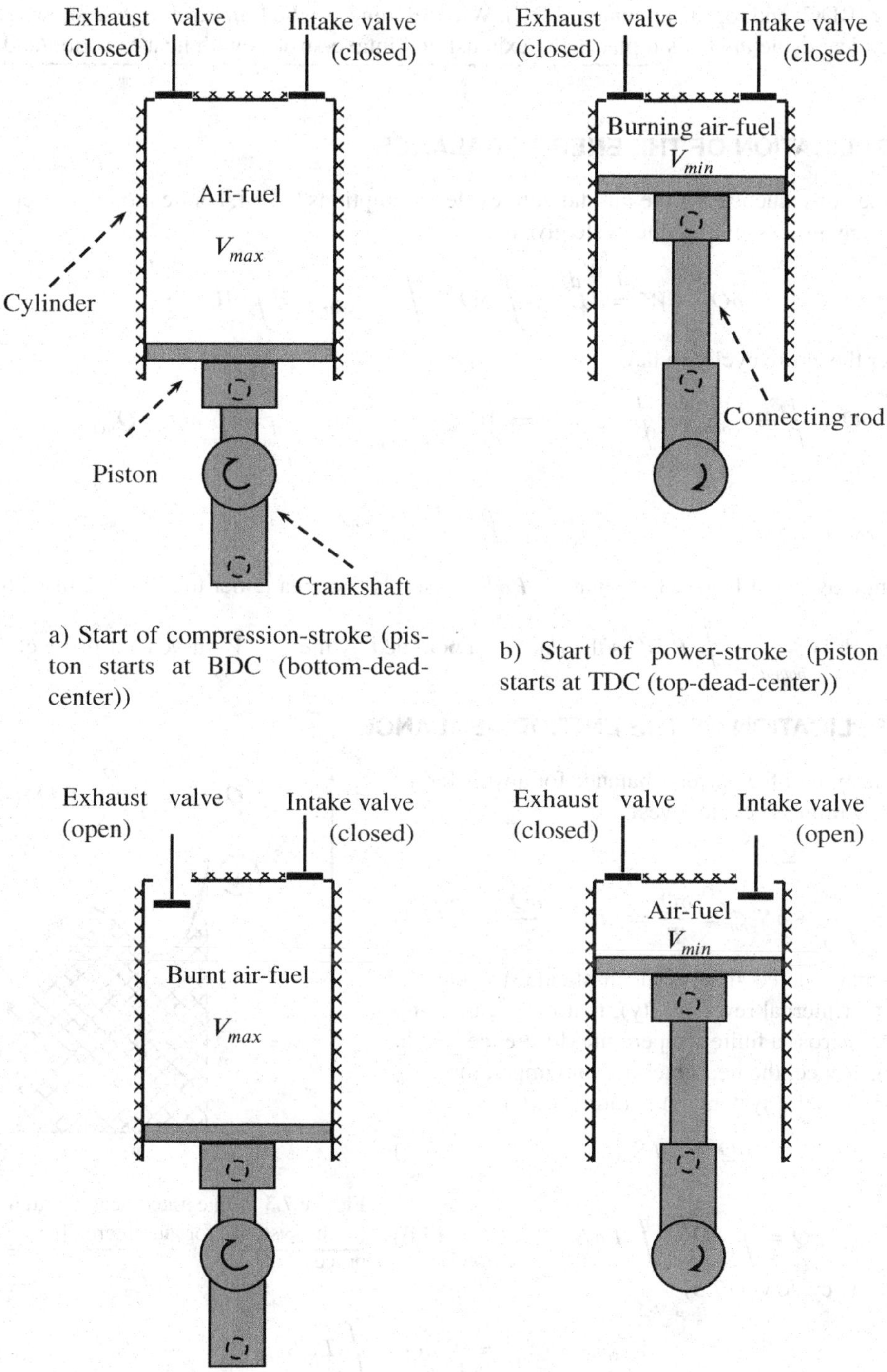

a) Start of compression-stroke (piston starts at BDC (bottom-dead-center))

b) Start of power-stroke (piston starts at TDC (top-dead-center))

c) Start of exhaust-stroke (piston starts at BDC (bottom-dead-center))

d) Start of intake-stroke (piston starts at TDC (top-dead-center))

Figure 7.2 Schematic and operation of a four-stroke internal combustion (IC) engine.

dead-center (BDC) and top-dead-center (TDC). With reference to the four-stroke cycle shown in the Figure, the ideal cycle analysis replaces the exhaust and intake-strokes with heat rejection/addition processes.

7.2.1 APPLICATION OF THE ENERGY-BALANCE

What are the consequences of the air-standard cycle assumptions? Well, application of the energy-balance for any process within the cycle gives

$$\delta Q - \delta W_{bound} = dU \Rightarrow \int \delta Q - \int \delta W_{bound} = \int dU \tag{7.1}$$

Indeed over the entire cycle we have

$$\oint \delta Q - \oint \delta W_{bound} = \oint dU = 0 \Rightarrow W_{\substack{cycle\\bound}} = \oint \delta W_{bound} = \oint \delta Q = Q_{net} \tag{7.2}$$

or

$$W_{\substack{cycle\\bound}} = \oint PdV = Q_{net} \tag{7.3}$$

The student may recall from §3.3.2 that $\int PdV$ is simply the area under the $P - V$ curve, hence the cycle work $W_{\substack{cycle\\bound}} = \oint PdV$ is the total area bounded by the $P - V$ curve over the cycle.

7.2.2 APPLICATION OF THE ENTROPY-BALANCE

Next, application of the entropy-balance for any differential process within the cycle gives

$$dS = \frac{\delta Q}{T} + \overset{0}{\delta S_{gen}} = \frac{\delta Q}{T} \Rightarrow dS = \frac{\delta Q}{T} \tag{7.4}$$

where we have noted that from air-standard cycle assumption #3 (internal reversibility), that there can be no heat transfer across a finite temperature difference within the system. Hence, the heat inter-action temperature T_{int} must be equal to the system temperature T. Or

$$\delta Q = TdS \tag{7.5}$$

and

$$Q = \int \delta Q = \int TdS \tag{7.6}$$

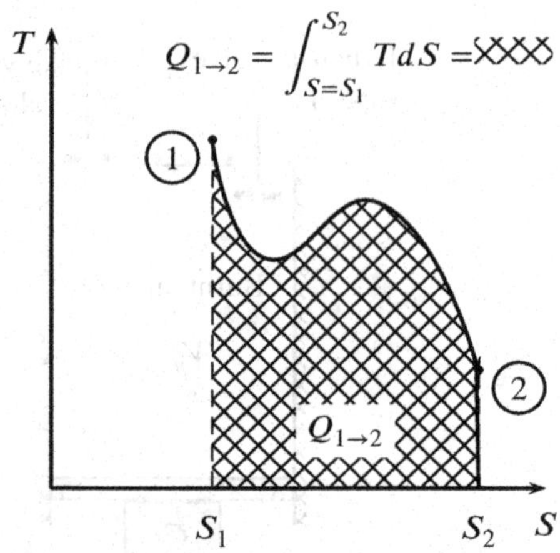

Figure 7.3 Integrated heat addition $Q_{1\to2}$ (to the system) for an internally reversible process.

Thus, over the cycle via (7.3)

$$W_{\substack{cycle\\bound}} = Q_{net} = \oint \delta Q = \oint TdS \tag{7.7}$$

Now (7.6) and (7.7) are both interesting and highly useful results since they indicate a graphical interpretation of the area under the $T - S$ curve as shown in Figure 7.3 (analogous to the area under the $P - V$ curve). Hence, careful examination of the cycle $T - S$ plot allows us to represent cycle work, heat addition, heat rejection and thus efficiency.

EXAMPLE 7.1 Cycle analysis using $T-S$ diagrams.

Problem Statement:
$T-S$ diagrams for two cycles A and B are given below. Using graphical methods, compare the cycle work, heat addition (not net heat addition) and thus efficiency of the cycles. Air-standard cycle assumptions are valid!

Given:
$T-S$ diagrams for cycles A and B,
air-standard cycle analysis

Find:
For both cycles A and B, graphically compare:
W_{cycle},
Q_{in},
η_{cycle}

Assumptions:
None

Schematic:

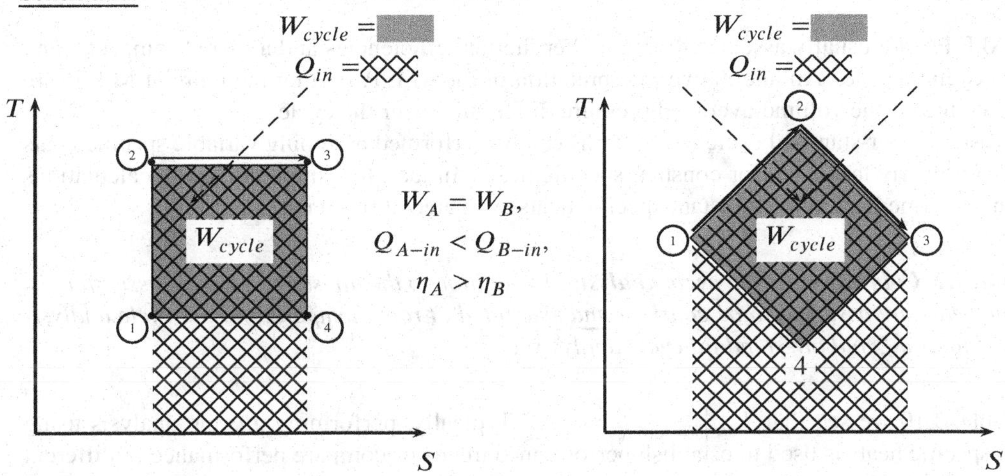

a) $T-S$ diagram:Cycle A

b) $T-S$ diagram:Cycle B

Analysis:

From (7.7) $W_{cycle} = \oint TdS$. Thus, from the $T-S$ diagrams we see that the area bound by the $T-S$ curves, the cycle work, for both cycles (in gray) are equal. Thus, $W_{cycle:A} = W_{cycle:B}$.

Now, from (7.5) $\delta Q = TdS$ and a heat addition process $\delta Q > 0$ occurs when $dS > 0$. Thus, integrating along the path where $dS > 0$ we have $Q_{in} = \int TdS$ as indicated by the crosshatched areas under the $T-S$ curve. By inspection

$$Q_{in:A} = Q_{A:2\to3} = \left[\int TdS\right]_A < \left[\int TdS\right]_B = Q_{B:1\to3} = Q_{in_B} \tag{7.8}$$

or $Q_{in_A} < Q_{in_B}$.

Finally, from the definition of efficiency $\eta = \frac{W_{cycle}}{Q_{in}}$ since the cycle works are equal, but $Q_{in:A} < Q_{in:B}$ this implies $\eta_A > \eta_B$.

$$\boxed{W_{cycle:A} = W_{cycle:B}, Q_{in:A} < Q_{in:B}, \eta_A > \eta_B}$$

□

Thus far, the only relevant quantities used to describe the performance of our cycle has been efficiency η and cycle work W_{cycle}. As engineers, one might also be interested in the cycle work produced relative to the *size* of the device. In what sense is the term *size* meant? Well, one way to do this is to normalize the cycle work by the total piston swept (or displaced) volume $\Delta V = V_{max} - V_{min}$, giving the M.E.P or "mean-effective-pressure" as defined below

Definition 7.1 ***Mean Effective Pressure*** ***M.E.P.:***

$$M.E.P. = \frac{\oint PdV}{\Delta V} = \frac{W_{cycle}}{V_{max} - V_{min}} \tag{7.9}$$

Higher M.E.P.'s are usually associated with higher thermal efficiencies and a more "compact" and powerful cyclic device. An alternative interpretation of the M.E.P. (noting the units of M.E.P. are that of pressure) is the volume-averaged pressure differential over the cycle.

As a last note, air-standard cycle calculations can be performed assuming variable specific heats (i.e. using property tables) or for constant specific heats. In fact, for simplicity, cycle calculations are often performed assuming constant specific heats evaluated at room temperature:

Definition 7.2 ***Cold air-standard cycle analysis:*** *In addition to the air-standard cycle assumptions, if specific heats are assumed to be constant and evaluated at room temperature, the cycle analysis is referred to as a "cold air-standard cycle analysis".*

From Table B.10 we see that $k_{air}|_{T=300°K} = 1.4$. Typically, performing a cycle analysis using constant specific heats is used to establish performance trends or compare performance of different cycles. Such an analysis will be performed in §7.3.

EXAMPLE 7.2 Detailed cycle analysis: Heat engine

Problem Statement:
A heat engine containing $0.002kg$ of working fluid (air) executes the following air standard cycle:
-Adiabatic compression from $305°K$ and $85kPa$ to $1/7.5$ of original volume.
-Isochoric heat addition to $960°K$
-Adiabatic expansion to 7.5 time original volume
-Isochoric heat rejection back to $305°K$ and $85kPa$

Determine the heat rejection, net cycle work, thermal efficiency and M.E.P.. Sketch $P - v$ and $T - s$ plots. Perform the calculations for:
a) variable specific heats (using property Tables B.11)
b) cold-air cycle ($k = 1.4$)

Given:
Air-standard cycle,
cycle processes:
$1 \to 2$ Adiabatic compression,
$2 \to 3$ Isochoric heat addition,
$3 \to 4$ Adiabatic expansion,
$4 \to 1$ Isochoric heat rejection,
$m = 0.002kg$,
$T_1 = 305°K$,
$P_1 = 85kPa$,
$r = \frac{V_1}{V_2} = \frac{V_4}{V_3} = 7.5$,
$T_3 = 960°K$

Schematic:

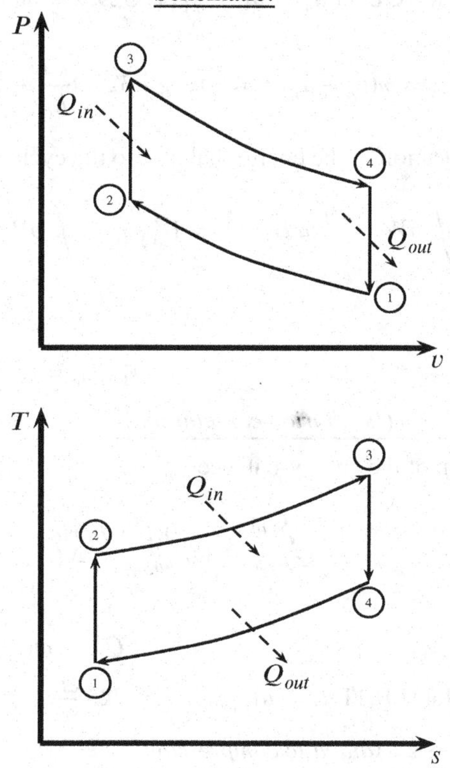

Find:
$Q_{out} = -Q_{4\to1}$,
W_{cycle},
η,
M.E.P.$=\frac{W_{cycle}}{V_{max}-V_{min}}$

Assumptions:
None

Analysis:

a) **Variable specific heats:**

Process $4 \to 1$ *(Isochoric heat rejection):*
Application of the energy-balance for an overall process

$$\overset{-Q_{out}}{\cancel{Q}_{4\to1}} - \overset{0,\,isochoric}{\cancel{W}_{bound_{4\to1}}} = \Delta E_{4\to1} = \Delta U_{4\to1} = m(u_1 - u_4) \qquad (7.10)$$

or

$$Q_{out} = -Q_{4\to1} = m(u_4 - u_1) \qquad (7.11)$$

From air Tables B.11 $u_1 = u|_{T_1=305°K} = 217.6\frac{kJ}{kg}$.

Process $3 \to 4$ *(Adiabatic compression):*
Application of S-balance gives

$$\Delta S_{3\to4} = \int_{3\to4} \frac{\overset{0,adia.}{\delta \cancel{Q}}}{T} + \overset{0,\,rev.}{\cancel{S}_{gen_{3\to4}}} = 0 \Rightarrow \Delta S_{3\to4} = 0 \qquad (7.12)$$

Thus, process $3 \to 4$ is isentropic and from (6.58) noting that $v_{r_3} = v_r|_{T_3=960°K} = 28.4$ via Table B.11

$$\frac{v_4}{v_3} = \frac{v_{r_4}}{v_{r_3}} \Rightarrow v_{r_4} = v_{r_3}\overset{r}{\cancel{\frac{v_4}{v_3}}} = v_{r_3} r = 28.4(7.5) = 213 \qquad (7.13)$$

Hence via Table B.11 $u_4 = u|_{v_{r_4}=213} = 329.9\frac{kJ}{kg}$ and (7.11)

$$Q_{out} = m(u_4 - u_1) = 0.002\cancel{kg}\left(329.9\frac{kJ}{\cancel{kg}} - 217.6\frac{kJ}{\cancel{kg}}\right) = 0.223kJ \xLeftarrow{Q_{out}} \tag{7.14}$$

Next, application of the energy-balance to the cycle

$$\oint \delta Q - \oint \delta W = \oint dE = 0 \Rightarrow W_{cycle} = \oint \delta W = \oint \delta Q = Q_{2\to3} + Q_{4\to1} = Q_{in} - Q_{out} \tag{7.15}$$

or

$$W_{cycle} = Q_{in} - Q_{out} \tag{7.16}$$

Process 2 → 3 (Isochoric heat addition):

Application of the energy-balance

$$\overset{Q_{in}}{\cancel{Q}_{2\to3}} - \overset{0,\,isochoric}{\cancel{W}_{boun_{2\to3}}} = \Delta E_{2\to3} = \Delta U_{2\to3} = m(u_3 - u_2) \tag{7.17}$$

or

$$Q_{in} = m(u_3 - u_2) \tag{7.18}$$

and from Table B.11 $u_3 = u|_{T_3=960^\circ K} = 725\frac{kJ}{kg}$.

Process 1 → 2 (Adiabatic compression):

Application of S-balance:

$$\Delta S_{1\to2} = \int_{1\to2} \frac{\overset{0,\,adia.}{\cancel{\delta Q}}}{T} + \overset{0,\,rev.}{\cancel{S}_{gen_{1\to2}}} = 0 \Rightarrow \Delta S_{1\to2} = 0 \tag{7.19}$$

Thus, process 1 → 2 is isentropic, $v_{r_1} = v_r|_{T_1=305^\circ K} = 596$ via Table B.11 and

$$\frac{v_1}{v_2} = \frac{v_{r_1}}{v_{r_2}} \Rightarrow v_{r_2} = v_{r_1}\overset{1/r}{\cancel{\frac{v_2}{v_1}}} = \frac{v_{r_1}}{r} = \frac{596}{7.5} = 79.5 \tag{7.20}$$

Hence $u_2 = u|_{v_{r_2}=79.5} = 486.7\frac{kJ}{kg}$ and via (7.18)

$$Q_{in} = m(u_3 - u_2) = 0.002\cancel{kg}\left(725\frac{kJ}{\cancel{kg}} - 486.7\frac{kJ}{\cancel{kg}}\right) = 0.476kJ \tag{7.21}$$

and hence via (7.16)

$$W_{cycle} = Q_{in} - Q_{out} = 0.476kJ - 0.223kJ = 0.253kJ \xLeftarrow{W_{cycle}} \tag{7.22}$$

From the definition of cycle efficiency (4.136)

$$\eta = \frac{W_{cycle}}{Q_{in}} = \frac{0.253\cancel{kJ}}{0.476\cancel{kJ}} = 0.529 = 53\% \xLeftarrow{\eta} \tag{7.23}$$

Finally via the definition of M.E.P. (7.9)

$$M.E.P. = \frac{W_{cycle}}{V_{max} - V_{min}} = \frac{W_{cycle}}{V_1 - V_2} = \frac{W_{cycle}}{V_1[1 - \frac{V_2}{V_1}]} = \frac{W_{cycle}}{V_1[1 - \frac{1}{r}]} \tag{7.24}$$

Now, via I.G.B.

$$P_1V_1 = mRT_1 \Rightarrow V_1 = \frac{mRT_1}{P_1} = \frac{0.002\cancel{kg}(0.287\frac{kJ}{\cancel{kg}-\cancel{°K}})305°\cancel{K}}{85kPa} = 2.05(10^{-3})m^3 \tag{7.25}$$

or returning to the expression for M.E.P. (7.24)

$$M.E.P. = \frac{W_{cycle}}{V_1[1 - \frac{1}{r}]} = \frac{0.253kJ}{2.05(10^{-3})m^3[1 - \frac{1}{7.5}]} = 142kPa \overset{M.E.P.}{\Longleftarrow} \tag{7.26}$$

$$\boxed{Q_{out} = 0.223kJ,\ W_{cycle} = 0.253kJ,\ \eta = 53\%,\ M.E.P. = 142kPa}$$

Comment: The calculated M.E.P. indicates a volume averaged pressure rise over the cycle is approximately one and a half atmospheres. Also, comparison of this cycle's efficiency with that of a reversible cycle operating between the same high and low temperature limits (960° K and 305° K) yields an efficiency of $\eta_{rev} = 1 - \left[\frac{Q_L}{Q_H}\right]_{rev} = 1 - \frac{T_L}{T_H} = 1 - \frac{305°\cancel{K}}{960°\cancel{K}} = 0.682 = 68.2\%$. As expected, the efficiency of the internally and externally reversible HE exceeds the calculated efficiency of our internally reversible HE herein.

b) **Cold-air (constant specific heats $k = 1.4$):**

Process 4 → 1 (Isochoric heat rejection):

Application of the energy-balance

$$Q_{out} = -Q_{4\to1} = m(u_4 - u_1) = mC_v|_{k=1.4}(T_4 - T_1) \tag{7.27}$$

Now via I.G.B. $C_P = C_v - R$ and

$$k = \frac{C_P}{C_v} = \frac{C_v + R}{C_v} = 1 + \frac{R}{C_v} \Rightarrow \frac{R}{C_v} = k - 1 \Rightarrow C_v = \frac{R}{k-1} = \frac{0.287\frac{kJ}{kg-°K}}{1.4 - 1} = 0.718\frac{kJ}{kg-°K} \tag{7.28}$$

Process 3 → 4 (Adiabatic compression):

Application of S-balance indicates that process 3 → 4 is isentropic. Hence via the isentropic property relationship (6.61)

$$\frac{T_4}{T_3} = \left(\cancelto{r}{\frac{v_4}{v_3}}\right)^{1-k} \Rightarrow T_4 = T_3 r^{1-k} = 960°K(7.5)^{1-1.4} = 428.8°K \tag{7.29}$$

Thus from (7.27) we have

$$Q_{out} = mC_v|_{k=1.4}(T_4 - T_1) = 0.002\cancel{kg}\left(0.718\frac{kJ}{\cancel{kg}-°\cancel{K}}\right)(428.8°\cancel{K} - 305°\cancel{K})$$

$$= 0.177kJ \overset{Q_{out}}{\Longleftarrow} \tag{7.30}$$

Next, application of the energy-balance to the cycle gives

$$W_{cycle} = Q_{in} - Q_{out} \tag{7.31}$$

Process 2 → 3 (Isochoric heat addition):
Application of the energy-balance gives

$$Q_{in} = Q_{2\to3} = m(u_3 - u_2) = mC_v\big|_{k=1.4}(T_3 - T_2) \tag{7.32}$$

Process 1 → 2: Adiabatic compression:

Application of S-balance indicates that process 1 → 2 is isentropic. Via the isentropic property relationship (6.61)

$$\frac{T_2}{T_1} = \frac{\not{v}_2}{\not{v}_1}^{1-k} \Rightarrow T_2 = T_1\left(\frac{1}{r}\right)^{1-k} = 305^\circ K\left(\frac{1}{7.5}\right)^{-0.4} = 682.8^\circ K \tag{7.33}$$

Hence from (7.32)

$$Q_{in} = mC_v\big|_{k=1.4}(T_3 - T_2) = 0.002\not{kg}\left(0.718\frac{kJ}{\not{kg} - ^\circ \not{K}}\right)(960^\circ\not{K} - 682.8^\circ\not{K}) = 0.397kJ \tag{7.34}$$

Thus via (7.31)

$$W_{cycle} = Q_{in} - Q_{out} = 0.397kJ - 0.177kJ = 0.22kJ \xLeftarrow{W_{cycle}} \tag{7.35}$$

Hence via (4.136) and (7.34)

$$\eta = \frac{W_{cycle}}{Q_{in}} = \frac{0.22\not{kJ}}{0.397\not{kJ}} = 0.55 = 55\% \xLeftarrow{\eta} \tag{7.36}$$

Finally, via the definition of M.E.P. (7.9)

$$M.E.P. = \frac{W_{cycle}}{V_{max} - V_{min}} = \frac{W_{cycle}}{V_1 - V_2} = \frac{W_{cycle}}{V_1[1 - \frac{V_2}{V_1}]} = \frac{W_{cycle}}{V_1[1 - \frac{1}{r}]} \tag{7.37}$$

where via I.G.B.

$$P_1V_1 = mRT_1 \Rightarrow V_1 = \frac{mRT_1}{P_1} = \frac{0.002\not{kg}(0.287\frac{kJ}{\not{kg}-^\circ\not{K}})305^\circ\not{K}}{85kPa} = 2.05(10^{-3})m^3 \tag{7.38}$$

Which gives from (7.38) and (7.37)

$$M.E.P. = \frac{W_{cycle}}{V_1[1 - \frac{1}{r}]} = \frac{0.22kJ}{2.05(10^{-3})m^3[1 - \frac{1}{7.5}]} = 124kPa \xLeftarrow{M.E.P.} \tag{7.39}$$

$$\boxed{Q_{out} = 0.177kJ, W_{cycle} = 0.22kJ, \eta = 55\%, M.E.P. = 124kPa}$$

Comment: Note the similarity between results for constant and variable specific heats.

□

7.3 TWO COMMON CLOSED SYSTEM CYCLES:

7.3.1 THE DIESEL CYCLE

The first of two common cycles to be analyzed is the air-standard Diesel cycle: an approximation for the cycles used commonly in marine, industrial and transport applications. This cycle is composed of the following processes:

Process $1 \rightarrow 2$**:** Adiabatic Compression

Process $2 \rightarrow 3$**:** Isobaric heat addition

Process $3 \rightarrow 4$**:** Adiabatic expansion

Process $4 \rightarrow 1$**:** Isochoric heat rejection

with $T - s$ and $P - v$ diagrams shown in Figure 7.4.

Now, for the purpose of illustrating this cycle, let us assume, as usual, an air-standard cycle but for a *perfect* gas (constant specific heats) and determine an expression for the cycle efficiency as a function of various parameters such as the compression ratio r defined as follows

$$r = \frac{V_{max}}{V_{min}}, \text{ (Compression ratio)} \tag{7.40}$$

amongst others. From the definition of efficiency we have

$$\eta_{diesel} = \frac{W_{cycle}}{Q_{in}} = \frac{\not{m} w_{cycle}}{\not{m} q_{in}} = \frac{w_{cycle}}{q_{in}} \tag{7.41}$$

Now, the cycle assumptions stipulate that the cycle work is due to boundary work over the cycle. From Figure 7.4 a) there are three process which have non-zero boundary-work, thus to minimize

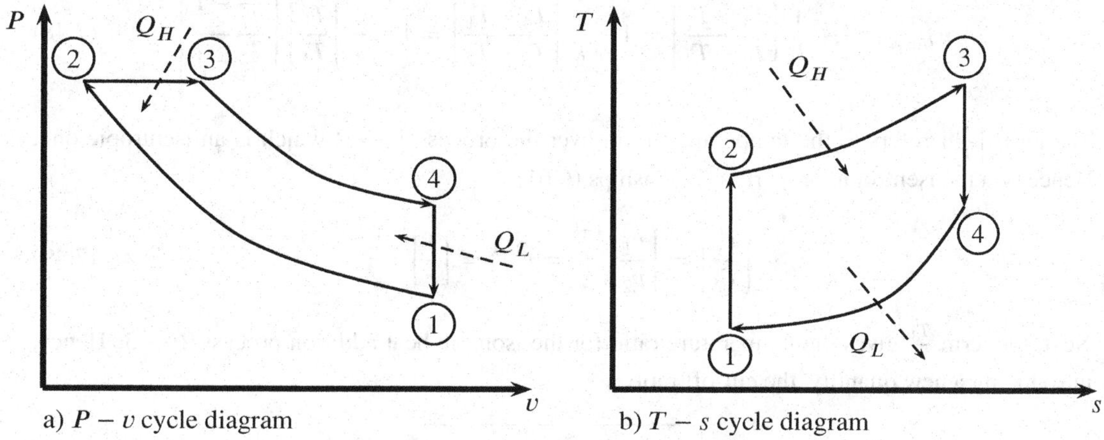

Figure 7.4 Diesel cycle diagrams.

effort let us integrate the energy-balance (for a closed system) over the cycle and substitute for the cycle work in (7.41).

$$\oint \delta q - \underbrace{\oint \delta w_{bound}}_{w_{cycle}} = \underbrace{\oint de}_{0} \Rightarrow w_{cycle} = \oint \delta q = q_{2\to 3} + q_{4\to 1} \tag{7.42}$$

where $q_{out} = -q_{4\to 1} > 0$. Noting that $q_{in} = q_{2\to 3}$ (7.41) becomes

$$\eta_{diesel} = \frac{q_{2\to 3} + q_{4\to 1}}{q_{2\to 3}} = 1 + \frac{q_{4\to 1}}{q_{2\to 3}} \tag{7.43}$$

Application of the energy-balance to the isochoric heat rejection process 4 → 1 and isobaric heat addition process 2 → 3 gives (for constant specific heats)

$$q_{4\to 1} - \underbrace{w_{bound_{4\to 1}}}_{0,\,isochoric} = \Delta e_{4\to 1} = \Delta u_{4\to 1} = C_v \Delta T_{4\to 1} \Rightarrow q_{4\to 1} = C_v \Delta T_{4\to 1} \tag{7.44}$$

and

$$q_{2\to 3} - \underbrace{w_{bound_{2\to 3}}}_{P\Delta v_{2\to 3},\,isobaric} = \Delta e_{2\to 3} = \Delta u_{2\to 3} \Rightarrow q_{2\to 3} = P\Delta v_{2\to 3} + \Delta u_{2\to 3} = \Delta h_{2\to 3} = C_P \Delta T_{2\to 3} \tag{7.45}$$

or

$$q_{2\to 3} = C_P \Delta T_{2\to 3} \tag{7.46}$$

Substituting these results (7.44) and (7.46) from the energy-balance into (7.43) and noting the definition of k gives

$$\eta_{diesel} = 1 + \frac{q_{4\to 1}}{q_{2\to 3}} = 1 + \frac{C_v \Delta T_{4\to 1}}{C_P \Delta T_{2\to 3}} = 1 + \underbrace{\frac{C_v}{C_P}}_{1/k} \left[\frac{T_4 - T_1}{T_2 - T_3}\right] = 1 + \frac{1}{k}\left[\frac{T_4 - T_1}{T_2 - T_3}\right] \tag{7.47}$$

Now, recall that our goal is to express (7.47) in terms of the compression ratio r, etc. Factoring out the term $\left[\frac{T_1}{T_2}\right]$ gives

$$\eta_{diesel} = 1 + \frac{1}{k}\left[\frac{T_4 - T_1}{T_2 - T_3}\right] = 1 - \frac{1}{k}\left[\frac{T_4 - T_1}{T_3 - T_2}\right] = 1 - \frac{1}{k}\left[\frac{T_1}{T_2}\right]\left[\frac{\frac{T_4}{T_1} - 1}{\frac{T_3}{T_2} - 1}\right] \tag{7.48}$$

The term $\left[\frac{T_1}{T_2}\right]$ refers to the temperature ratio over the process 1 → 2 which is an isentropic one: Hence, via the isentropic property relationships (6.61)

$$\left[\frac{T_1}{T_2}\right] = \left[\frac{v_1}{v_2}\right]^{1-k} = r^{1-k} = \left[\frac{1}{r}\right]^{k-1} \tag{7.49}$$

Next, the term $\frac{T_3}{T_2}$ gives the temperature ratio for the isobaric heat addition process 2 → 3. Hence, introducing a new quantity, the cut-off ratio

$$\boxed{r_c = \frac{v_3}{v_2}, \text{ (Cut-off ratio)}} \tag{7.50}$$

and, via the ideal gas law we have

$$\frac{T_3}{T_2} = \frac{\frac{P_3 v_3}{R_{air}}}{\frac{P_2 v_2}{R_{air}}} = \frac{v_3}{v_2} = r_c \Rightarrow \frac{T_3}{T_2} = r_c \tag{7.51}$$

Finally, to address the term $\frac{T_4}{T_1}$, we note from the $T - s$ diagram in Figure 7.4 b) that $\Delta s_{2\to3} = -\Delta s_{4\to1}$. Hence, via the property relationships (6.33) for constant specific heats we get

$$\Delta s_{2\to3} = \left[\int_{T_2}^{T_3} \frac{C_P(T)}{T} dT - Rln\left(\frac{P_3}{P_2}\right)\right] = \left[C_P ln\left(\frac{T_3}{T_2}\right) - Rln\left(\frac{P_3}{P_2}\right)\right]$$

$$= -\left[C_v ln\left(\frac{T_4}{T_1}\right) + Rln\left(\frac{v_4}{v_1}\right)\right] = -\left[\int_{T_4}^{T_1} \frac{C_v(T)}{T} dT + Rln\frac{v_4}{v_1}\right] = -\Delta s_{4\to1} \tag{7.52}$$

or

$$\left[C_P ln\left(\frac{T_3}{T_2}\right) - Rln\left(\frac{P_3}{P_2}\right)\right] = -\left[C_v ln\left(\frac{T_4}{T_1}\right) + Rln\left(\frac{v_4}{v_1}\right)\right]$$

$$\Rightarrow C_P ln\left(\frac{T_3}{T_2}\right) = -C_v ln\left(\frac{T_4}{T_1}\right) \Rightarrow \frac{T_4}{T_1} = \left[\frac{T_2}{T_3}\right]^k \tag{7.53}$$

Now, via (7.51) $\frac{T_3}{T_2} = r_c$ thus from (7.53) we have

$$\frac{T_4}{T_1} = \left[\frac{T_3}{T_2}\right]^k = r_c^k \tag{7.54}$$

Hence, combining (7.48), (7.49), (7.51) and (7.54) gives our final result

$$\eta_{diesel} = 1 - \frac{1}{k}\left[\frac{1}{r^{k-1}}\right]\left[\frac{r_c^k - 1}{r_c - 1}\right] \tag{7.55}$$

The performance implications of this expression are as follows: as the compression ratio r increase (cycle minimum volume decreases relative to the maximum volume) the cycle efficiency η_{diesel} increases. This can be seen by inspection of the $T - s$ diagram in Figure 7.4 b): Decreasing the minimum volume increases the temperature at state 2 and thus the total area bound by the $T - s$ curve (the specific cycle work) *for the same heat rejection* $q_{4\to1}$. Typical compression ratios range from $\approx$ 15 for low-speed Diesels to as high as $\approx$ 25 for modern high-speed Diesels. What about the effect of increasing or decreasing the heat addition $q_{2\to3}$ *for a fixed compression ratio* r? Well, increasing the heat addition results in a higher cut-off ratio (i.e. as $q_{2\to3} \uparrow \Rightarrow r_c \uparrow$). By inspection of (7.3) we see that (since $k > 1$ and $r_c > 1$) as the cut-off ratio r_c increases, the fraction $\left[\frac{r_c^k-1}{r_c-1}\right]$, increases as well, resulting in a reduction in efficiency. Thus, a higher heat addition results in a lower efficiency for the Diesel cycle: For maximum efficiency, don't "floor your Diesel".

EXAMPLE 7.3 Detailed cycle analysis: Diesel cycle

Problem Statement:
A piston-cylinder device executes an air-standard Diesel cycle with a compression and cut-off ratio of 18 and 2 respectively. The temperature and pressure at bottom-dead-center prior to the adiabatic compression process are $300°K$ and $100kPa$. Assuming variable specific heats determine the temperatures and pressures at all states, thermal efficiency and the M.E.P.. Sketch the $P-v$ and $T-s$ diagrams.

Schematic:

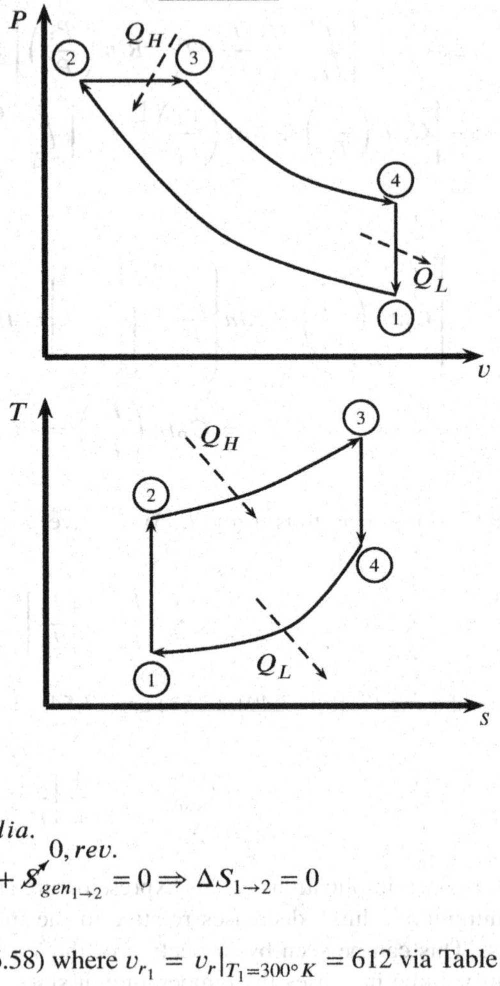

Given:
Air-standard Diesel cycle,
$T_1 = 300°K$,
$P_1 = 100kPa$,
$r = 18$,
$r_c = 2$

Find:
T_2, P_2,
T_3, P_3,
T_4, P_4,
η,
M.E.P.

Assumptions:
I.G.B. property Tables B.11

Analysis:

Process 1 → 2 (Adiabatic compression):
Application of entropy-balance gives

$$\Delta S_{1\rightarrow 2} = \int_{1\rightarrow 2} \overbrace{\frac{\delta Q}{T}}^{0,\, adia.} + \overbrace{S_{gen_{1\rightarrow 2}}}^{0,\, rev.} = 0 \Rightarrow \Delta S_{1\rightarrow 2} = 0 \tag{7.56}$$

Thus process 1 → 2 is isentropic and from (6.58) where $v_{r_1} = v_r|_{T_1=300°K} = 612$ via Table B.11

$$\frac{v_1}{v_2} = \frac{v_{r_1}}{v_{r_2}} \Rightarrow v_{r_2} = v_{r_1} \overbrace{\frac{v_2}{v_1}}^{\frac{1}{r}} = \frac{v_{r_1}}{r} = \frac{612}{18} = 34.5 \tag{7.57}$$

Hence via Table B.11

$$T_2 = T|_{v_{r_2}=34.5} = 898.3°K \xLeftarrow{T_2} \tag{7.58}$$

Also, via I.G.B. and the definition of the compression ratio r (7.40)

$$R_{air} = \frac{P_1 v_1}{T_1} = \frac{P_2 v_2}{T_2} \Rightarrow \frac{P_2 v_2}{T_2} = \frac{P_1 v_1}{T_1} \Rightarrow P_2 = P_1 \overset{r}{\frac{v_1}{v_2}} \frac{T_2}{T_1}$$
$$= P_1 r \frac{T_2}{T_1} = 100kPa(18)\frac{898.3°K}{300°K} = 5390kPa \overset{P_2}{\Longleftarrow} \tag{7.59}$$

<u>*Process 2 → 3 (Isobaric heat addition):*</u>

If process is isobaric then

$$P_3 = P_2 = 5390kPa \overset{P_3}{\Longleftarrow} \tag{7.60}$$

Now, from I.G.B. and the definition of the cut-off ratio we have

$$R_{air} = \frac{P_2 v_2}{T_2} = \frac{P_3 v_3}{T_3} \Rightarrow \frac{\cancel{P_2} v_2}{T_2} = \frac{\cancel{P_3} v_3}{T_3} \Rightarrow T_3 = T_2 \overset{r_c}{\frac{v_3}{v_2}} = 898.3°K(2) = 1796°K \overset{T_3}{\Longleftarrow} \tag{7.61}$$

<u>*Process 3 → 4 (Adiabatic expansion):*</u>

Application of entropy-balance gives

$$\Delta S_{3\to4} = \int_{3\to4} \overset{0,\,adia.}{\frac{\delta Q}{T}} + \overset{0,\,rev.}{S_{gen_{3\to4}}} = 0 \Rightarrow \Delta S_{3\to4} = 0 \tag{7.62}$$

Thus, process 3 → 4 is isentropic and from (6.58) we have $T_4 = T|_{v_{r4}}$. From Table B.11 $v_{r_3} = v_r|_{T_3=1796°K} = 3.97$ and

$$\frac{v_4}{v_3} = \frac{v_{r_4}}{v_{r_3}} \Rightarrow v_{r_4} = v_{r_3}\left[\frac{v_4}{v_3}\right] = v_{r_3}\left[\frac{\overset{v_1}{\cancel{v_4}}}{v_3}\right]\left[\frac{v_2}{v_2}\right] = v_{r_3}\left[\overset{r}{\frac{v_1}{v_2}}\right]\left[\overset{1/r_c}{\frac{v_2}{v_3}}\right] = 3.97\frac{18}{2} = 35.7 \tag{7.63}$$

Thus,

$$T_4 = T|_{v_{r_4}=35.7} = 887.9°K \overset{T_4}{\Longleftarrow} \tag{7.64}$$

<u>*Process 4 → 1 (Isochoric heat rejection):*</u>

From I.G.B. (noting $v_4 = v_1$) we have

$$R_{air} = \frac{P_1 v_1}{T_1} = \frac{P_4 v_4}{T_4} \Rightarrow \frac{P_4 \cancel{v_4}}{T_4} = \frac{P_1 \cancel{v_1}}{T_1} \Rightarrow P_4 = P_1 \frac{T_4}{T_1} = 100kPa\frac{887.9°K}{300°K} = 296kPa \overset{P_4}{\Longleftarrow} \tag{7.65}$$

Now via the definition of efficiency for a work producing device (4.136)

$$\eta = \frac{W_{cycle}}{Q_{in}} = \frac{\cancel{m} w_{cycle}}{\cancel{m} q_{in}} = \frac{w_{cycle}}{q_{in}} \tag{7.66}$$

where application of the energy-balance to the cycle gives

$$\oint \delta q - \oint \delta w = \oint de = 0 \Rightarrow w_{cycle} = \oint \delta w = \oint \delta q = q_{2\to3} + q_{4\to1} \quad (7.67)$$

or

$$w_{cycle} = q_{2\to3} + q_{4\to1} \quad (7.68)$$

$$q_{2\to3} - \overbrace{W_{boun_{2\to3}}}^{P\Delta v_{2\to3},\, isobaric} = \Delta e_{2\to3} = \Delta u_{2\to3} \Rightarrow q_{2\to3} = P\Delta v_{2\to3} + \Delta u_{2\to3} = \Delta h_{2\to3} \Rightarrow q_{2\to3} = \Delta h_{2\to3} \quad (7.69)$$

While for the heat rejection process $4 \to 1$ we have

$$q_{4\to1} - \overbrace{W_{boun_{4\to1}}}^{0,\, isochoric} = \Delta e_{4\to1} = \Delta u_{4\to1} \Rightarrow q_{4\to1} = \Delta u_{4\to1} \quad (7.70)$$

Thus, combining (7.66), (7.68), (7.69) and (7.70) gives

$$\eta = \frac{w_{cycle}}{q_{2\to3}} = \frac{q_{2\to3} + q_{4\to1}}{q_{2\to3}} = 1 + \frac{q_{4\to1}}{q_{2\to3}} = 1 + \frac{\Delta u_{4\to1}}{\Delta h_{2\to3}} = 1 + \frac{u_1 - u_4}{h_3 - h_2} \quad (7.71)$$

or

$$\eta = 1 + \frac{u_1 - u_4}{h_3 - h_2} \quad (7.72)$$

From air Tables B.11 $u_4 = u|_{T_4=887.9^\circ K} = 664kJ/kg$, $u_1 = u|_{T_1=300^\circ K} = 214kJ/kg$, $h_2 = h|_{T_2=898.3^\circ K} = 931kJ/kg$ and $h_3 = h|_{T_3=1796^\circ K} = 1999kJ/kg$ we get

$$\eta = 1 + \frac{u_1 - u_4}{h_3 - h_2} = 1 + \frac{214\frac{kJ}{kg} - 664\frac{kJ}{kg}}{1999\frac{kJ}{kg} - 931\frac{kJ}{kg}} = 0.578 = 57.8\% \xLeftarrow{\eta} \quad (7.73)$$

Finally, via the definition of M.E.P. (7.9)

$$M.E.P. = \frac{W_{cycle}}{V_{max} - V_{min}} = \frac{m w_{cycle}}{m v_1 - m v_2} = \frac{w_{cycle}}{v_1[1 - \frac{v_2}{v_1}]} = \frac{q_{2\to3} + q_{4\to1}}{v_1[1 - \frac{v_2}{v_1}]} = \frac{\Delta h_{2\to3} + \Delta u_{4\to1}}{\frac{R_{air}T_1}{P_1}[1 - \frac{1}{r}]}$$

$$= \frac{h_3 - h_2 + u_1 - u_4}{\frac{R_{air}T_1}{P_1}[1 - \frac{1}{r}]} = \frac{1998\frac{kJ}{kg} - 931\frac{kJ}{kg} + 214\frac{kJ}{kg} - 664\frac{kJ}{kg}}{\frac{0.287\frac{kJ}{kg-^\circ K}300^\circ K}{100kPa}[1 - \frac{1}{18}]} = 759kPa \xLeftarrow{M.E.P.} \quad (7.74)$$

$T_2 = 898.3^\circ K$, $P_2 = 5390kPa$, $T_3 = 1796^\circ K$, $P_3 = 5390kPa$, $T_4 = 887.9^\circ K$, $P_4 = 296kPa$, $\eta = 57.8\%$, $M.E.P. = 759kPa$

Comment: We can compare the calculated efficiency (for variable specific heats) with the result for constant specific heats for $k = 1.4$ (as a check) via (7.3):

$$\eta_{diesel} = 1 - \frac{1}{k}\left[\frac{1}{r^{k-1}}\right]\left[\frac{r_c^k - 1}{r_c - 1}\right] = 1 - \frac{1}{1.4}\left[\frac{1}{18^{1.4-1}}\right]\left[\frac{2^{1.4} - 1}{2 - 1}\right] = 63\% \quad (7.75)$$

This calculation is within $\approx 6\%$ of the calculated value based on variable specific heats and lends justification to our previous results.

□

7.3.2 THE OTTO CYCLE

The Otto cycle, a model for the commonly used gasoline engine, is composed of the following processes:

Process 1 → 2: Adiabatic Compression

Process 2 → 3: Isochoric heat addition

Process 3 → 4: Adiabatic expansion

Process 4 → 1: Isochoric heat rejection

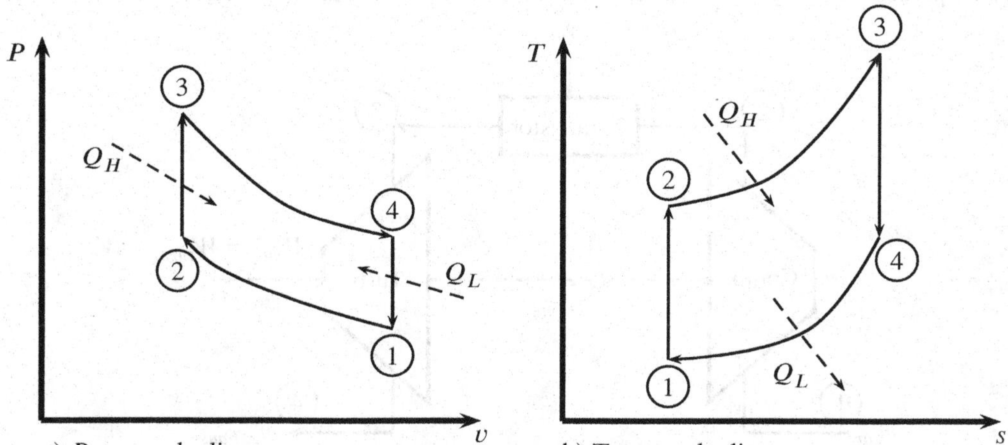

a) $P - v$ cycle diagram b) $T - s$ cycle diagram

Figure 7.5 Otto cycle diagrams.

where the $P - v$ and $T - s$ diagrams are shown in Figure 7.5. Note that the difference between the Otto and Diesel cycle is the heat addition process: In the Diesel cycle the isobaric heat addition process is due to a relatively slow, metered injection and burning of fuel in the combustion chamber which initially contains air. Conversely, for the Otto cycle, the fuel and air are pre-mixed and combustion (burning) is initiated instantaneously due to application of a spark. Thus, the heating of the air-fuel charge for the Otto-cycle is approximately isochoric.

Now, if you refer back to Example 7.2, you will see that this cycle was, in fact, an Otto cycle. Using an approach similar to that in the previous section, the efficiency of an Otto cycle (assuming constant specific heats) is found to be

$$\eta_{Otto} = 1 - \left[\frac{1}{r^{k-1}}\right] \tag{7.76}$$

Typical compression ratios r for gasoline engines (Otto cycles) are low compared with Diesel engines (due to the need to avoid mechanically damaging "pre-ignition" or "pinging") caused by explosive buring of the fuel, and range from 8 to 10.5. Hence, efficiencies of Diesel engines are higher than Otto (gasoline) engines.

As a last word, given that there are theoretically an infinite number of cycles, in a very real sense, once you've analyzed one cycle, you've analyzed them all. Hence, analysis of exotic cycles and cycles exhibiting regeneration, are reserved for the homework.

7.4 OPEN-SYSTEM WORK PRODUCING CYCLES

7.4.1 THE IDEAL BRAYTON CYCLE

The previous sections dealt with close system work producing devices (cycles). Continuing the approach in the chapters on the energy and entropy-balance, we will now investigate open-system work producing devices. Figure 7.6 shows a simplified schematic of a gas-turbine work producing device composed of three distinct sub-devices: an adiabatic compressor which compresses the inlet air, a combustor where fuel is added and burnt (releasing heat), and an adiabatic turbine where the hot pressurized exhaust gas is expanded producing power. This device typically produces shaft-power W_{net} to drive a generator (producing electricity) or a fan (to produce thrust).

Before we proceed to analyze this device, let us make a few points at this juncture:

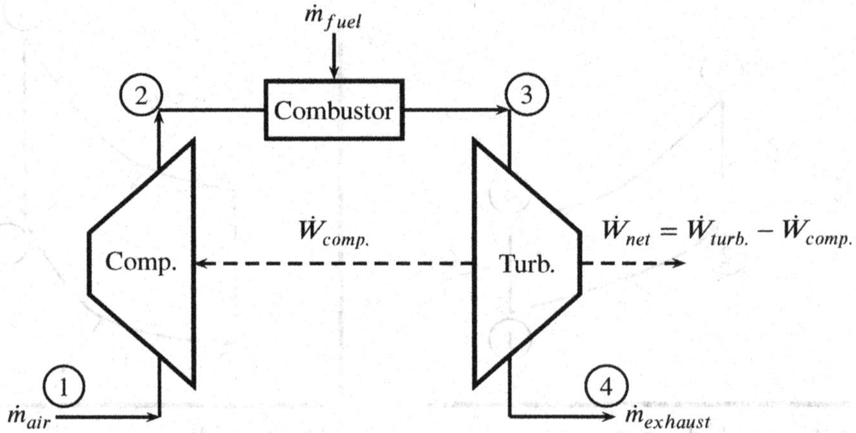

Figure 7.6 Schematic of a simplified gas-turbine power-plant

1. Typical air mass-flow rates greatly exceed fuel flow rates into the combustor ($\dot{m}_{air} \gg \dot{m}_{fuel}$). Hence, the inlet air and exhaust flow are approximately equal: $\dot{m}_{air} \approx \dot{m}_{exhaust}$.
2. Physically, the exhaust rejection at location/state 4 is to the surrounding environment composed of air. The dispersion of this (heated) exhaust gas and the excess energy carried with it occurs isobarically within the environment. Thus, there is a heat rejection process ($\dot{Q}_{out} = \dot{Q}_{1\to4}$) occurring isobarically within the environment.
3. Air is composed primarily of Nitrogen N_2 as is the case of the exhaust gases. Hence, the air and exhaust thermodynamic properties (specific heats, gas constant, etc.) are *similar* and assumed to be the same: i.e that of air.

If we take these three points together, we can recast and model the gas-turbine as a cyclic open-system device with heat rejection $\dot{Q}_{out}$ occurring isobarically within a heat-exchanger. Hence, we will model the gas turbine shown in Figure 7.6 via the Brayton cycle device given in Figure 7.7 a) with $T-s$ diagram shown in Figure 7.7 b) which will be justified shortly. Let us now make the following assumptions as to the operation of an ideal Brayton cycle power producing device:

Open-system air-standard ideal Brayton cycle assumptions:

1. Device working fluid is air

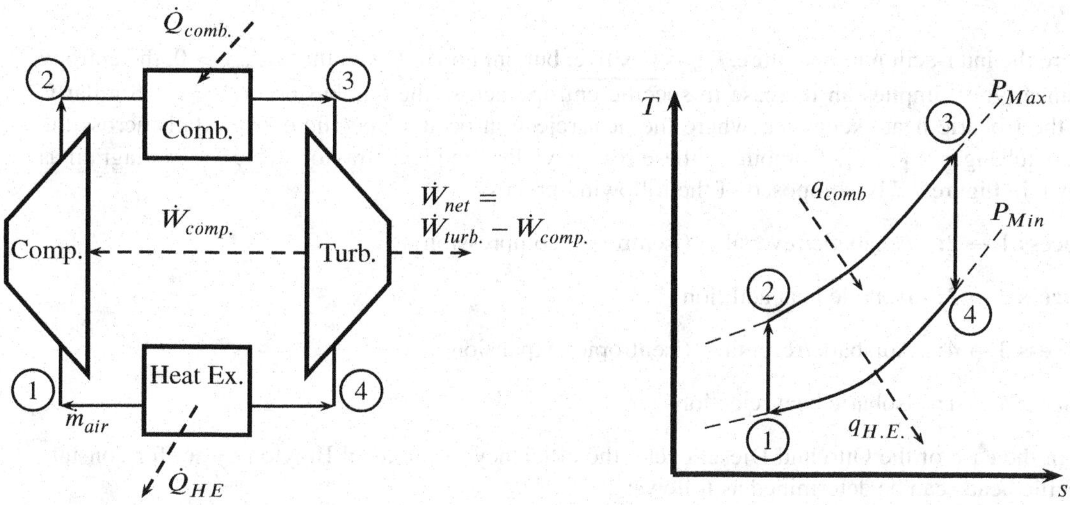

a) Device schematic

b) $T - s$ diagram

Figure 7.7 Schematic of an ideal gas-turbine power-cycle (ideal Brayton cycle)

2. Component devices (ideally) reversible
3. Turbine $3 \rightarrow 4$ and compressor $1 \rightarrow 2$ (typically) adiabatic
4. Combustion $2 \rightarrow 3$ and exhaust $4 \rightarrow 1$ processes $\Rightarrow$ Heat addition/rejection processes (ideally) isobaric
5. $\Delta pe = 0$ for all processes
6. $\dot{W}_{electric} = \dot{W}_{bound} = 0$ for all processes
7. Steady-state, steady-flow operation of all components

For the ideal Brayton cycle, application of open-system entropy-balance to, for example, the adiabatic turbine (process $3 \rightarrow 4$) gives

$$\sum \underbrace{\left(\frac{\dot{Q}}{T_{int}}\right)_{system}}_{0,\,adia.} = \underbrace{\frac{dS_{CV}}{dt}}_{0,\,S.S.} + \sum \underbrace{(\dot{m}s)_{outlets}}_{(\dot{m}s)_4} - \sum \underbrace{(\dot{m}s)_{inlets}}_{(\dot{m}s)_3} - \underbrace{\dot{S}_{gen}}_{0,\,rev.} \Rightarrow (\dot{m}s)_4 = (\dot{m}s)_3 \tag{7.77}$$

Application of the mass-balance to the turbine gives $\dot{m}_3 = \dot{m}_4 = \dot{m}$ which combined with (7.77) gives $s_3 = s_4$. Hence, for the <u>ideal</u> Brayton cycle, the turbine operates adiabatically, reversibly and hence, isentropically. The same applies to the compressor. Application of the entropy-balance to the isobaric combustor (modeled as an isobaric reversible heat-exchanger) gives

$$\sum \underbrace{\left(\frac{\dot{Q}}{T_{int}}\right)_{system}}_{\dot{Q}_{comb}/T_{int}} = \underbrace{\frac{dS_{CV}}{dt}}_{0,\,S.S.} + \sum \underbrace{(\dot{m}s)_{outlets}}_{(\dot{m}s)_3} - \sum \underbrace{(\dot{m}s)_{inlets}}_{(\dot{m}s)_2} - \underbrace{\dot{S}_{gen}}_{0,\,rev.} \Rightarrow \frac{\dot{Q}_{comb}}{T_{int}} = (\dot{m}s)_3 - (\dot{m}s)_2 \tag{7.78}$$

Application of the mass-balance to the combustor gives $\dot{m}_2 = \dot{m}_3 = \dot{m}$. Thus, via (7.78)

$$\frac{\dot{Q}_{comb}}{T_{int}} = \dot{m}(s_3 - s_2) \tag{7.79}$$

where the inter-action temperature T_{int} is positive, but unknown. Given that $\dot{Q}_{comb} > 0$, the entropy balance (7.79) implies an increase in specific entropy across the combustor: $s_3 > s_2$. Similarly, for the isobaric heat-exchanger, where the heat rejection occurs, specific entropy falls across the heat-exchanger: $s_4 > s_1$. Combining these results yields the ideal Brayton cycle $T - s$ diagram as shown in Figure 7.7 b) composed of the following process:

Process $1 \rightarrow 2$**:** Adiabatic/reversible (isentropic) compression

Process $2 \rightarrow 3$**:** Isobaric heat addition

Process $3 \rightarrow 4$**:** Adiabatic/reversible (isentropic) expansion

Process $4 \rightarrow 1$**:** Isobaric heat rejection

As in the case of the Otto and Diesel cycle, the efficiency of the ideal Brayton cycle, for constant specific heats, can be determined as follows:

$$\eta_{Brayton} = \frac{\dot{W}_{net}}{\dot{Q}_{in}} = \frac{\dot{W}_{net}}{\dot{Q}_{comb}} \tag{7.80}$$

Application of the energy-balance to the overall gas-turbine device (a closed system operating in steady-state) gives

$$\cancelto{\dot{Q}_{comb} - \dot{Q}_{H.E.}}{\dot{Q}} - \dot{W}_{net} = \cancelto{0,\, S.S.}{\dot{E}} \Rightarrow \dot{W}_{net} = \dot{Q}_{comb} - \dot{Q}_{H.E.} \tag{7.81}$$

where, as shown in Figure 7.6 a), $\dot{Q}_{H.E.}$ is positive for heat rejection from the system! Thus, from the general definition of efficiency (7.80)

$$\eta_{Brayton} = \frac{\dot{W}_{cycle}}{\dot{Q}_{comb}} = \frac{\dot{Q}_{comb} - \dot{Q}_{H.E.}}{\dot{Q}_{comb}} = 1 - \frac{\dot{Q}_{H.E.}}{\dot{Q}_{comb}} = 1 - \frac{\cancel{\dot{m}} q_{H.E.}}{\cancel{\dot{m}} q_{comb}} = 1 - \frac{q_{H.E.}}{q_{comb}} \tag{7.82}$$

or

$$\eta_{Brayton} = 1 - \frac{q_{H.E.}}{q_{comb}} \tag{7.83}$$

Application of the energy-balance to combustor, an open-system (assuming constant specific heats, neglecting changes in kinetic and potential energy) gives

$$\cancelto{\dot{Q}_{comb}}{\dot{Q}} - \cancelto{0}{\dot{W}} = \cancelto{0,\, S.S.}{\frac{dE_{CV}}{dt}} + \sum \cancelto{\dot{m}h_3}{(\dot{m}(h + ke + pe))_{outlets}} - \sum \cancelto{\dot{m}h_2}{(\dot{m}(h + ke + pe))_{inlets}} \tag{7.84}$$

or

$$q_{comb} = \frac{\dot{Q}_{comb}}{\dot{m}} = h_3 - h_2 \tag{7.85}$$

Similarly, for the heat rejection (to the environment) at the heat exchanger

$$q_{H.E.} = \frac{\dot{Q}_{H.E.}}{\dot{m}} = h_4 - h_1 \tag{7.86}$$

Integrating the property relationships (4.39) assuming constant specific heats we get

$$\eta_{Brayton} = 1 - \frac{q_{H.E.}}{q_{comb}} = 1 - \frac{h_4 - h_1}{h_3 - h_2} = 1 - \frac{\Delta h_{1\rightarrow 4}}{\Delta h_{2\rightarrow 3}} = 1 - \frac{\cancel{c_P} \Delta T_{1\rightarrow 4}}{\cancel{c_P} \Delta T_{2\rightarrow 3}} = 1 - \frac{T_4 - T_1}{T_3 - T_2} \tag{7.87}$$

or, factoring out the temperature ratio T_1/T_2

$$\eta_{Brayton} = 1 - \frac{T_4 - T_1}{T_3 - T_2} = 1 - \left[\frac{T_1}{T_2}\right]\left[\frac{T_4/T_1 - 1}{T_3/T_2 - 1}\right] \tag{7.88}$$

Now, inspection of the Brayton cycle $T-s$ diagram in Figure 7.7 b) indicates that $\Delta s_{2\to 3} = -\Delta s_{4\to 1}$. From the I.G.B. property relationships (6.33) for constant specific heats, noting that both process $2 \to 3$ and $4 \to 1$ are isobaric, we have

$$\Delta s_{2\to 3} = C_p ln\left[\frac{T_3}{T_2}\right] - Rln\underbrace{\left[\frac{P_3}{P_2}\right]}_{0,\ isobaric} = C_p ln\left[\frac{T_3}{T_2}\right]$$

$$\Delta s_{4\to 1} = C_p ln\left[\frac{T_1}{T_4}\right] - Rln\underbrace{\left[\frac{P_1}{P_4}\right]}_{0,\ isobaric} = C_p ln\left[\frac{T_1}{T_4}\right] \tag{7.89}$$

Hence,

$$\Rightarrow \Delta s_{2\to 3} = C_p ln\left[\frac{T_3}{T_2}\right] = -C_p ln\left[\frac{T_1}{T_4}\right] = -\Delta s_{4\to 1} \Rightarrow$$

$$C_p ln\left[\frac{T_3}{T_2}\right] = -C_p ln\left[\frac{T_1}{T_4}\right] \Rightarrow ln\left[\frac{T_3}{T_2}\right] = ln\left[\frac{T_4}{T_1}\right] \Rightarrow \left[\frac{T_3}{T_2}\right] = \left[\frac{T_4}{T_1}\right] \tag{7.90}$$

or $\left[\frac{T_3}{T_2}\right] = \left[\frac{T_4}{T_1}\right]$. Combining this result with the last expression for efficiency (7.88) gives

$$\eta_{Brayton} = 1 - \left[\frac{T_1}{T_2}\right]\left[\frac{T_4/T_1 - 1}{T_3/T_2 - 1}\right] = 1 - \left[\frac{T_1}{T_2}\right]\left[\frac{T_3/T_2 - 1}{T_3/T_2 - 1}\right] = 1 - \left[\frac{T_1}{T_2}\right] \tag{7.91}$$

Noting that process $1 \to 2$ is isentropic, we use the isentropic property relationship for constant specific heats (6.63)

$$\eta_{Brayton} = 1 - \left[\frac{T_1}{T_2}\right] = 1 - \frac{1}{\left[\frac{T_1}{T_2}\right]} = 1 - \frac{1}{\left(\frac{P_2}{P_1}\right)^{\frac{k-1}{k}}} \tag{7.92}$$

Finally, introducing the compressor pressure ratio

$$r_p = \frac{P_{max}}{P_{min}}, \text{ (Compressor pressure ratio)} \tag{7.93}$$

we have via (7.92)

$$\eta_{Brayton} = 1 - \frac{1}{r_p^{\frac{k-1}{k}}} \tag{7.94}$$

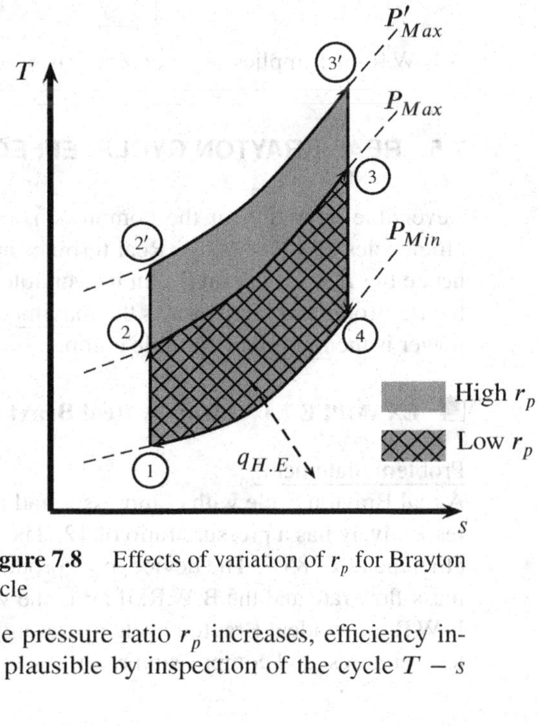

Figure 7.8 Effects of variation of r_p for Brayton cycle

Inspection of (7.94) indicates that as the Brayton cycle pressure ratio r_p increases, efficiency increases and tends towards 1. This conclusion is made plausible by inspection of the cycle $T - s$

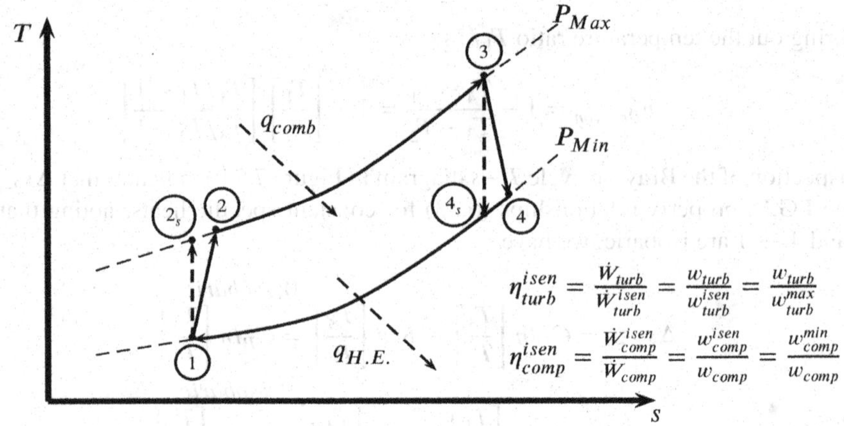

Figure 7.9 $T-s$ diagram of real vs. ideal Brayton cycle (– – – – –)

diagram for two cycles as shown in Figure 7.8: A high peak cycle pressure (pressure ratio) cycle $1 \to 2' \to 3' \to 4 \to 1$ and a low peak cycle pressure (pressure ratio) cycle $1 \to 2 \to 3 \to 4 \to 1$ both of which have the same specific heat rejection $q_{H.E.}$. Clearly, the area bound by the $T-s$ curve (specific cycle work) for the high peak cycle pressure cycle (in gray) is higher than the low peak cycle pressure cycle (in cross-hatched). Since the specific heat rejection is the same for both cycles, the efficiency of the high-pressure ratio cycle must exceed the low pressure cycle.

Finally, aside from efficiency, the effectiveness of a open-system work producing steady-state device can be characterized by a quantity referred to as the Back-Work-Ratio (B.W.R.):

Definition 7.3 ***Back-Work-Ratio (B.W.R.)*** *Ratio of compressor(s)-to-turbine(s) power*

$$B.W.R. = \frac{\dot{W}_{comp}}{\dot{W}_{turb}}, \text{ (Back-Work-Ratio)} \tag{7.95}$$

A B.W.R. of 1 implies *zero net power* is produced, while low B.W.R's imply high net power devices.

7.5 REAL BRAYTON CYCLE: EFFECTS OF IRREVERSIBILITIES

Reversible operation of the compressor and turbine in the ideal Brayton cycle imply isentropic efficiencies of 1 (see §6.8). Real turbines and compressors, on the other hand, operate irreversibly, hence the $T-s$ diagram for an irreversible or real Brayton cycle is shown in Figure 7.9. Note that the isentropic turbine power is the maximum possible turbine power, and the isentropic compressor power is the minimum possible compressor power.

EXAMPLE 7.4 Ideal vs. Real Brayton cycle

Problem Statement:

A real Brayton cycle with compressor and turbine inlet temperatures 300°K and 1000°K respectively has a pressure ratio of 12. The compressor and turbine both have isentropic efficiencies of 85%. The device must produce power at a rate of $70MW$. Determine the required mass-flow rate and the B.W.R. if air is the working fluid. Compare this with the mass-flow rate and B.W.R. of an ideal Brayton cycle operating under the same turbine and compressor inlet temperatures, and compression ratio. Assume I.G.B. and neglect flow kinetic energy.

Schematic:

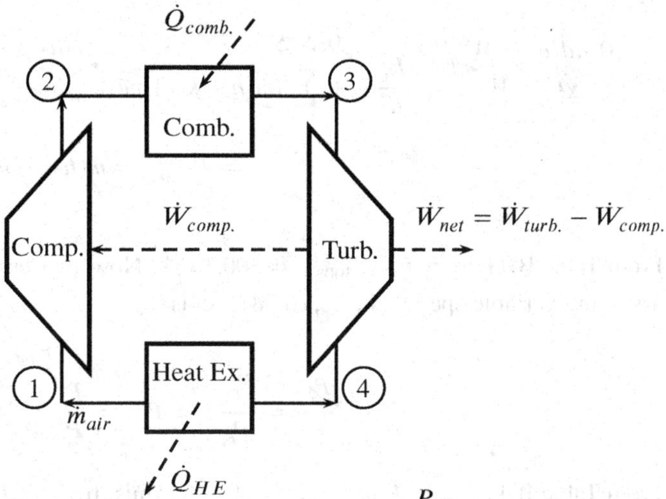

Given:
Brayton cycle,
$T_1 = 300°K$,
$T_3 = 1000°K$,
$\dot{W}_{req} = 70MW$,
I.G.B.,
$\eta_{turb} = \eta_{comp} = 0.85$,
$r_p = 12$

Find:
$\dot{m}$,
$B.W.R.$,
$\dot{m}_{ideal}$,
$B.W.R._{ideal}$

Assumptions:
I.G.B. property Tables B.11

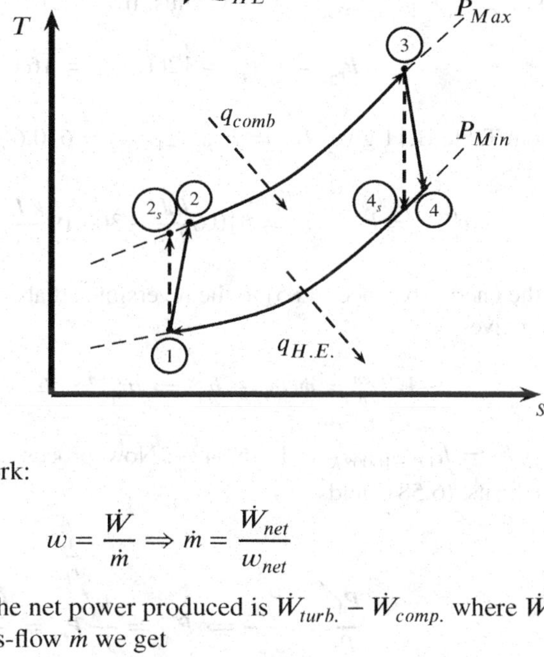

Analysis:

From the definition of specific work:

$$w = \frac{\dot{W}}{\dot{m}} \Rightarrow \dot{m} = \frac{\dot{W}_{net}}{w_{net}} \tag{7.96}$$

Now, from the above schematic, the net power produced is $\dot{W}_{turb.} - \dot{W}_{comp.}$ where $\dot{W}_{comp.}$ is positive. Dividing though by mass-flow $\dot{m}$ we get

$$w_{net} = w_{turb} - w_{comp} \tag{7.97}$$

Note, from the definition of isentropic efficiency for a work consuming device (6.83) we have for the compressor

$$\eta_{comp}^{isen} = \frac{w_{comp}^{isen}}{w_{comp}} \Rightarrow w_{comp} = \frac{w_{comp}^{isen}}{\eta_{comp}^{isen}} \tag{7.98}$$

and for a work producing device (6.82) we have for the turbine

$$\eta_{turb}^{isen} = \frac{w_{turb}}{w_{turb}^{isen}} \Rightarrow w_{turb} = \eta_{turb}^{isen} w_{turb}^{isen} \tag{7.99}$$

Hence, to determine w_{turb} and w_{comp} (and thus $\dot{m}$ via (7.96)) we must determine w_{comp}^{isen} and w_{turb}^{isen}. Application of the energy-balance (4.85) to the reversible adiabatic (i.e. isentropic) compressor (process $1 \rightarrow 2_s$) gives

$$\overset{0,\, adia.}{\cancel{\dot{Q}}} - \overset{-\dot{W}_{comp}^{isen}}{\cancel{\dot{W}}} = \overset{0,\, S.S.}{\cancel{\frac{dE_{CV}}{dt}}} + \sum \overset{\dot{m}h_{2_s}}{\cancel{(\dot{m}(h + ke + pe))_{outlets}}} - \sum \overset{\dot{m}h_1}{\cancel{(\dot{m}(h + ke + pe))_{inlets}}}$$

$$\Rightarrow \dot{W}_{comp}^{isen} = \dot{m}(h_{2s} - h_1) \Rightarrow w_{comp}^{isen} = h_{2s} - h_1 \tag{7.100}$$

From Table B.11 $h_1 = h|_{T_1=300°K} = 300.19\frac{kJ}{kg}$. Now, process $1 \rightarrow 2_s$ is isentropic so for I.G.B., assuming variable specific heats, (6.58) yields

$$\frac{\overset{P_2}{\cancel{P_{2_s}}}}{P_1} = \frac{P_{r_{2_s}}}{P_{r_1}} \Rightarrow P_{r_{2_s}} = \overset{r_p}{\cancel{\frac{P_2}{P_1}}} P_{r_1} = r_p P_{r_1} \tag{7.101}$$

From Table B.11 $P_{r_1} = P_r|_{T_1=300°K} = 1.386$. Thus, from (7.101) we have

$$P_{r_{2_s}} = r_p P_{r_1} = 12(1.386) = 16.632 \tag{7.102}$$

Interpolating from Table B.11 gives $h_{2s} = h|_{P_{r_{2_s}}=16.632} = 610.6\frac{kJ}{kg}$ and from (7.100)

$$w_{comp}^{isen} = h_{2s} - h_1 = 610.6\frac{kJ}{kg} - 300.19\frac{kJ}{kg} = 310.41\frac{kJ}{kg} \tag{7.103}$$

Application of the energy-balance (4.85) to the reversible adiabatic (i.e. isentropic) turbine (process $3 \rightarrow 4_s$) gives

$$-\dot{W}_{turb}^{isen} = \dot{m}(h_{4_s} - h_3) \Rightarrow w_{turb}^{isen} = h_3 - h_{4_s} \tag{7.104}$$

From Table B.11 $h_3 = h|_{T_3=1000°K} = 1046.04\frac{kJ}{kg}$. Now, process $3 \rightarrow 4_s$ is isentropic so for I.G.B., variable specific heats, (6.58) yields

$$\frac{\overset{P_4}{\cancel{P_{4_s}}}}{P_3} = \frac{P_{r_{4_s}}}{P_{r_3}} \Rightarrow P_{r_{4_s}} = \overset{\frac{1}{r_p}}{\cancel{\frac{P_4}{P_3}}} P_{r_3} = \frac{P_{r_3}}{r_p} \tag{7.105}$$

Now $P_{r_3} = P_r|_{T_3=1000°K} = 114.0$ and thus from (7.105) we have

$$P_{r_{4_s}} = \frac{1}{r_p} P_{r_3} = \frac{114}{12} = 9.5 \tag{7.106}$$

Interpolating from Table B.11 gives $h_{4s} = h|_{P_r=9.5} = 520.7\frac{kJ}{kg}$ and from (7.100)

$$w_{turb}^{isen} = h_3 - h_{4s} = 1046.04\frac{kJ}{kg} - 520.7\frac{kJ}{kg} = 525.34\frac{kJ}{kg} \tag{7.107}$$

Hence, from the energy-balance applied to the isentropic processes (7.107), (7.100) along with the definition of isentropic efficiencies (7.98) and (7.99), we have

$$w_{comp} = \frac{w^{isen}_{comp}}{\eta^{isen}_{comp}} = \frac{310.41\frac{kJ}{kg}}{0.85} = 365.18\frac{kJ}{kg},$$

$$w_{turb} = \eta^{isen}_{turb} w^{isen}_{turb} = 0.85\left(525.34\frac{kJ}{kg}\right) = 446.54\frac{kJ}{kg} \quad (7.108)$$

Combining these results with the expression for $\dot{m}$ (7.96) gives

$$\dot{m} = \frac{\dot{W}_{req}}{w} = \frac{\dot{W}_{req}}{w_{turb} - w_{comp}} = \frac{70MW}{(446.54 - 365.18)\frac{kJ}{kg}} = 0.86\frac{\cancel{MJ}}{s}\frac{kg}{\cancel{kJ}}\frac{1000\cancel{kJ}}{\cancel{MJ}} = 860.0\frac{kg}{s} \xLeftarrow{\dot{m}} \quad (7.109)$$

The back-work-ratio for the actual gas-turbine is

$$B.W.R. = \frac{\dot{W}_{comp}}{\dot{W}_{turb}} = \frac{w_{comp}}{w_{turb}} = \frac{365.18\frac{\cancel{kJ}}{\cancel{kg}}}{446.54\frac{\cancel{kJ}}{\cancel{kg}}} = 0.817 \xLeftarrow{B.W.R.} \quad (7.110)$$

Finally, we can recalculate $\dot{m}$ and the $B.W.R.$ for the reversible cycle from the reversible (isentropic) specific works calculated previously (7.108) giving

$$\dot{m}^{isen} = \frac{\dot{W}_{req}}{w^{isen}_{turb} - w^{isen}_{comp}} = \frac{70MW}{(525.34 - 310.41)\frac{kJ}{kg}} = 0.3256\frac{\cancel{MJ}}{s}\frac{kg}{\cancel{kJ}}\frac{1000\cancel{kJ}}{\cancel{MJ}} = 325.6\frac{kg}{s} \xLeftarrow{\dot{m}^{isen}} \quad (7.111)$$

with a back-work-ratio

$$B.W.R.^{isen} = \frac{\dot{W}^{isen}_{comp}}{\dot{W}^{isen}_{turb}} = \frac{w^{isen}_{comp}}{w^{isen}_{turb}} = \frac{310.41\frac{\cancel{kJ}}{\cancel{kg}}}{525.34\frac{\cancel{kJ}}{\cancel{kg}}} = 0.591 \xLeftarrow{B.W.R.^{isen}} \quad (7.112)$$

$$\boxed{\dot{m} = 860.0\tfrac{kg}{s},\ B.W.R. = 0.817,\ \dot{m}^{isen} = 325.6\tfrac{kg}{s},\ B.W.R.^{isen} = 0.591}$$

Comment: If the gas turbine where internally reversible ($\eta^{isen}_{turb}, \eta^{isen}_{comp} = 1$), the required mass-flow rate would be more than 50% lower than the real, internally irreversible device.

□

7.6 MODIFICATIONS TO THE BRAYTON CYCLE

Finally, there are many strategies for increasing cycle specific work and/or efficiency of the Brayton cycle. These strategies include inter-cooling, re-heating and regeneration all of which will be investigated forthwith.

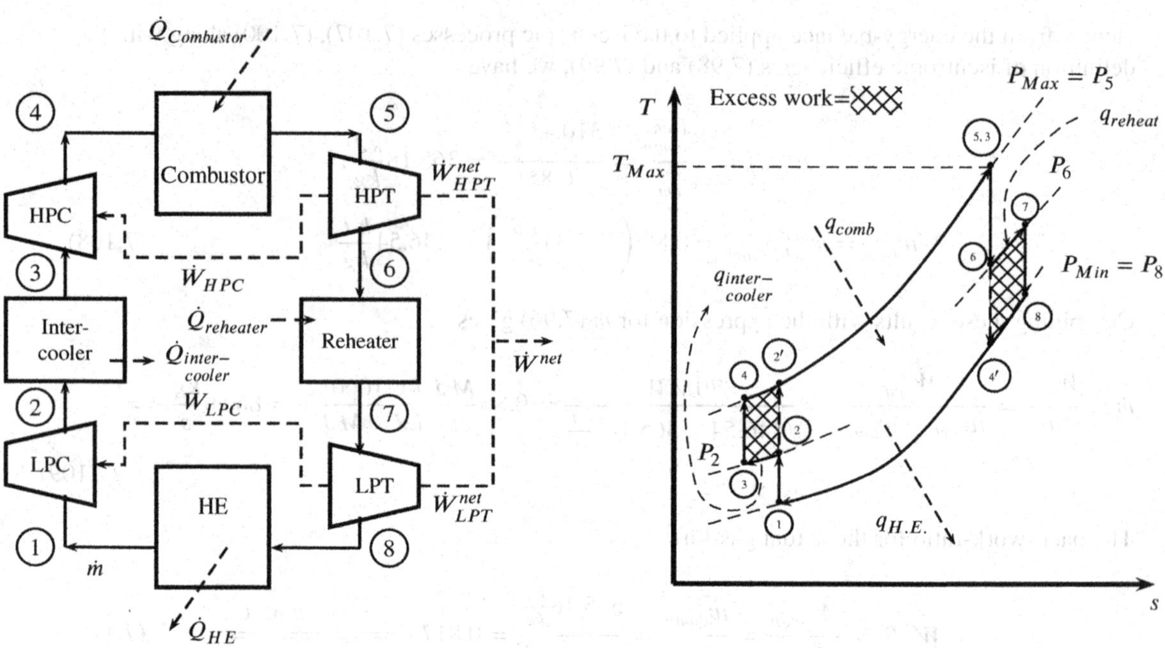

a) Device schematic

b) $T - s$ diagram

Figure 7.10 Schematic of an ideal multi-stage gas-turbine power-cycle (reheat and intercooling)

7.6.1 MULTI-STAGE COMPRESSION AND EXPANSION

Let us concern ourselves for the moment with a simple Brayton cycle operating within the pressure limits P_{max} and P_{min} subject to a maximum cycle temperature T_{max}. These pressure and temperature limits are imposed due to material or mechanical constraints on the devices (e.g. melting temperature of turbine components, etc.). This Brayton cycle is shown over the process $1 \rightarrow 2' \rightarrow 5 \rightarrow 4' \rightarrow 1$ in Figure 7.10 b). Now, let us split the compression and expansion processes across a high and low pressure compressor (HPC and LPC) and a high and low pressure turbine (HPT and LPT): This device now exhibits two-stage compression and expansion. Next, we install an iso-baric heat exchanger referred to as an "inter-cooler" between the low and high pressure compressors (LPC and HPC) from which heat is rejected as a rate of $\dot{Q}_{\substack{inter-\\cooler}}$. Similarly, we install a iso-baric heat exchanger referred to as a "reheater" between he low and high pressure turbines (LPT and HPT) to which heat is added at a rate of $\dot{Q}_{reheater}$. The new cycle shown in Figure 7.10 b) executes the cycle $1 \rightarrow 2 \rightarrow 3 \rightarrow 4 \rightarrow 5 \rightarrow 6 \rightarrow 7 \rightarrow 8 \rightarrow 1$. Comparing the area bound by the $T - s$ curve (which is w_{cycle}) for the multi-stage vs. the single-stage cycle we see that the two-stage cycle specific work is larger by the amount corresponding to the cross-hatched area: The multi-stage cycle produces higher cycle power per-unit mass-flow! In fact, it can be shown that the maximum additional specific work is achieved when turbine and compressor pressure ratios are equal (i.e. $P_2 = P_6$ and $P_2/P_1 = P_4/P_3 = P_5/P_6 = P_7/P_8 = \sqrt{r_P}$).

EXAMPLE 7.5 Real multi-stage gas turbine with inter-cooling and reheat

Problem Statement:

A real gas turbine (air is the working fluid) with two-stages of compressor and expansion operates between the temperature limits 300° K and 1200° K. Specifically, the LPC and HPC inlet temperatures are 300° K while the LPT and HPT inlet temperatures are 1200°. Each compressor and turbine have a pressure ratio of 3. The compressors and turbines have isentropic efficiencies of 80% and 85% respectively. Determine the back-work-ratio and thermal efficiency of the device. Assume I.G.B..

Schematic:

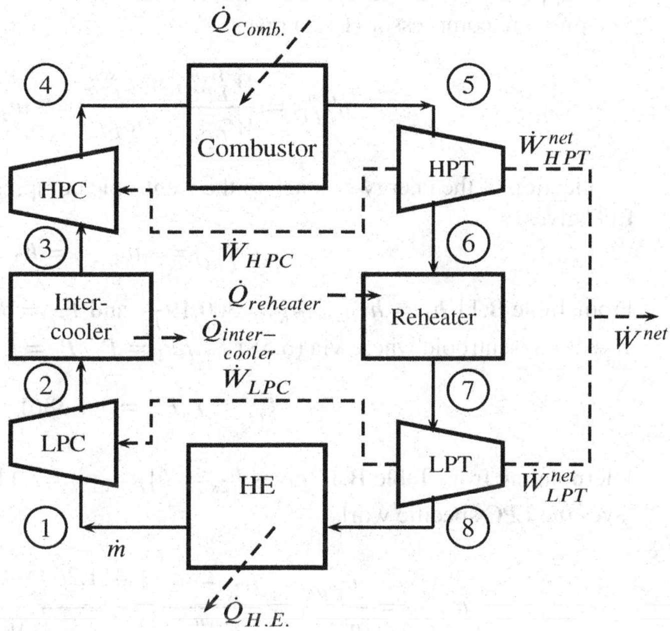

Given:

Brayton cycle with multi-stage compression and expansion,
$T_1 = T_3 = 300°K$,
$T_5 = T_7 = 1200°K$,
I.G.B.,
$\eta_{turb} = 0.85$,
$\eta_{comp} = 0.80$,
$r_p = 3$ (for each turbine/compressor)

Find:

$B.W.R.$,
η

Assumptions:

Property Tables B.11,
$\Delta pe = \Delta ke = 0$,
isobaric inter-cooler,
isobaric reheater

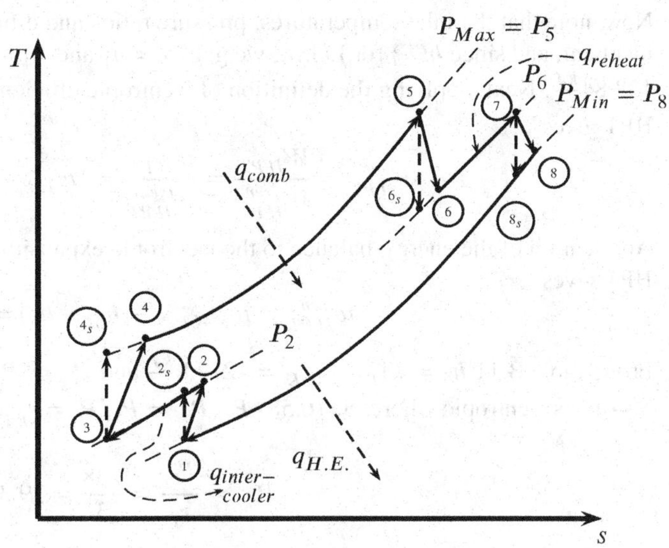

Analysis:

From the definition of the back-work ratio (noting that $\dot{m}$ is identical through all devices)

$$B.W.R. = \frac{\dot{W}_{comp}}{\dot{W}_{turb}} = \frac{\dot{W}_{HPC} + \dot{W}_{LPC}}{\dot{W}_{HPT} + \dot{W}_{LPT}} = \frac{\cancel{\dot{m}} w_{HPC} + \cancel{\dot{m}} w_{LPC}}{\cancel{\dot{m}} w_{HPT} + \cancel{\dot{m}} w_{LPT}} = \frac{w_{HPC} + w_{LPC}}{w_{HPT} + w_{LPT}} \tag{7.113}$$

where we have noted that the total turbine and compressor power are

$$\dot{W}_{turb} = \dot{W}_{HPT} + \dot{W}_{LPT}, \dot{W}_{comp} = \dot{W}_{HPC} + \dot{W}_{HPC} \tag{7.114}$$

Now, applying the definition of isentropic efficiency for a work consuming device to the low-pressure compressor (LPC) gives

$$\eta_{LPC}^{isen} = \frac{\dot{W}_{LPC}^{isen}}{\dot{W}_{LPC}} = \frac{w_{LPC}^{isen}}{w_{LPC}} \Rightarrow w_{LPC} = \frac{w_{LPC}^{isen}}{\eta_{LPC}^{isen}} \tag{7.115}$$

Application of the energy-balance to the isentropic compression process $1 \rightarrow 2_s$ for the isentropic LPC gives

$$w_{LPC}^{isen} = -w_{1\rightarrow 2_s} = h_{2_s} - h_1 \tag{7.116}$$

From Table B.11 $h_1 = h|_{T=300°K} = 300.19\frac{kJ}{kg}$ and $P_{r_1} = P_r|_{T_1=300°K} = 1.386$. Given that process $1 \rightarrow 2_s$ is isentropic where via (6.58) $P_{2_s}/P_1 = P_2/P_1 = r_p$ and we have

$$P_{r_{2_s}} = r_p P_{r_1} = 3(1.386) = 4.158 \tag{7.117}$$

Interpolating from Table B.11 gives $h_{2s} = h|_{P_r=4.158} = 411.26\frac{kJ}{kg}$. Combining (7.116) and (7.115) gives the LPC specific work

$$w_{LPC} = \frac{w_{LPC}^{isen}}{\eta_{LPC}^{isen}} = \frac{h_{2_s} - h_1}{\eta_{LPC}^{isen}} = \frac{411.26\frac{kJ}{kg} - 300.19\frac{kJ}{kg}}{0.8} = 138.84\frac{kJ}{kg} \tag{7.118}$$

Now, note that the inlet temperatures, pressure ratios and efficiencies for the LPC and HPC are identical, and since $h(T)$ for I.G.B., we get $h_3 = h_1$ and $h_4 = h_2$. Hence, $w_{HPC} = w_{LPC} = 138.84\frac{kJ}{kg}$. Next, applying the definition of isentropic efficiency for a work producing device to the HPT gives

$$\eta_{HPT}^{isen} = \frac{\dot{W}_{HPT}}{\dot{W}_{HPT}^{isen}} = \frac{w_{HPT}}{w_{HPT}^{isen}} \Rightarrow w_{HPT} = \eta_{HPT}^{isen} w_{HPT}^{isen} \tag{7.119}$$

Application of the energy-balance to the isentropic expansion process $5 \rightarrow 6_s$ for the isentropic HPT gives

$$w_{HPT}^{isen} = w_{5\rightarrow 6_s} = -(h_{6_s} - h_5) = h_5 - h_{6_s} \tag{7.120}$$

From Table B.11 $h_5 = h|_{T=1200°K} = 1277.8\frac{kJ}{kg}$ and $P_{r_5} = P_r|_{T_1=1200°K} = 238$. Given that process $5 \rightarrow 6_s$ is isentropic where via (6.58) $P_5/P_{6_s} = P_5/P_6 = r_p$ and we have

$$P_{r_{6_s}} = \frac{P_{r_5}}{r_p} = \frac{238}{3} = 79.333 \tag{7.121}$$

Interpolating from Table B.11 gives $h_{6s} = h|_{P_r=79.333} = 946.36\frac{kJ}{kg}$. Combining (7.119) and (7.130) gives the HPT specific work

$$w_{HPT} = \eta_{HPT}^{isen} w_{HPT}^{isen} = \eta_{HPT}^{isen}(h_5 - h_{6_s}) = 0.85\left(1277.8\frac{kJ}{kg} - 946.36\frac{kJ}{kg}\right) = 281.724\frac{kJ}{kg} \tag{7.122}$$

Noting that the inlet temperatures, pressure ratios and efficiencies for the HPT and LPT are identical, and since $h(T)$ for I.G.B., we get $h_5 = h_7$ and $h_6 = h_8$. Hence, $w_{HPT} = w_{LPT}$. Returning to the expression for the B.W.R. (7.113)

$$B.W.R. = \frac{w_{HPC} + w_{LPC}}{w_{HPT} + w_{LPT}} = \frac{\cancel{2} w_{HPC}}{\cancel{2} w_{HPT}} = \frac{w_{HPC}}{w_{HPT}} = \frac{138.84\frac{k\cancel{J}}{\cancel{kg}}}{281.724\frac{k\cancel{J}}{\cancel{kg}}} = 0.493 \overset{B.W.R.}{\Longleftarrow} \tag{7.123}$$

or 49.3%. As for the thermal efficiency

$$\eta = \frac{\dot{W}_{net}}{\dot{Q}_{in}} = \frac{\dot{W}_{HPT} + \overset{\dot{W}_{HPT}}{\cancel{\dot{W}_{LPT}}} - (\dot{W}_{HPC} + \overset{\dot{W}_{HPC}}{\cancel{\dot{W}_{LPC}}})}{\dot{Q}_{comb.} + \dot{Q}_{reheat}} = 2\left[\frac{w_{HPT} - w_{HPC}}{q_{comb.} + q_{reheat}}\right] \tag{7.124}$$

or

$$\eta = 2\frac{w_{LPT} - w_{HPC}}{q_{comb.} + q_{reheat}} \tag{7.125}$$

Application of the energy-balance to the combustor gives

$$q_{comb} = q_{4\to5} = h_5 - h_4 \tag{7.126}$$

Since h_5 is known we turn our attention to h_4: Application of the energy-balance to the HPC (process $3 \to 4$) gives

$$w_{HPC} = -w_{3\to4} = h_4 - h_3 \Rightarrow$$

$$h_4 = \overset{h_1}{\cancel{h_3}} + w_{HPC} = h_1 + w_{HPC} = 300.19\frac{kJ}{kg} + 138.84\frac{kJ}{kg} = 439.03\frac{kJ}{kg} \tag{7.127}$$

or $h_4 = 439.03\frac{kJ}{kg}$. This gives via (7.126)

$$q_{comb} = h_5 - h_4 = 1277.79\frac{kJ}{kg} - 439.03\frac{kJ}{kg} = 838.76\frac{kJ}{kg} \tag{7.128}$$

Next, we apply the energy-balance to the reheater

$$q_{reheater} = q_{6\to7} = h_7 - h_6 \tag{7.129}$$

Since h_7 is known we turn our attention to h_6: Application of the energy-balance to the HPT (process $5 \to 6$) gives

$$w_{HPT} = w_{5\to6} = h_5 - h_6 \Rightarrow h_6 = h_5 - w_{HPT} = 1277.79\frac{kJ}{kg} - 281.724\frac{kJ}{kg} = 996.06\frac{kJ}{kg} \tag{7.130}$$

or $h_6 = 996.06\frac{kJ}{kg}$. This gives via (7.129)

$$q_{reheater} = h_7 - h_6 = 1277.79\frac{kJ}{kg} - 996.06\frac{kJ}{kg} = 281.724\frac{kJ}{kg} \tag{7.131}$$

Combining (7.129), (7.131) and (7.125) gives

$$\eta = 2\frac{w_{LPT} - w_{HPC}}{q_{comb.} + q_{reheat}} = 2\left[\frac{281.724\frac{kJ}{kg} - 138.84\frac{kJ}{kg}}{838.76\frac{kJ}{kg} + 281.724\frac{kJ}{kg}}\right] = 25.5\% \overset{\eta}{\Longleftarrow} \tag{7.132}$$

$$\boxed{B.W.R. = 49.3\%, \eta = 25.5\%}$$

Comment: The compressor and turbine outlet temperatures (which may be of interest) are $T_4 = T_2 = 437°K$ and $T_8 = T_6 = 956°K$.

□

7.6.2 REGENERATION

The previous example (multi-stage gas-turbine) exhibited compressor and turbine exit temperatures $T_4 = T_2 = 437°K$ and $T_8 = T_6 = 956°K$ respectively. Hence, inspection of the $T-s$ diagram reveals that for a substantial portion of the heat-addition process in the combustor, the turbine exhaust is rejecting heat via the heat exchanger (to the environment) at a higher temperature. Thus, it should be possible to harness some of the rejected (waste) heat to the environment to pre-heat the combustor inlet air. To see this graphically, Figure 7.11 gives the $T-s$ diagram for the cycle in the previous example with all superfluous information removed: For additional clarity, the combustor inlet temperature T_4 and LPT exit temperature T_8 are give their own isotherms. Let us now ask ourselves the question: "For the cross-hatched regions of the cycle, just after the HPC and LPT exit states, will heat move spontaneously (from hot to cold) from the later to the former?" The answer is since the HPC exit state is at a temperature of $\approx T_4$, while the LPT exit state is at $\approx T_8$, where $T_8 > T_4$, heat would flow from the region just after the LPT exit to the HPC exit provided the mass-flows are in thermal contact: Clearly the device begs for the addition of a heat-exchange, called a "regenerator", which does just that! Figure 7.12 a) shows a multi-stage Brayton cycle with a regenerator installed. Note, the regenerator exit state prior to the combustor inlet is denoted by x as shown in the device schematic Figure 7.12 a) with $T-s$ diagram 7.12 b).

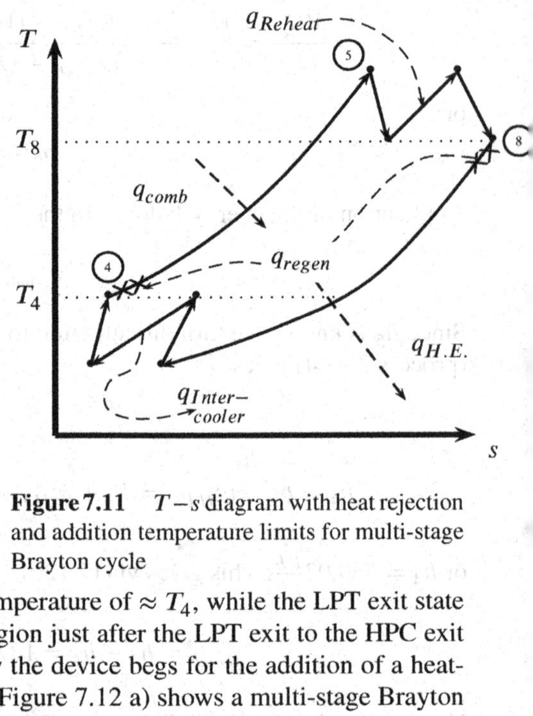

Figure 7.11 $T-s$ diagram with heat rejection and addition temperature limits for multi-stage Brayton cycle

Next question: "What is the maximum amount of heat that can be transferred from the hot LPT exhaust to the relatively cool HPC outlet flow?" Well, Figure 7.12 c) gives an exploded view of the regenerator complete with mass-flows, heat transfer between the flow streams $\dot{Q}_{regen.}$ and state labels. As the Figure shows, heat will continue to be transferred from the hot LPT exhaust to the cooler combustor inlet flow up to the point where their temperatures are equal: $T_x = T_8$. Hence, applying the energy-balance to the dashed control volume shown in Figure 7.12 c) we have the regenerator heat transfer rate

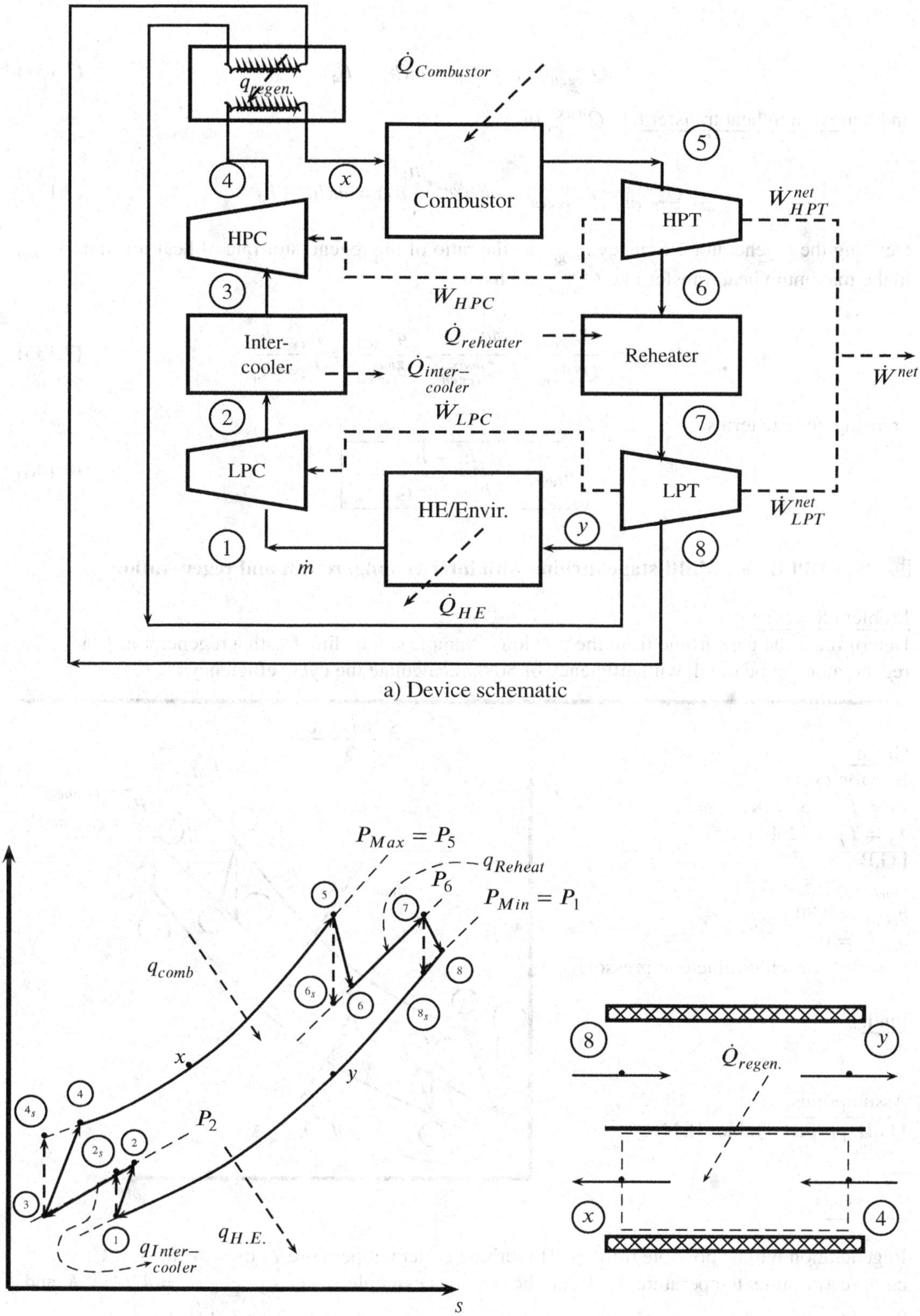

a) Device schematic

b) $T - s$ diagram

c) Regenerator

Figure 7.12 Schematic of an multi-stage gas-turbine with regeneration a), $T - s$ diagram b) and regenerator shown c).

$$\dot{Q}_{regen} = \dot{m}q_{regen} = \dot{m}(h_x - h_4) \tag{7.133}$$

and a maximum heat transfer rate $\dot{Q}^{max}_{regen}$ of

$$\dot{Q}^{max}_{regen} = \dot{m}q^{max}_{regen} = \dot{m}(h^{max}_x - h_4) = \dot{m}(h_8 - h_4) \tag{7.134}$$

Defining the regenerator efficiency η_{regen} as the ratio of the regenerator rate of heat transfer $\dot{Q}_{regen}$ to the maximum heat transfer rate $\dot{Q}^{max}_{regen}$ we have

$$\eta_{regen} = \frac{\dot{Q}_{regen}}{\dot{Q}^{max}_{regen}} = \frac{\dot{m}q_{regen}}{\dot{m}q^{max}_{regen}} = \frac{q_{regen}}{q^{max}_{regen}} = \frac{h_x - h_4}{h_8 - h_4} \tag{7.135}$$

or, using generic terms

$$\boxed{\eta_{regen} = \frac{h_x - h_{\text{HPC exit}}}{h_{\text{LPT exit}} - h_{\text{HPC exit}}}} \tag{7.136}$$

EXAMPLE 7.6 Multi-stage turbine with inter-cooling, reheat and regeneration

Problem Statement:
Determine if the gas turbine from the previous example can be fitted with a regenerator. If a regenerator can be fitted, with efficiency of 80%, recalculate the cycle efficiency.

Given:
Brayton cycle,
$T_1 = T_3 = 300°K$,
$T_5 = T_7 = 1200°K$,
I.G.B.,
$\eta_{turb} = 0.85$,
$\eta_{comp} = 0.80$,
$\eta_{regen} = 0.8$,
$r_p = 3$ (for each turbine/compressor)

Find:
η

Assumptions:
I.G.B. property Tables B.11

Schematic:

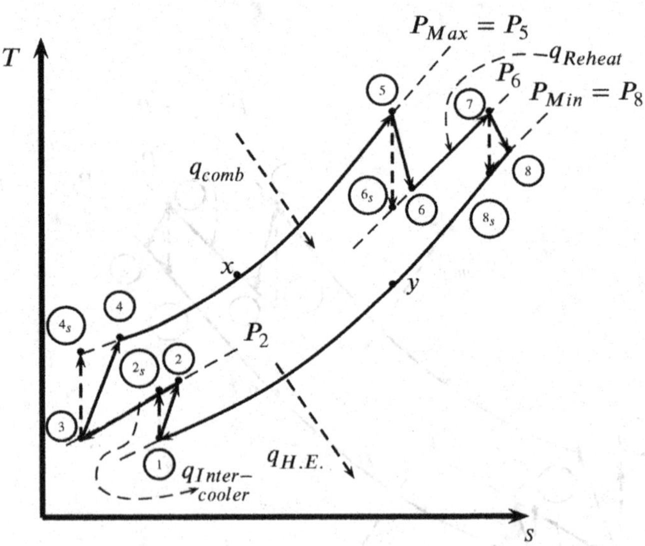

Analysis:

Regeneration will be possible if the (LPT) turbine outlet temperature T_8 exceeds the (HPC) compressor outlet temperature T_4. From the previous example: $T_4 = T|_{h_4=439\frac{kJ}{kg}} = 436.54°K$ and $T_8 = T_6 = T|_{h_6=996.06\frac{kJ}{kg}} = 956.03°K$ thus $T_4 < T_8$ and regeneration is possible. From the definition of regenerator efficiency

$$\eta_{regen} = \frac{h_x - h_{\text{HPC exit}}}{h_{\text{LPT exit}} - h_{\text{HPC exit}}} = \frac{h_x - h_4}{h_8 - h_4} \Rightarrow h_x = h_4 + \eta_{regen}(h_8 - h_4)$$

$$= 439\frac{kJ}{kg} + 0.8\left(996.06\frac{kJ}{kg} - 439\frac{kJ}{kg}\right) = 884.65\frac{kJ}{kg} \tag{7.137}$$

Restating the expression for the combustor heat addition with combustor inlet state x (instead of 4) and recalculating q_{comb}

$$q_{comb} = h_5 - h_x = 1277.79\frac{kJ}{kg} - 884.65\frac{kJ}{kg} = 393.14\frac{kJ}{kg} \tag{7.138}$$

Recalculating the efficiency

$$\eta = 2\frac{w_{LPT} - w_{HPC}}{q_{comb} + q_{reheat}} = 2\left[\frac{281.724\frac{kJ}{kg} - 138.84\frac{kJ}{kg}}{393.14\frac{kJ}{kg} + 281.724\frac{kJ}{kg}}\right] = 42.3\% \stackrel{\eta}{\Longleftarrow} \tag{7.139}$$

$$\boxed{\eta = 42.3\%}$$

Comment: Addition of the regenerator increased the efficiency from 24.5% to 42.3%.

□

7.7 AEROSPACE APPLICATIONS: TURBO-JET PROPULSION

To finish our investigation of the Brayton cycle, we now turn our attention to the Brayton-based jet propulsion cycle: The turbo-jet engine. As shown in Figure 7.13 a) and b), inlet air flow passes through a diffuser and is thus decelerated and pressurized. The pressurized air is now pressurized further in a compressor after-which it is heated isobarically via a combustion process with fuel in a combustion chamber. The pressurized, high temperature products of combustion then expand in the turbine producing power which is used to drive the compressor (i.e. $\dot{W}_{Compressor} = \dot{W}_{Turbine}$). Finally, as the exhaust gases pass through the nozzle, the exhaust is accelerated and expanded further producing thrust. The cycle process and $T - s$ diagram are shown in Figure 7.13 c) and d) respectively. How do we calculate the thrust produced by such a device? Well, from Table 4.2 we have the conservation equation for momentum $\sum F_{i:\substack{closed\\system}} = \frac{d}{dt}\beta_{\substack{closed\\system}}$, $\beta = mV_i$ and $b = V_i$ where i refers to the x, y or z direction. From the Reynold's Transport Theorem (4.69) we get

$$\sum F_{i:\substack{closed\\system}} = \frac{d(mV_i)_{CV}}{dt} + \sum(\dot{m}V_i)_{outlets} - \sum(\dot{m}V_i)_{inlets} \tag{7.140}$$

Noting that in the limit, the open and closed-system occupy the same volume, $\sum F_{i:\substack{closed\\system}} \rightarrow \sum F_{i:\substack{open\\system}}$, yielding the open-system momentum conservation equation

$$\sum F_{i:\substack{open\\system}} = \frac{d(mV_i)_{CV}}{dt} + \sum(\dot{m}V_i)_{outlets} - \sum(\dot{m}V_i)_{inlets} \tag{7.141}$$

Now, with respect to the jet engine application shown with control volume – – –– in Figure 7.13 a), we have a single inlet/outlet flow device (neglecting fuel mass-flow delivered to the combustor $\dot{m}_{fuel} \ll \dot{m}_{air}$). Thus, for steady-state, steady-flow operation, neglecting all pressure forces on the control volume, (7.141) yields

$$\sum \cancel{F_{i,closed\ system}}^{F_{i:restraint}} = \cancel{\frac{d(mV_i)_{CV}}{dt}}^{0,\ S.S.} + \sum \cancel{(\dot{m}V_i)_{outlets}}^{\dot{m}_{exhaust} V_{i:exhaust}} - \sum \cancel{(\dot{m}V_i)_{inlets}}^{\dot{m}_{inlet} V_{i:inlet}}$$

$$\Rightarrow F_{i:restraint} = (\dot{m}V)_{i:exhaust} - (\dot{m}V)_{i:inlet} \tag{7.142}$$

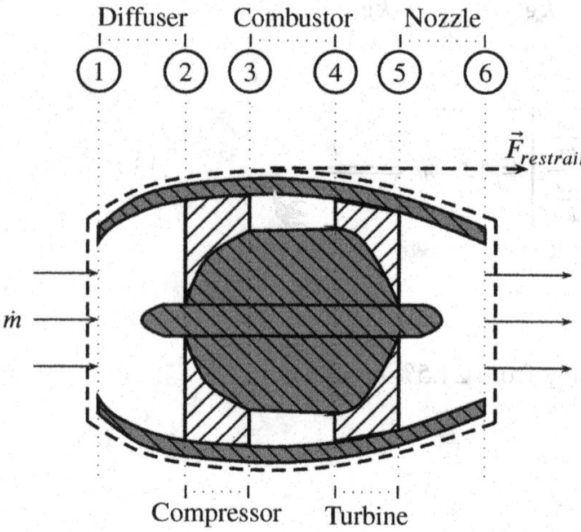

a) Jet engine (cut-section). Control volume – – ––.

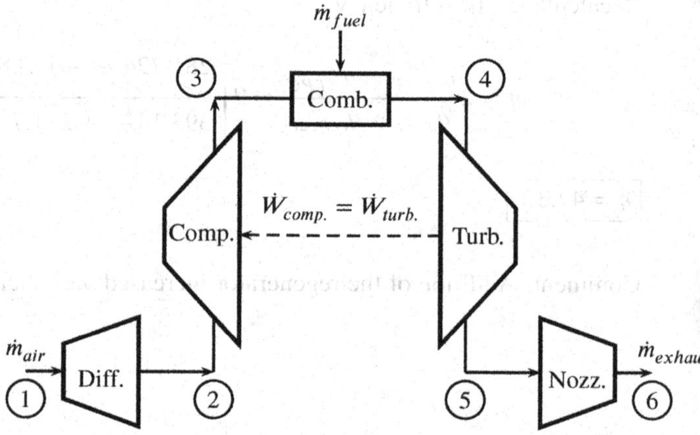

b) Device schematic

Process 1 → 2 Adiabatic reversible deceleration in diffuser $ke_1 \gg ke_2$

Process 2 → 3 Adiabatic compression in compressor

Process 3 → 4 Isobaric heat addition in combustor

Process 4 → 5 Adiabatic expansion in turbine

Process 5 → 6 Adiabatic reversible acceleration in nozzle $ke_6 \gg ke_5$

c) Cycle description

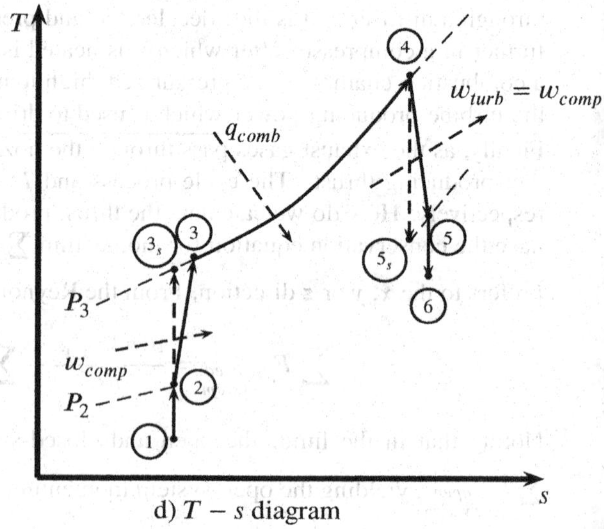

d) $T - s$ diagram

Figure 7.13 Jet engine based on Brayton power cycle

where, from the mass-balance, we note that since $\dot{m}_{air} \gg \dot{m}_{fuel} \Rightarrow \dot{m} = \dot{m}_{inlet} \approx \dot{m}_{exhaust}$. The restraining force $F_{i:restraint}$, we see from Figure 7.13 a) is simply equal and opposite to the thrust $F_{i:thrust}$. Hence via (7.142)

$$\boxed{F_{i:thrust} = \dot{m}(V_{i:inlet} - V_{i:exhaust})} \tag{7.143}$$

As a final reminder, the turbo-jet propulsion cycle uses the power output from the turbine to drive the compressor: Thus, the constraint on the cycle is $\dot{W}_{comp} = \dot{W}_{turb}$. Now, recall that the standard Brayton cycle can be specified via the compressor (and turbine) pressure ratios. The turbo-jet cycle, on the other hand, is specified via the compressor pressure ratio r_p while the turbine pressure ratio is dictated via the turbine-to-compressor power constraint $\dot{W}_{comp} = \dot{W}_{turb}$: This implies that the turbine and compressor pressure ratios are not generally the same!

Finally, as discussed in §6.9, nozzles and diffusers may operate reversibly or irreversibly.

EXAMPLE 7.7 Jet propulsion cycle

Problem Statement:

A turbo-jet engine based on the Brayton cycle is mounted to a plane traveling at a speed of $300\frac{m}{s}$ where the atmospheric conditions are at $300° K$ and $80kPa$. The diffuser inlet diameter is $1m$, the compressor pressure ratio is 5 and the combustor outlet temperature is $750° K$. The compressor and turbine isentropic efficiencies are 0.85 and 0.9 respectively. Determine the diffuser outlet pressure, compressor outlet temperature and pressure, mass-flow rate of the air, the rate of heat addition in the combustor, the turbine outlet pressure and the thrust produced. Assume I.G.B. as well as adiabatic/reversible nozzle and diffuser. In addition, the nozzle exit pressure is atmospheric. You may neglect flow kinetic energy everywhere except at the diffuser inlet and nozzle exit. In addition, the mass-flow rate of the fuel injected into the combustor is very small compared to the air-flow.

Schematic:

Given:
Turbo-jet cycle,
$Vel_1 = 300\frac{m}{s}$,
$T_1 = 300° K$,
$P_1 = P_6 = 80kPa$,
$D_1 = 1m$,
$T_4 = 750° K$,
$\eta_{comp} = 0.85$,
$\eta_{turb} = 0.9$,
$r_{comp} = 5$,
I.G.B.

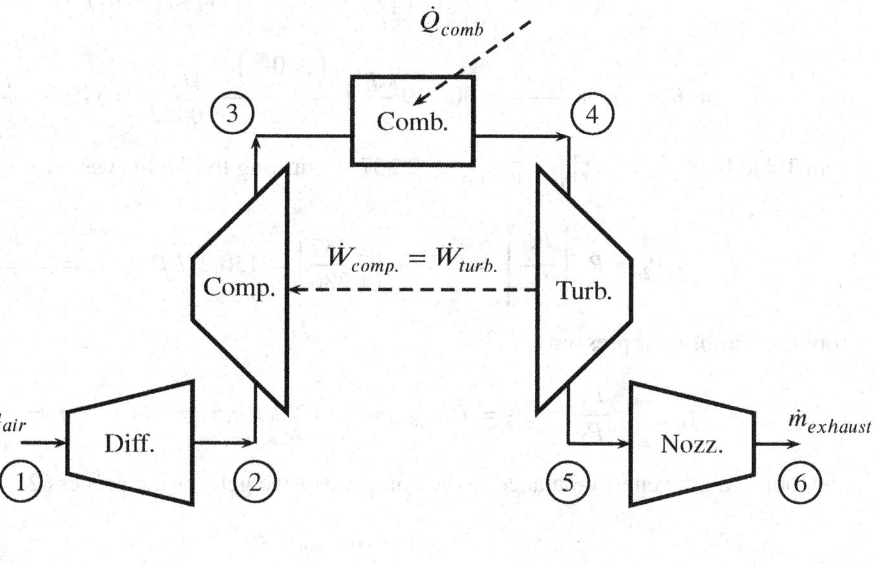

Find:
P_2,
P_3,
T_3,
$\dot{m}$,
$\dot{Q}_{comb}$,
P_5,
F_{thrust}

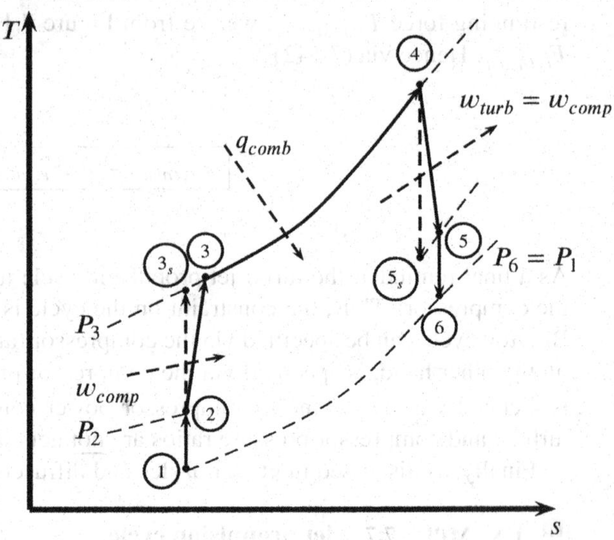

Assumptions:
I.G.B. property Tables B.11,
adiabatic and reversible nozzle and diffuser,
$\Delta pe = 0$,
isobaric combustion ($P_3 = P_4$),
$ke \approx 0$ except at diffuser inlet and nozzle exit,
mass-flow is the same through all devices.

Analysis:
Application of the entropy-balance to the adiabatic reversible diffuser yields $s_2 = s_1$, hence the process $1 \rightarrow 2$ is isentropic. From (6.58)

$$\frac{P_2}{P_1} = \frac{P_{r_2}}{P_{r_1}} \Rightarrow P_2 = P_1 \left[\frac{P_{r_2}}{P_{r_1}}\right] \tag{7.144}$$

where $P_{r_1} = P_r|_{T_1=300°K} = 1.386$. Application of the energy-balance to the diffuser where $h_1 = h|_{T_1=300°K} = 300.19\frac{kJ}{kg}$ yields (neglecting the outlet flow kinetic energy)

$$0 = (h_2 - h_1) + (\overset{\approx 0}{\cancel{ke_2}} - ke_1) \approx (h_2 - h_1) - \frac{Vel_1^2}{2}$$

$$\Rightarrow h_2 = h_1 + \frac{Vel^2}{2} = 300.19\frac{kJ}{kg} + \frac{\left(300\frac{m}{s}\right)^2}{2}\frac{1kJ}{1000J} = 345.19\frac{kJ}{kg} \tag{7.145}$$

From Table B.11 $P_{r_2} = P_r|_{h_2=345.19\frac{kJ}{kg}} = 2.257$. Returning to (7.144) we get

$$P_2 = P_1\left[\frac{P_{r_2}}{P_{r_1}}\right] = 80kPa\left[\frac{2.257}{1.386}\right] = 130.27kPa \overset{P_2}{\Longleftarrow\!=\!=} \tag{7.146}$$

From the compressor pressure ratio

$$r_{comp} = \frac{P_3}{P_2} \Rightarrow P_3 = P_2 r_{comp} = 130.27kPa(5) = 651.4kPa \overset{P_3}{\Longleftarrow\!=\!=} \tag{7.147}$$

Next, applying the energy-balance to the compressor we get, for the process $2 \rightarrow 3$

$$h_3 = w_{comp} + h_2 \tag{7.148}$$

where the compressor outlet temperature is found via $T_3 = T|_{h_3}$. Now, from the definition of isentropic efficiency

$$\eta_{comp} = \frac{\dot{W}^{isen}_{comp}}{\dot{W}_{comp}} = \frac{w^{isen}_{comp}}{w_{comp}} \Rightarrow w_{comp} = \frac{w^{isen}_{comp}}{\eta_{comp}} \tag{7.149}$$

where the isentropic work over process $2 \to 3_s$ is (via the energy-balance applied to the isentropic compressor) $w^{isen} = h_{3_s} - h_2$ and h_{3_s} is found via (6.58)

$$\frac{P_3}{P_2} = \frac{P_{r_{3s}}}{P_{r_2}} \Rightarrow P_{r_{3s}} = P_{r_2}\left[\frac{P_3}{P_2}\right] = P_{r_2} r_{comp} = 2.257(5) = 11.285 \tag{7.150}$$

giving $h_{3_s} = h|_{P_{r_{3s}}=11.285} = 546.9\frac{kJ}{kg}$. Combining (7.148) and (7.149) gives

$$h_3 = w_{comp} + h_2 = \frac{w^{isen}_{comp}}{\eta_{comp}} + h_2 = \frac{h_{3_s} - h_2}{\eta_{comp}} + h_2$$

$$= \frac{546.9\frac{kJ}{kg} - 345.19\frac{kJ}{kg}}{0.85} + 345.19\frac{kJ}{kg} = 582.5\frac{kJ}{kg} \tag{7.151}$$

Thus,

$$T_3 = T|_{h_3=582.5\frac{kJ}{kg}} = 576.6°K \overset{T_3}{\Longleftarrow\!=\!=} \tag{7.152}$$

To determine the mass-flow rate we simply calculate the air flow at the diffuser inlet (all of which is assumed to ultimately flow into the compressor)

$$\dot{m} = \dot{m}_{diffuser} = \dot{m}_1 = \pi\left(\frac{D_1}{2}\right)^2 Vel_1 \rho_1 = \pi\left(\frac{D_1}{2}\right)^2 Vel_1\left(\frac{P_1}{R_{air}T_1}\right)$$

$$= \pi\left(\frac{1m}{2}\right)^2 300\frac{m}{s}\left(\frac{80kPa}{0.287\frac{kJ}{kg-°K}300°K}\right) = 218.9\frac{kg}{s} \overset{\dot{m}}{\Longleftarrow\!=\!=} \tag{7.153}$$

To find the combustor heat addition $\dot{Q}_{comb}$, we apply the energy-balance to the combustor, where $h_4 = h|_{T_4=750°K} = 767.3\frac{kJ}{kg}$, giving

$$\dot{Q}_{comb} = \dot{m}(h_4 - h_3) = 218.9\frac{kg}{s}\left(767.3\frac{kJ}{kg} - 582.5\frac{kJ}{kg}\right)$$

$$= 40452kW\frac{1MW}{1000kW} = 40.45MW \overset{\dot{Q}_{comb}}{\Longleftarrow\!=\!=} \tag{7.154}$$

Now, to find the turbine exit pressure we note $P_5 = P_{5_s}$ from the definition of isentropic efficiency. The process $4 \to 5_s$ is isentropic, thus

$$\frac{P_{5_s}}{P_4} = \frac{P_{r_{5_s}}}{P_{r_4}} \Rightarrow P_{5_s} = P_4\left[\frac{P_{r_{5_s}}}{P_{r_4}}\right] \tag{7.155}$$

where $P_{r_{5_s}} = P_r|_{h_{5_s}}$. Enthalpy at state 5_s can be found via the energy-balance applied to the turbine for the isentropic process $4 \to 5_s$

$$w^{isen}_{turb} = h_4 - h_{5_s} \Rightarrow h_{5_s} = h_4 - w^{isen}_{turb} \tag{7.156}$$

in conjunction with the definition of isentropic efficiency

$$\eta_{turb} = \frac{\dot{W}_{turb}}{\dot{W}^{isen}_{turb}} = \frac{w_{turb}}{w^{isen}_{turb}} \Rightarrow w^{isen}_{turb} = \frac{w_{turb}}{\eta_{turb}} \quad (7.157)$$

Combining (7.156) and (7.157), noting that the turbine drives the compressor $\dot{W}_{comp} = \dot{W}_{turb} \Rightarrow w_{comp} = w_{turb}$ gives

$$h_{5_s} = h_4 - w^{isen}_{turb} = h_4 - \frac{\overset{w_{comp}}{\cancel{w_{turb}}}}{\eta_{turb}} = h_4 - \frac{w_{comp}}{\eta_{turb}}$$

$$= h_4 - \frac{h_3 - h_2}{\eta_{turb}} = 767.6\frac{kJ}{kg} - \frac{582.5\frac{kJ}{kg} - 345.19\frac{kJ}{kg}}{0.9} = 503.9\frac{kJ}{kg} \quad (7.158)$$

Thus, interpolating from Table B.11 $P_{r_{5_s}} = P_r|_{h_{5_s}=503.9\frac{kJ}{kg}} = 8.42$, $P_{r_4} = P_r|_{T_4=750^\circ K} = 37.35$ and via (6.58)

$$P_5 = P_{5_s} = \left[\frac{P_{r_{5_s}}}{P_{r_4}}\frac{P_3}{\cancel{P_4}}\right] = \left[\frac{P_{r_{5_s}}}{P_{r_4}}\right] P_3 = 651.4kPa\left[\frac{8.42}{37.35}\right] = 146.85kPa \overset{P_5}{\Longleftarrow} \quad (7.159)$$

Finally, to determine the thrust in the horizontal direction (flow direction is positive) we have from (7.143)

$$F_{thrust} = \dot{m}(Vel_{inlet} - Vel_{exhaust}) \Rightarrow F_{thrust} = \dot{m}(Vel_1 - Vel_6) \quad (7.160)$$

To determine Vel_6 we apply the energy-balance to the nozzle neglecting the inlet kinetic energy term

$$0 = (h_6 - h_5) + (ke_6 - \overset{\approx 0}{\cancel{ke_5}}) \approx (h_6 - h_5) + \frac{Vel_6^2}{2} \Rightarrow Vel_6 \approx \sqrt{2(h_5 - h_6)} \quad (7.161)$$

Noting that the acceleration in the nozzle is adiabatic and reversible, the entropy-balance gives $s_6 = s_5$ and the process $5 \to 6$ is isentropic. Now $P_{r_5} = P_r|_{h_5}$ where, via the energy-balance on the turbine, h_5 is equal to

$$h_5 = h_4 - \overset{w_{comp}}{\cancel{w_{turb}}} = h_4 - w_{comp} = h_4 - (h_3 - h_2)$$

$$= 767.3\frac{kJ}{kg} - \left(582.5\frac{kJ}{kg} - 345.19\frac{kJ}{kg}\right) = 529.9\frac{kJ}{kg} \quad (7.162)$$

giving $P_{r_5} = P_r|_{h_5=529.9\frac{kJ}{kg}} = 10.37$. Thus, from (6.58)

$$\frac{P_6}{P_5} = \frac{P_{r_6}}{P_{r_5}} \Rightarrow P_{r_6} = P_{r_5}\left[\frac{P_6}{P_5}\right] = 10.37\left[\frac{80\cancel{kPa}}{146.85\cancel{kPa}}\right] = 5.65 \quad (7.163)$$

and via Table B.11 $h_6 = h|_{P_{r_6}=5.65} = 448.9\frac{kJ}{kg}$. Returning to (7.161) gives

$$Vel_6 \approx \sqrt{2(h_5 - h_6)} = \sqrt{2(529.9\frac{\cancel{kJ}}{kg} - 448.9\frac{\cancel{kJ}}{kg})\frac{1000J}{1\cancel{kJ}}} = 402.5\frac{m}{s} \quad (7.164)$$

Returning to the expression for thrust (7.160)

$$F_{thrust} = \dot{m}(Vel_1 - Vel_6) = 218.9\frac{kg}{s}(300 - 402.5)\frac{m}{s} = -22437\frac{kg-m}{s^2} = -22.44kN \overset{F_{thrust}}{\Longleftarrow} \tag{7.165}$$

$P_2 = 130.27kPa,\ P_3 = 651kPa,\ T_3 = 576.6°K,\ \dot{m} = 218.9\frac{kg}{s},\ \dot{Q}_{comb} = 40.45MW,$ $P_5 = 146.85kPa,\ F_{thrust} = -22.44kN$

Comment: Thus, the thrust force is in the opposite direction of the flow. Hence, the thrust force points to the left in the schematic and has a magnitude of $22.44kN$.

□

7.7.1 INCREASING THRUST: AFTER-BURN

As in the multi-stage Brayton power producing cycle investigated in §7.6.1, we can increase thrust (previously power) by adding heat after the expansion process in the turbine (the LPT in the multi-stage device). Physically, this increases the internal energy, and hence enthalpy, of the flow entering the nozzle, thus increasing the exit velocity and thrust as shown in the following example. Such a device is referred to as an "after-burner".

EXAMPLE 7.8 Jet propulsion cycle with afterburn

Problem Statement:
The turbo-jet engine based on the Brayton cycle analyzed previously is now fitted with a secondary combustor, an after-burner (A.B.), installed between the turbine and nozzle. Heat is added isobarically via the after-burner giving a nozzle inlet temperature of $700°K$. Determine the thrust produced and the volume flow-rate of fuel consumed in the combustor and after-burner if the heating value and density of the fuel are $46MJ/kg_{fuel}$ and $800kg/m^3$ respectively. Assume I.G.B.. In addition, the nozzle exit pressure is atmospheric.

Given:
Turbo-jet cycle,
$Vel_1 = 300\frac{m}{s}$,
$T_1 = 300°K$,
$P_1 = P_7 = 80kPa$,
$D_1 = 1m$,
$T_4 = 750°K$,
$\eta_{comp} = 0.85$,
$\eta_{turb} = 0.9$,
$r_{comp} = 5$,
$T_6 = 700°K$,
$\frac{\dot{Q}_{fuel}}{\dot{m}_{fuel}} = 46\frac{MJ}{kg_{fuel}}$,
$\rho_{fuel} = 800\frac{kg}{m^3}$,
I.G.B.

Schematic:

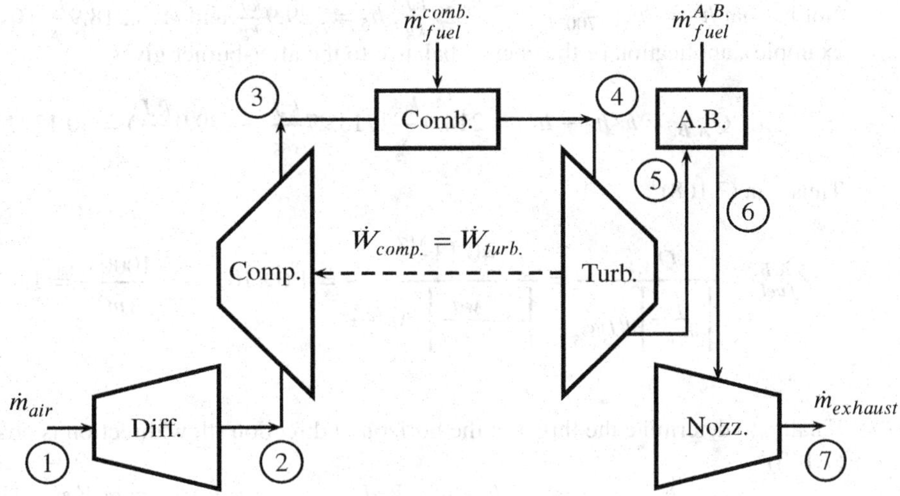

Find:

F_{thrust},
$\dot{V}^{comb}_{fuel}$, $\dot{V}^{A.B.}_{fuel}$

Assumptions:

I.G.B. property Tables B.11,
adiabatic and reversible nozzle and diffuser,
$\Delta pe = 0$,
isobaric combustion ($P_3 = P_4$),
isobaric afterburner ($P_5 = P_6$),
ke negligible except at diffuser inlet and nozzle exit.

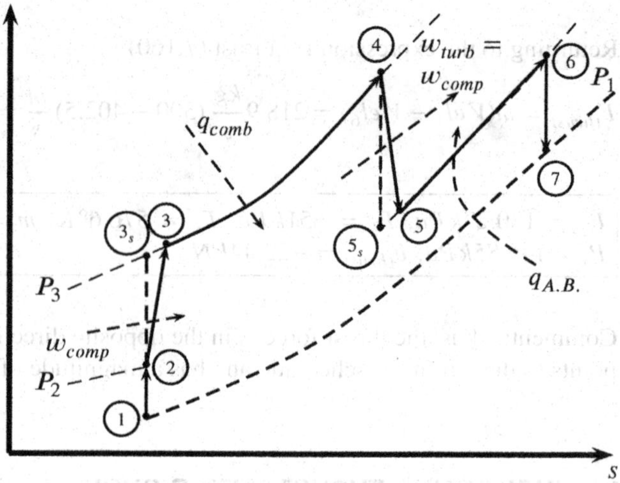

Analysis:

To determine the combustor fuel volume flow-rate, note from the previous example $\dot{Q}_{comb} = 66.596MW$ thus

$$\dot{Q}_{comb} = \left[\frac{\dot{Q}_{fuel}}{\dot{m}_{fuel}}\right] \dot{m}^{comb}_{fuel} = \left[\frac{\dot{Q}_{fuel}}{\dot{m}^{comb}_{fuel}}\right] \dot{V}^{comb}_{fuel}\, \rho_{fuel} \tag{7.166}$$

or

$$\dot{V}^{comb}_{fuel} = \frac{\dot{Q}_{comb}}{\left[\frac{\dot{Q}_{fuel}}{\dot{m}_{fuel}}\right]\rho_{fuel}} = \frac{40.45\frac{MJ}{s}}{\left[46\frac{MJ}{kg_{fuel}}\right] 800\frac{kg}{m^3}} = 1.1(10^{-3})\frac{m^3}{s}\frac{1000L}{1m^3} = 1.1\frac{L}{s} \overset{\dot{V}^{comb}_{fuel}}{\Longleftarrow} \tag{7.167}$$

With respect to the after-burner (A.B.), we modify (7.166) giving

$$\dot{Q}_{A.B.} = \left[\frac{\dot{Q}_{fuel}}{\dot{m}_{fuel}}\right] \dot{m}^{A.B.}_{fuel} = \left[\frac{\dot{Q}_{fuel}}{\dot{m}^{comb}_{fuel}}\right] \dot{V}^{A.B.}_{fuel}\, \rho_{fuel} \Rightarrow \dot{V}^{A.B.}_{fuel} = \frac{\dot{Q}_{A.B.}}{\left[\frac{\dot{Q}_{fuel}}{\dot{m}_{fuel}}\right]\rho_{fuel}} \tag{7.168}$$

Noting that $h_6 = h|_{T_6=700^\circ K} = 713.27\frac{kJ}{kg}$, $h_5 = 529.9\frac{kJ}{kg}$ and $\dot{m} = 218.9\frac{kg}{s}$ (from the previous example), application of the energy-balance to the after-burner gives

$$\dot{Q}_{A.B.} = \dot{m}(h_6 - h_5) = 218.9\frac{kg}{s}(713.27\frac{kJ}{kg} - 529.9\frac{kJ}{kg}) = 40.13MW \tag{7.169}$$

Thus, via (7.168)

$$\dot{V}^{A.B.}_{fuel} = \frac{\dot{Q}_{A.B.}}{\left[\frac{\dot{Q}_{fuel}}{\dot{m}_{fuel}}\right]\rho_{fuel}} = \frac{40.13\frac{MJ}{s}}{\left[46\frac{MJ}{kg_{fuel}}\right] 800\frac{kg}{m^3}} = 1.09(10^{-3})\frac{m^3}{s}\frac{1000L}{1m^3} = 1.09\frac{L}{s} \overset{\dot{V}^{A.B.}_{fuel}}{\Longleftarrow} \tag{7.170}$$

Finally, to determine the thrust in the horizontal direction (flow direction is positive) we have from (7.143)

$$F_{i:thrust} = \dot{m}(Vel_{i:inlet} - Vel_{i:exhaust}) \Rightarrow F_{thrust} = \dot{m}(Vel_1 - Vel_7) \tag{7.171}$$

To determine Vel_7 we apply the energy-balance to the adiabatic reversible nozzle neglecting the inlet kinetic energy term

$$Vel_7 \approx \sqrt{2(h_6 - h_7)} \tag{7.172}$$

Noting that the acceleration in the nozzle is adiabatic and reversible, the entropy-balance gives $s_7 = s_6$ and the process $6 \rightarrow 7$ is isentropic. From Table B.11 $P_{r_6} = P_r\big|_{h_6=713.27\frac{kJ}{kg}} = 28.8$, and since process $5 \rightarrow 6$ is isobaric (with $P_5 = 146.85kPa$ from the previous example)

$$\frac{P_7}{P_6} = \frac{P_{r_7}}{P_{r_6}} \Rightarrow P_{r_7} = P_{r_6}\left[\frac{P_7}{P_6}\right] = P_{r_6}\left[\frac{P_7}{P_5}\right] = 28.8\left[\frac{80\cancel{kPa}}{146.85\cancel{kPa}}\right] = 15.69 \tag{7.173}$$

and $h_7 = h|_{P_{r_7}=15.69} = 600.9\frac{kJ}{kg}$. Returning to (7.172) gives the nozzle outlet velocity Vel_7

$$Vel_7 \approx \sqrt{2(h_5 - h_6)} = \sqrt{2(713.27\frac{\cancel{kJ}}{kg} - 600.9\frac{\cancel{kJ}}{kg})\frac{1000J}{1\cancel{kJ}}} = 474.0\frac{m}{s} \tag{7.174}$$

which from (7.171) we have the engine thrust F_{thrust}

$$F_{thrust} = \dot{m}(Vel_1 - Vel_7) = 218.9\frac{kg}{s}(300 - 474.0)\frac{m}{s} = -38.1\frac{kg-m}{s^2} = -38.1kN \xleftarrow{F_{thrust}} \tag{7.175}$$

Thus, the thrust force is in the opposite direction of the flow. Hence, the thrust force points to the left in the schematic and has a magnitude of $38.1kN$.

$\boxed{\dot{V}^{comb}_{fuel} = 1.1\frac{L}{s},\ \dot{V}^{A.B.}_{fuel} = 1.09\frac{L}{s},\ F_{thrust} = 38.1kN}$

Comment: The increase in thrust from $24.32kN$ to $38.1kN$, an increase of approximately 50%, due to installation of an after-burner is at the cost doubling fuel consumption. Hence, after-burners are used for only short periods compared to the total flight duration. Finally, the total fuel mass-flow rate is found to be $1.752\frac{kg}{s}$ giving an "air-fuel ratio" $\frac{\dot{m}_{fuel}}{\dot{m}_{air}} = \frac{1.752\frac{\cancel{kg}}{\cancel{s}}}{218.9\frac{\cancel{kg}}{\cancel{s}}} = 0.008 \ll 1$. Hence, our original assumption $\dot{m}_{fuel} \ll \dot{m}_{air}$ is justified!

□

7.8 LIQUID-VAPOR POWER CYCLES

7.8.1 MOTIVATION: CARNOT CYCLE REVISITED

Recall from §5.5 that the Carnot heat engine is internally and externally reversible: All heat interactions are isothermal and occur at the heat source/sink temperture, while all processes exclude friction and un-restrained expansion. Figure 7.14 a) shows an open-system steady-flow device composed of a reversible adiabatic turbine and compressor/pump along with a frictionless iso-baric boiler and condenser. As we can see from $T - s$ diagram in Figure 7.14 b), this device is equivalent to a Carnot cycle if the compressor exit and turbine inlet states are saturated liquid and vapor respectively: The heat addition and rejection process occur within the dome and are isobaric, and hence, isothermal.

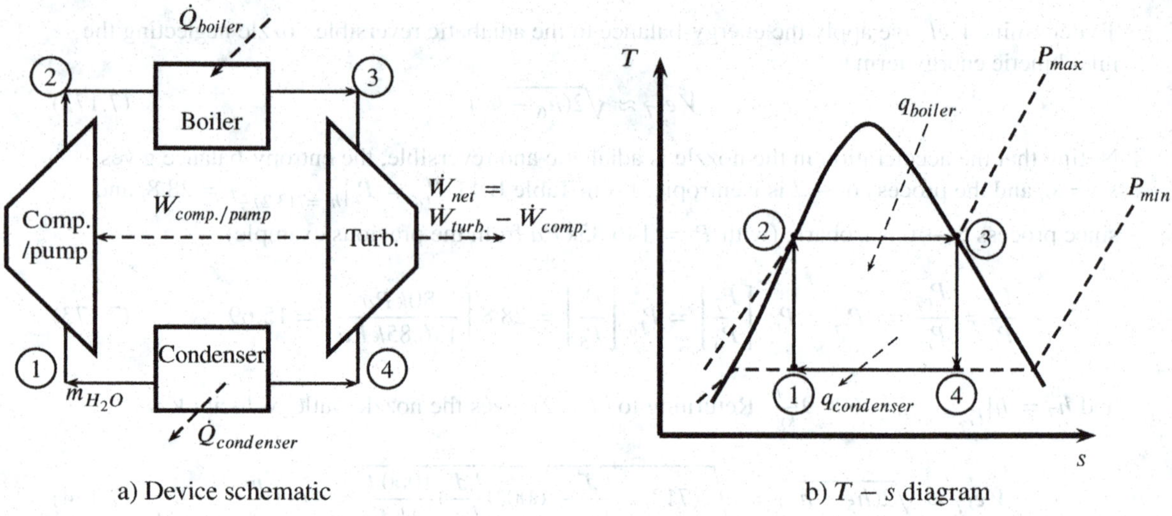

Figure 7.14 Schematic of liquid-vapor Carnot power-cycle device and $T - s$ diagram

7.8.2 IDEAL RANKINE CYCLE

As with the closed system Carnot cycle (see §7.1) there are several objections, on technical grounds, to an open-system Carnot cycle: First, the expansion process $3 \rightarrow 4$ in the turbine results in a significant mass fraction of the working fluid condensing to a liquid within the turbine. The presence of a significant amount of liquid (traveling as high speed through the turbine) would rapidly erode internal components (blades, baffles, endwalls, etc.) resulting in a short turbine life. The same limitation applies to the compression process $1 \rightarrow 2$ within the compressor/pump. To address the rapid erosion in the compressor, we choose to reject additional heat from the condenser, thus reducing the condenser outlet state to a saturated liquid. Since the inlet state 1 to the compressor

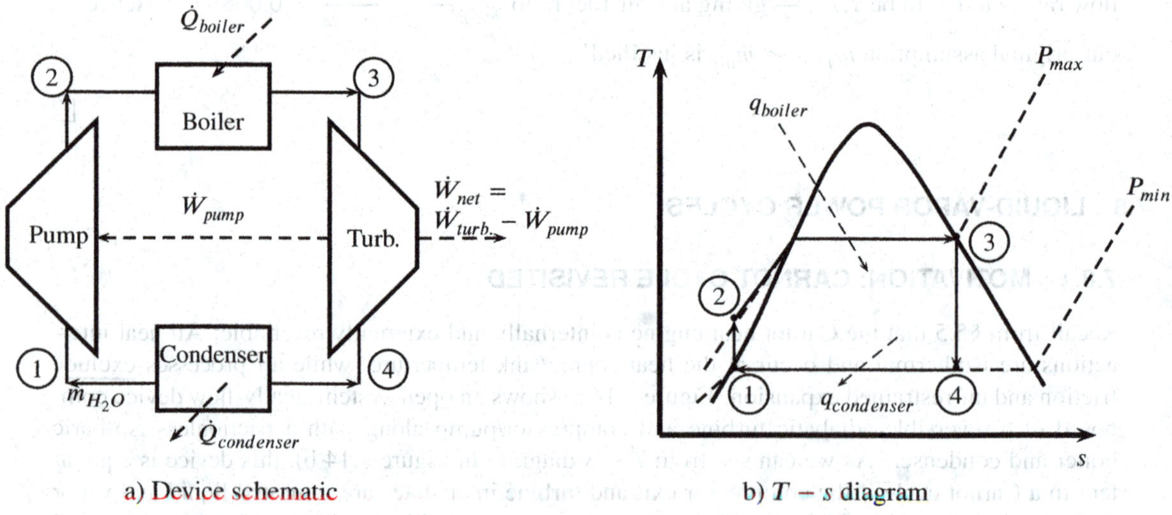

Figure 7.15 Schematic of an Ideal Rankine power-cycle device and $T - s$ diagram

is now a saturated liquid, we simply replace the compressor with an adiabatic pump. The revised cycle, the ideal Rankine cycle, is composed of the following processes as shown in Figure 7.15

Process 1 → 2**:** Adiabatic and reversible (isentropic) compression via a pump from saturated to super-cooled liquid

Process 2 → 3**:** Isobaric heat addition in the boiler/steam-generator to a saturated vapor

Process 3 → 4**:** Expansion to a saturated mixture via a reversible adiabatic (isentropic) turbine

Process 4 → 1**:** Isobaric heat rejection via the condenser to a saturated liquid

7.8.3 REVERSIBLE PUMP-WORK

We could now proceed to analyze the ideal Rankine cycle by application of the energy and entropy-balances to arrive at the cycle efficiency and specific work, etc. Analysis of the adiabatic reversible pump presents no problem; we have at our disposal the approximation for specific enthalpy of a super-cooled liquid (2.22). However, there exists a very simple expression for reversible adiabatic (i.e. isentropic) specific pump work which will greatly simplify our analysis. Application of the mass-balance (assuming steady-state/steady-flow operation) to our pump shown in Figure 7.16 gives $\dot{m}_1 = \dot{m}_2$, while application of the entropy-balance gives $s_1 = s_2$. Now, if we mark and follow a particle of liquid (a closed system) passing through the pump (an open-system) we have, from the Tds property relationships (6.25)

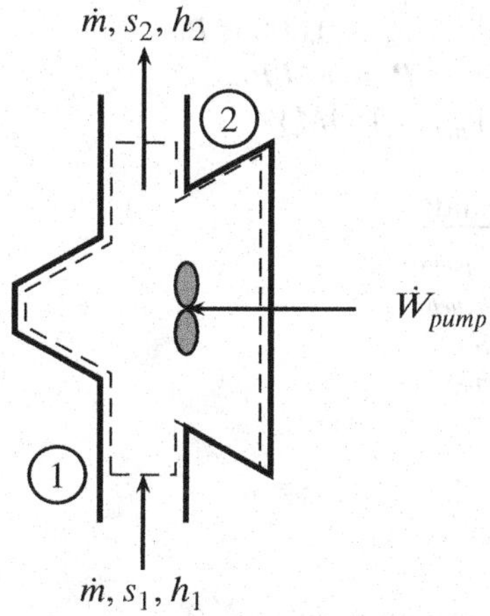

Figure 7.16 Adiabatic pump with inlet and outlet states 1 and 2.

$$\overset{0,\,isen.}{T\cancel{ds}} = du + \overset{0,\,incompressible}{P\cancel{dv}} \Rightarrow du = 0 \tag{7.176}$$

Thus, since $du = 0$ over the process $1 \rightarrow 2$, we have from the property relationship for an incompressible substance (4.34) $du = C_v dT = 0 \Rightarrow dT = 0$, hence $\Delta u_{1\rightarrow 2} = 0$ and $\Delta T_{1\rightarrow 2} = 0$. Applying the open-system energy-balance (4.85) to the pump yields

$$\overset{0,\,adia.}{\cancel{\dot{Q}}} - \overset{-\dot{W}_{pump}}{\cancel{\dot{W}}} = \overset{0,\,S.S.}{\cancel{\frac{dE_{CV}}{dt}}} + \sum \overset{(\dot{m}(h+ke+pe))_2}{\cancel{(\dot{m}(h+ke+pe))_{outlets}}} - \sum \overset{(\dot{m}(h+ke+pe))_1}{\cancel{(\dot{m}(h+ke+pe))_{inlets}}} \tag{7.177}$$

Or, neglecting changes in kinetic and potential energy

$$\dot{W}_{pump_{1\rightarrow 2}} = \dot{m}(h + \cancel{ke + pe})_2 - \dot{m}(h + \cancel{ke + pe})_1$$

$$\Rightarrow w_{pump} = h_2 - h_2 = \overset{\approx 0}{\cancel{\Delta u_{1\rightarrow 2}}} + \Delta(Pv)_{1\rightarrow 2} \approx \Delta(Pv)_{1\rightarrow 2} \tag{7.178}$$

Finally, from (2.19) we have for super-cooled liquid $v|_{T,P} \approx v^{sat}_{liq.}\big|_T$ and (7.178) becomes

$$w_{pump_{1\rightarrow 2}} \approx (P\overset{v^{sat}_{liq.}|_{T_1}}{\cancel{v}})_{1\rightarrow 2} \approx v^{sat}_{liq.}\Big|_{T_1} \Delta P_{1\rightarrow 2} \tag{7.179}$$

or

$$\boxed{w^{isen}_{pump_{1\rightarrow 2}} \approx v^{sat}_{liq}\Big|_{T_1} \Delta P_{1\rightarrow 2}, \text{ (Reversible pump-work)}} \tag{7.180}$$

EXAMPLE 7.9 Ideal Rankin cycle

Problem Statement:

An ideal Rankine cycle (with H_2O as the working fluid) has min and max cycle pressures of $0.008MPa$ and $8MPa$ respectively. The cycle produces net power at a rate of $100MW$. Determine the turbine and pump specific work, the cycle efficiency and mass-flow rate.

Schematic:

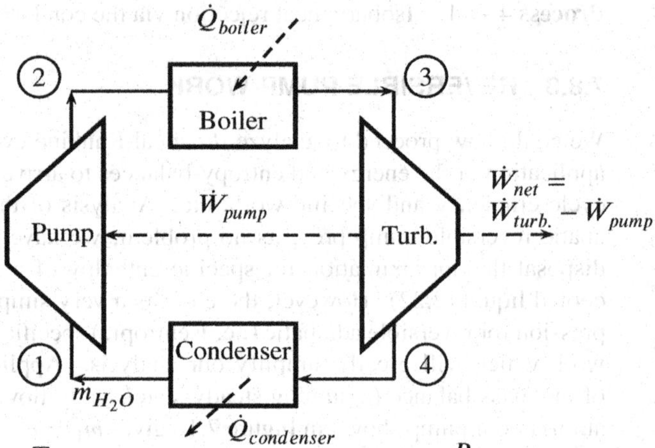

Given:

Ideal Rankine cycle,

H_2O,

$P_1 = P_4 = 0.008MPa$,

$P_2 = P_3 = 8MPa$,

$\dot{W}_{net} = 100MW$

Find:

w_{pump},

w_{turb},

η,

$\dot{m}$

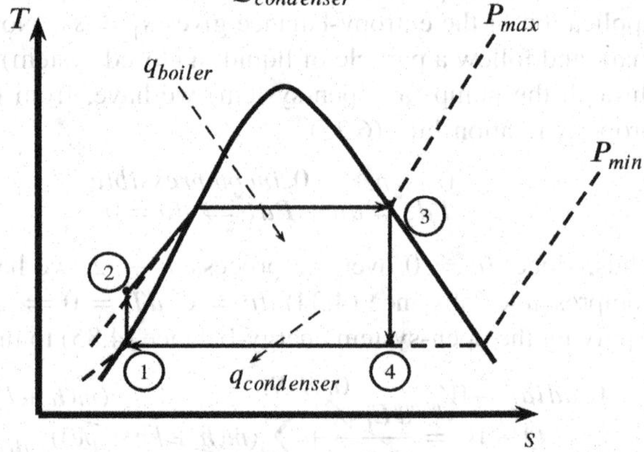

Assumptions:

H_2O property tables,

reversible/adiabatic pump and turbine,

$\Delta ke, pe$ negligible,

steady-state (S.S.)/steady-flow operation

Analysis:

Process 1 → 2 (Adiabatic and reversible pump work):

The pump is adiabatic and reversible, hence from the entropy-balance isentropic and (7.180) is valid. Thus, since $v_1 = v_{liq.}^{sat}\Big|_{P_1=0.008MPa} = 0.001\frac{m^3}{kg}$ via saturation pressure Table B.3, (7.180) gives

$$w_{pump_{1\to2}} \approx v_{liq.}^{sat}\Big|_{T_1} \Delta P_{1\to2} = 0.001\frac{m^3}{kg}(8\cancel{MPa} - 0.008\cancel{MPa})\frac{1000kPa}{1\cancel{MPa}} = 7.992\frac{kJ}{kg} \overset{w_{pump}}{\Longleftarrow} \tag{7.181}$$

Process 3 → 4 (Adiabatic and reversible turbine work):

Application of the energy-balance and mass-balance to the turbine yields

$$w_{turb} = w_{3\to4} = h_3 - h_4 \tag{7.182}$$

From saturation Table B.3 $h_3 = h^{sat}_{vapor}\Big|_{P_3=8MPa} = 2758\frac{kJ}{kg}$. The turbine is adiabatic and reversible, and hence from the entropy-balance isentropic (i.e. $s_4 = s_3$). From the saturation Tables $s_3 = s^{sat}_{vapor}\Big|_{P_3=8MPa} = 5.74\frac{kJ}{kg-°K}$, $s^{sat}_{liq}\Big|_{P_4=0.008MPa} = 0.5926\frac{kJ}{kg-°K}$ and $s^{sat}_{vapor}\Big|_{P_4=0.008MPa} = 8.228\frac{kJ}{kg-°K}$. Since $s^{sat}_{liq}\Big|_{P_4} < s_4 < s^{sat}_{vapor}\Big|_{P_4}$ state 4 is a saturated mixture with quality

$$x_4 = \frac{s_4 - s^{sat}_{liq.}\Big|_{P_4}}{s^{sat}_{vapor}\Big|_{P_4} - s^{sat}_{liq.}\Big|_{P_4}} = \frac{5.74 - 0.5926}{8.228 - 0.5926}\frac{\cancel{\frac{kJ}{kg-°K}}}{\cancel{\frac{kJ}{kg-°K}}} = 0.67 \tag{7.183}$$

and hence the enthalpy at state 4 is

$$h_4 = h^{sat}_{mix}\Big|_{\substack{P_4=0.008MPa,\\ x_4=0.67}} = h^{sat}_{liq}\Big|_{P_4=0.008MPa} + x_4\left(h^{sat}_{vapor} - h^{sat}_{liq}\right)\Big|_{P_4=0.008MPa} =$$

$$173.8\frac{kJ}{kg} + 0.67\left(2577.9\frac{kJ}{kg} - 173.8\frac{kJ}{kg}\right) = 1783.9\frac{kJ}{kg} \tag{7.184}$$

Returning to (7.182) we have

$$w_{turb} = h_3 - h_4 = 2758\frac{kJ}{kg} - 1783.9\frac{kJ}{kg} = 974.1\frac{kJ}{kg} \overset{w_{turb}}{\Longleftarrow} \tag{7.185}$$

Next, we can find $\dot{m}$ from the definition of specific work $\dot{W}_{net} = \dot{m}w_{net}$

$$\dot{m} = \frac{\dot{W}_{net}}{w_{net}} = \frac{\dot{W}_{req}}{w_{turb} - w_{pump}} = \frac{100\cancel{MW}}{974.1\frac{kJ}{kg} - 7.992\frac{kJ}{kg}}\frac{1000kW}{1\cancel{MW}} = 103.5\frac{kg}{s} \overset{\dot{m}}{\Longleftarrow} \tag{7.186}$$

Finally, from the definition of efficiency for a work producing cycle device (4.136)

$$\eta = \frac{\dot{W}_{net}}{\dot{Q}_{in}} = \frac{\cancel{\dot{m}}w_{net}}{\cancel{\dot{m}}q_{in}} = \frac{w_{net}}{q_{2\to3}} = \frac{w_{turb} - w_{pump}}{q_{2\to3}} \tag{7.187}$$

Process 2 → 3 (Reversible, isobaric heat addition in boiler/steam generator):

Application of the energy-balance and mass-balance to the steam-generator/boiler yields

$$q_{2\to3} = h_3 - h_2 \tag{7.188}$$

Enthalpy at state 2 can be found via application of the energy-balance to the pump (noting via saturation pressure Tables B.2 $h_1 = h^{sat}_{liquid}\Big|_{P_1=0.008MPa} = 173.88\frac{kJ}{kg}$)

$$w_{pump} = -w_{1\to2} = h_2 - h_1 \Rightarrow h_2 = h_1 + w_{pump} = 173.88\frac{kJ}{kg} + 7.992\frac{kJ}{kg} = 181.872\frac{kJ}{kg} \tag{7.189}$$

Combining (7.187) with (7.188)

$$\eta = \frac{w_{turb} - w_{pump}}{q_{2\to3}} = \frac{w_{turb} - w_{pump}}{h_3 - h_2} = \frac{963.2\cancel{\frac{kJ}{kg}} - 7.992\cancel{\frac{kJ}{kg}}}{2758\cancel{\frac{kJ}{kg}} - 181.872\cancel{\frac{kJ}{kg}}} = 37.5\% \overset{\eta}{\Longleftarrow} \tag{7.190}$$

$$\boxed{w_{pump} = 7.992\frac{kJ}{kg},\ w_{turb} = 974.1\frac{kJ}{kg},\ \dot{m} = 104.7\frac{kg}{s},\ \eta = 37.5\%}$$

Comment: Comparing the pump and turbine specific work indicates a very low $B.W.R. = \frac{w_{pump}}{w_{turb}}$ which is typical of Rankine cycle applications: Pump work is dwarfed by turbine work. Secondly, the turbine exit quality is low (0.67) and will cause rapid erosion of internal turbine components. Finally, comparing the Rankine vs. the Carnot cycle efficiency operating between the same temperature limits ($T_L = T^{sat}\big|_{P_1} = 314.5°K$ and $T_H = T^{sat}\big|_{P_3} = 568.1°K$) we find $\eta_{Carnot} = 1 - T_L/T_H = 44.6\% > 37.5\% = \eta$, lending plausibility to our calculated cycle efficiency. The lower efficiency of the Rankin compared to the Carnot cycle is due to the entropy generation between the furnace (a constant temperature heat source) and the boiler.

□

7.8.4 IMPROVING RANKINE CYCLE PERFORMANCE

The damaging effects of low turbine exit flow quality can be mitigated while increasing cycle specific work by super-heating the turbine inlet flow. An alternative is to expand the steam through *two* turbines: a high-pressure turbine (HPT) followed by a heat addition process in a reheater and further expansion in a second low-pressure turbine (LPT). Both strategies, super-heating and mult-stage expansion, are utilized in the following example.

EXAMPLE 7.10 Real Rankine cycle with super-heat and reheat

Problem Statement:
A real Rankine cycle (with H_2O as the working fluid) has min and max cycle pressures of $0.006MPa$ and $8MPa$ respectively. Heat is added, within the boiler/steam generator, till the H_2O is a super-heated vapor with a steam generator exit temperature of $873°K$. The working fluid is expanded through a high-pressure turbine (HPT) with an isentropic efficiency of 90% and pressure ratio of 8. Heat is then added to the working fluid within an isobaric reheater to a temperature of $773°K$. After reheating, the working fluid is further expanded to the minimum pressure through a low-pressure turbine (LPT) with an isentropic efficiency of 85%. The device produces net power at a rate of $100MW$. The pump isentropic efficiency is 85%. Determine the high and low pressure turbine and pump specific work, the LPT outlet quality, the cycle efficiency and mass-flow rate.

Given:
Real Rankine cycle w/reheat,
H_2O,
$P_1 = P_6 = 0.006MPa$,
$P_2 = P_3 = 8MPa$ (isobaric heat addition),
$\dot{W}_{req} = 100MW$,
$\eta_{pump} = 85\%$,
$\eta_{HPT} = 90\%$,
$\eta_{LPT} = 85\%$,
$r_{p:HPT} = \frac{P_3}{P_4} = 8$,
$T_3 = 873°K$,
$T_5 = 773°K$,
$P_4 = P_5$ (isobaric reheat)

Schematic:

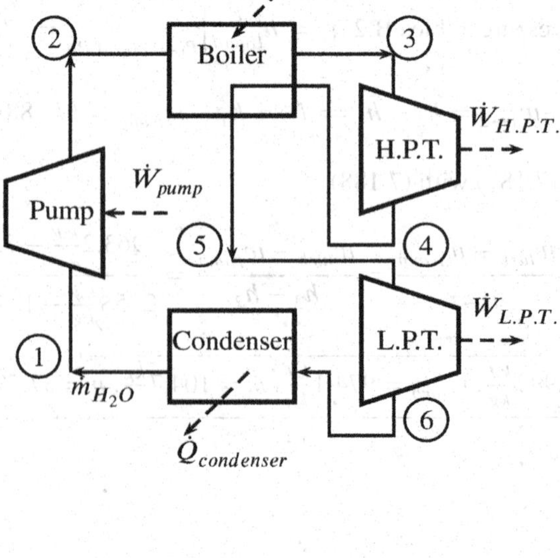

Find:
w_{pump},
w_{HPT},
w_{LPT},
x_6,
η,
$\dot{m}$

Assumptions:
H_2O property Tables,
steady-state (S.S.), steady-flow,
ke, *pe* negligible

Analysis:

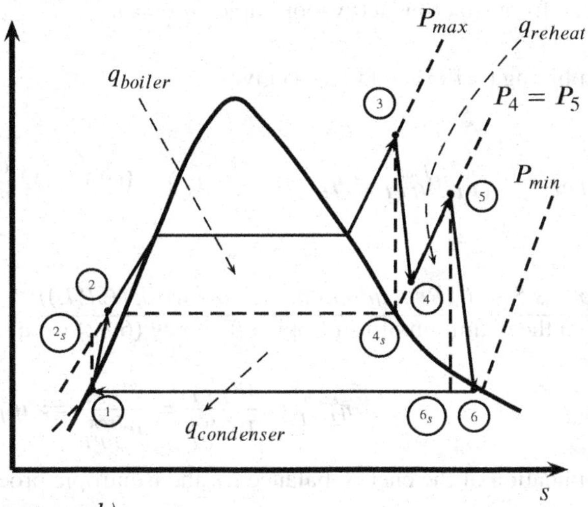

Process 1 → 2 (Adiabatic and reversible pump work):
From the definition of isentropic efficiency for a work consuming device (6.83)

$$\eta_{pump}^{isen} = \frac{\dot{W}_{pump}^{isen}}{\dot{W}_{pump}} = \frac{w_{pump}^{isen}}{w_{pump}} \Rightarrow w_{pump} = \frac{w_{pump}^{isen}}{\eta_{pump}^{isen}} \tag{7.191}$$

For the reversible and adiabatic (hence, isentropic) pump (7.180) is valid. Thus, since $v_1 = v_{liq}^{sat}\Big|_{P_1=0.006MPa} = 1.00(10^{-3})\frac{m^3}{kg}$ via saturation Tables B.3 (7.180) and (7.191) gives

$$w_{pump} = \frac{w_{pump}^{isen}}{\eta_{pump}^{isen}} \approx \frac{v_{liq}^{sat}\Big|_{P_1} \Delta P_{1\to 2}}{\eta_{pump}^{isen}} = \frac{0.001\frac{m^3}{kg}(8\cancel{MPa} - 0.006\cancel{MPa})\frac{1000kPa}{1\cancel{MPa}}}{0.85} = 9.4\frac{kJ}{kg} \overset{w_{pump}}{\Longleftarrow} \tag{7.192}$$

Process 3 → 4 (Real adiabatic turbine work (HPT)):
Application of the definition of isentropic efficiency (6.82) for a work producing device to the HPT yields

$$\eta_{HPT}^{isen} = \frac{\dot{W}_{HPT}}{\dot{W}_{HPT}^{isen}} = \frac{w_{HPT}}{w_{HPT}^{isen}} \Rightarrow w_{HPT} = \eta_{HPT}^{isen} w_{HPT}^{isen} \tag{7.193}$$

Application of the energy-balance and mass-balance for the isentropic process 3 → 4_s (the isentropic HPT) gives

$$w_{HPT}^{isen} = w_{3\to 4_s} = h_3 - h_{4_s} \tag{7.194}$$

From super-heated vapor Tables B.4 $h_3 = h_{vapor}^{super-heated}\Big|_{P_3=8MPa,\ T_3=873^\circ K} = 3642\frac{kJ}{kg}$ and $s_3 = s_{vapor}^{super-heated}\Big|_{P_3=8MPa,\ T_3=873^\circ K} = 7.02\frac{kJ}{kg-^\circ K}$. From the definition of the HPT pressure ratio $r_{P_{HPT}} = P_3/P_4$ we have

$$P_4 = P_3/r_{P_{HPT}} = 8MPa/8 = 1MPa \tag{7.195}$$

Thus, from super-heated vapor Tables $h_{4_s} = h_{vapor}^{super-heated}\Big|_{\substack{P_4=1MPa,\\ s_4=s_3=7.02\frac{kJ}{kg-{}^\circ K}}} = 2994\frac{kJ}{kg}$, hence

combining (7.193) and (7.194) gives

$$w_{HPT} = \eta_{HPT}^{isen} w_{HPT}^{isen} = \eta_{HPT}^{isen}(h_3 - h_{4_s}) = 0.9\left(3642\frac{kJ}{kg} - 2994\frac{kJ}{kg}\right) = 583.2\frac{kJ}{kg} \overset{w_{HPT}}{\Longleftarrow} \tag{7.196}$$

Process 5 → 6 (Real adiabatic turbine work (L.P.T.)):
From the definition of isentropic efficiency (6.82) for a work producing device

$$\eta_{LPT}^{isen} = \frac{\dot{W}_{LPT}}{\dot{W}_{LPT}^{isen}} = \frac{w_{LPT}}{w_{LPT}^{isen}} \Rightarrow w_{LPT} = \eta_{LPT}^{isen} w_{LPT}^{isen} \tag{7.197}$$

Application of the energy-balance for the isentropic process 5 → 6_s gives

$$w_{LPT}^{isen} = w_{5\to 6_s} = h_5 - h_{6_s} \tag{7.198}$$

From Table B.4 $h_5 = h_{vapor}^{super-heated}\Big|_{\substack{P_5=1MPa,\\ T_5=773^\circ K}} = 3478.5\frac{kJ}{kg}$, $s_5 = s_{vapor}^{super-heated}\Big|_{\substack{P_5=1MPa,\\ T_5=773^\circ K}} = 7.76\frac{kJ}{kg-{}^\circ K}$ and $h_{6_s} = h|_{\substack{P_{6_s}=P_6=0.006MPa,\\ s_{6_s}=s_5=7.76\frac{kJ}{kg-{}^\circ K}}}$. In addition, we also have $s_{liq}^{sat}\Big|_{P_{6_s}=P_6=0.006MPa} = 0.5210\frac{kJ}{kg-{}^\circ K}$ and $s_{vapor}^{sat}\Big|_{P_{6_s}=P_6=0.006MPa} = 8.3304\frac{kJ}{kg-{}^\circ K}$, state 6_s is a saturated mixture. Thus,

$$x_{6_s} = \frac{s_{6_s} - s_{liq}^{sat}\Big|_{P=0.006MPa}}{s_{vapor}^{sat}\Big|_{P=0.006MPa} - s_{liq}^{sat}\Big|_{P=0.006MPa}} = \frac{7.76 - 0.5210 \cancel{\frac{kJ}{kg-{}^\circ K}}}{8.3304 - 0.5210 \cancel{\frac{kJ}{kg-{}^\circ K}}} = 0.927 \tag{7.199}$$

and

$$\begin{aligned} h_{6_s} &= h_{liq}^{sat}\Big|_{P_{6_s}=0.006MPa} + x_{6_s}\left(h_{vapor}^{sat}\Big|_{P_{6_s}=0.006MPa} - h_{liq}^{sat}\Big|_{P_{6_s}=0.006MPa}\right) \\ &= 151.53\frac{kJ}{kg} + 0.927\left(2567.4\frac{kJ}{kg} - 151.53\frac{kJ}{kg}\right) = 2391\frac{kJ}{kg} \end{aligned} \tag{7.200}$$

Returning to the application of the energy-balance to the isentropic LPT (7.198) and the definition of isentropic efficiency (7.197) gives

$$w_{LPT} = \eta_{LPT}^{isen} w_{LPT}^{isen} = \eta_{LPT}^{isen}(h_5 - h_{6_s}) = 0.85\left(3478.5\frac{kJ}{kg} - 2391\frac{kJ}{kg}\right) = 924.3\frac{kJ}{kg} \overset{w_{LPT}}{\Longleftarrow} \tag{7.201}$$

Application of the energy-balance and mass-balance to the LPT gives

$$w_{LPT} = h_5 - h_6 \Rightarrow h_6 = h_5 - w_{LPT} = 3478.5\frac{kJ}{kg} - 924.3\frac{kJ}{kg} = 2554.2\frac{kJ}{kg} \tag{7.202}$$

State 6 is also found to be a saturated mixture

$$x_6 = \frac{h_6 - h_{liq}^{sat}\Big|_{P_6=0.006MPa}}{(h_{vapor}^{sat} - h_{liq}^{sat})\Big|_{P_6=0.006MPa}} = \frac{2554.1\frac{kJ}{kg} - 151.53\frac{kJ}{kg}}{2567.4\frac{kJ}{kg} - 151.53\frac{kJ}{kg}} = 0.994 \xleftarrow{x_6} \quad (7.203)$$

Hence, the outlet state for the LPT contains very little liquid water. Next, from the definition of efficiency of a work producing device (heat engine) (4.138), the definition of specific work and the fact that the mass-flow rate is identical through all components within the device

$$\eta = \frac{\dot{W}}{\dot{Q}_{in}} = \frac{w}{q_{in}} = \frac{w_{H.P.T.} + w_{L.P.T.} - w_{pump}}{q_{boiler} + q_{reheater}} \quad (7.204)$$

Application of the energy-balance to the boiler and reheater gives

$$q_{boiler} = h_3 - h_2 = 3642\frac{kJ}{kg} - 160.93\frac{kJ}{kg} = 3481.0\frac{kJ}{kg} \quad (7.205)$$

$$q_{reheater} = h_5 - h_4 = 3478.5\frac{kJ}{kg} - 3071.4\frac{kJ}{kg} = 407.1\frac{kJ}{kg} \quad (7.206)$$

where h_2 and h_4 are found via application of the energy-balance to the pump and HPT

$$w_{pump} = h_2 - h_1 \Rightarrow h_2 = h_1 + w_{pump} = 151.53\frac{kJ}{kg} + 9.4\frac{kJ}{kg} = 160.93\frac{kJ}{kg} \quad (7.207)$$

$$w_{HPT} = h_3 - h_4 \Rightarrow h_4 = h_3 - w_{HPT} = 3642\frac{kJ}{kg} - 570.6\frac{kJ}{kg} = 3071.4\frac{kJ}{kg} \quad (7.208)$$

Thus, returning to (7.204) gives

$$\eta = \frac{w_{HPT} + w_{LPT} + w_{pump}}{q_{boiler} + q_{reheater}} = \frac{583.2\frac{kJ}{kg} + 924.3\frac{kJ}{kg} - 9.4\frac{kJ}{kg}}{3481.0\frac{kJ}{kg} + 407.1\frac{kJ}{kg}} = 0.385 = 38.5\% \xleftarrow{\eta} \quad (7.209)$$

Finally, from the definition of specific work $\dot{m}$ is found to be

$$\dot{W} = \dot{m}w_{cycle} \Rightarrow \dot{m} = \frac{\dot{W}_{req}}{w_{cycle}} = \frac{\dot{W}_{req}}{w_{HPT} + w_{LPT} + w_{pump}}$$

$$= \frac{100MW}{583.2\frac{kJ}{kg} + 924.3\frac{kJ}{kg} - 9.4\frac{kJ}{kg}} \frac{1000\frac{kJ}{s}}{1MW} = 66.75\frac{kg}{s} \xleftarrow{\dot{m}} \quad (7.210)$$

$w_{pump} = 9.4\frac{kJ}{kg}$, $w_{H.P.T.} = 583.2\frac{kJ}{kg}$, $w_{L.P.T.} = 924.3\frac{kJ}{kg}$, $x_6 = 0.994$, $\eta = 38.5\%$, $\dot{m} = 66.75\frac{kg}{s}$

Comment: From the $T - s$ diagram, the reheat process (splitting the expansion process between a high and low pressure turbine with reheat in-between), results in a significant increase in turbine exit quality and hence turbine longevity. Specifically, setting $\dot{Q}_{reheat} = 0$ gives an LPT exit quality of 0.90. In addition, cycle specific work is reduced resulting in a higher required mass-flow rate: $76.61 \frac{kg}{s}$ for the cycle without reheat vs. $\dot{m} = 66.75 \frac{kg}{s}$ with reheat.

□

Finally, as in the Brayton cycle, strategies for increasing efficiency include the transfer of heat which would have been rejected from the condenser to the working fluid prior to heating in the steam-generator/boiler via regeneration. The recovery of lost heat occurs in one of two types of devices: Open or closed water feed-heaters the former of which is investigated in the following example.

EXAMPLE 7.11 Multi-stage Rankine cycle with open-water feed heating (OWFH)

Problem Statement:
A Rankine like cycle operates between a min and max pressure of $0.06bar$ and $100bar$, utilizes two stages of compression (low-pressure pump (LPP) and high-pressure pump (HPP)) and two stages of expansion (in a low-pressure turbine (LPT) and high-pressure turbine (HPT)). The turbine and pump isentropic efficiencies are all 0.8 and the LPP inlet state is a saturated liquid. To boost cycle efficiency a regenerator in the form of an open-water feed heater (OWFH) is installed between the LPP and HPP: The purpose of the OWFH is to bleed off a precise amount of steam from the HPT outlet to heat the working fluid to a saturated liquid state at the HPP inlet. Thus, recycling some of energy which would have been rejected via the condenser. The OWFH is an isobaric mixing chamber operating at $7bar$. If the boiler (steam-generator) outlet temperature is $480°C$, determine the cycle efficiency and the bleed mass-flow rate from the HPT exit as a fraction of the steam-generator/boiler mass-flow $\gamma = \dot{m}_{bleed} / \dot{m}_{steam-gen.}$.

Given:
Multi-stage Rankine cycle with OWFH,
$P_1 = 0.06bar$,
$P_4 = 100bar$,
$\eta_{LPT} = \eta_{HPT} = 0.8$,
$\eta_{LPP} = \eta_{HPP} = 0.8$,
$x_1 = x_3 = 0$,
$T_5 = 753.2°K$,
isobaric condenser and steam-generator,
OWFH is isobaric ($P_2 = P_3 = P_6 = P_7 = P_9 = 7bar$)

Schematic:

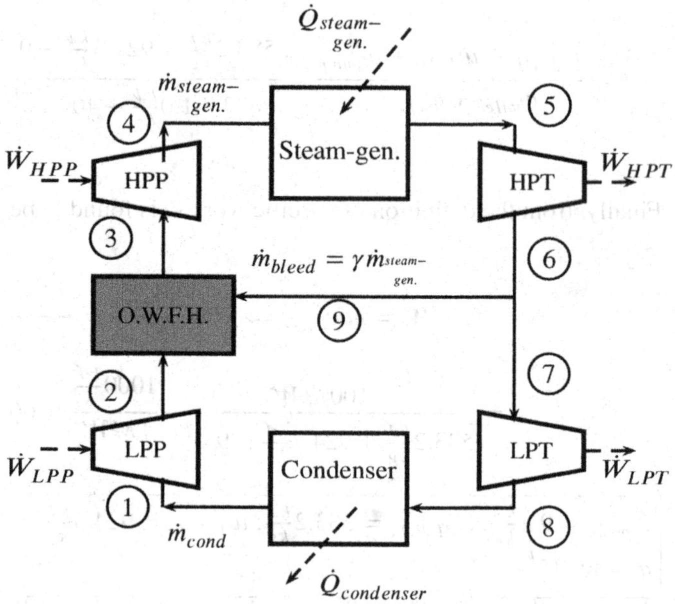

Find:

η,

$\gamma = \dfrac{\dot{m}_{bleed}}{\dot{m}_{steam-gen.}}$

Assumptions:

H_2O property Tables,
steady-state (S.S.),
steady-flow,
ke, pe negligible

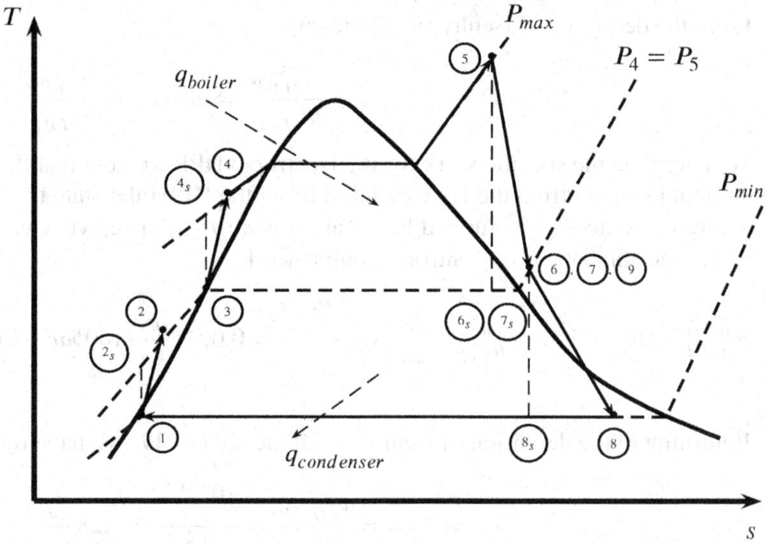

Analysis:

From the definition of cycle efficiency

$$\eta = \frac{\dot{W}_{net}}{\dot{Q}_{steam-gen.}} = \frac{\dot{W}_{HPT} + \dot{W}_{LPT} - (\dot{W}_{HPP} + \dot{W}_{LPP})}{\dot{Q}_{steam-gen.}} \tag{7.211}$$

To proceed, we begin by calculating $\dot{W}_{LPP}$ via the specific pump-work. Noting that, from the device diagram, the mass-flow through the LPP is $(1-\gamma)\dot{m}_{steam-gen.}$, we have

$$\dot{W}_{LPP} = (1-\gamma) w_{LPP} \dot{m}_{steam-gen.} \tag{7.212}$$

From the definition of isentropic efficiency

$$\eta_{LPP} = \frac{w^{isen}_{LPP}}{w_{LPP}} \Rightarrow w_{LPP} = \frac{w^{isen}_{LPP}}{\eta_{LPP}} \tag{7.213}$$

Noting that the process $1 \rightarrow 2_s$ is an isentropic low pressure compression process, we have via the expression for reversible adiabatic (i.e. isentropic) pump-work (7.180)

$$w^{isen}_{LPP} = h_{2_s} - h_1 \approx v^{sat}_{liq}\Big|_{P_1=0.06bar} (\overset{P_2}{\cancel{P_{2_s}}} - P_1) = 0.001\frac{m^3}{kg}(7\cancel{bar} - 0.06\cancel{bar})\frac{100kPa}{1\cancel{bar}} = 0.694\frac{kJ}{kg} \tag{7.214}$$

Returning to (7.213) w_{LPP} is found to be

$$w_{LPP} = \frac{w^{isen}_{LPP}}{\eta_{LPP}} = \frac{0.694\frac{kJ}{kg}}{0.8} = 0.87\frac{kJ}{kg} \tag{7.215}$$

Now, to find $\dot{W}_{HPP}$, noting that the mass-flow through the HPP is $\dot{m}_{steam-gen.}$ we have

$$\dot{W}_{HPP} = w_{HPP} \dot{m}_{steam-gen.} \tag{7.216}$$

From the definition of isentropic efficiency

$$\eta_{HPP} = \frac{w_{HPP}^{isen}}{w_{HPP}} \Rightarrow w_{HPP} = \frac{w_{HPP}^{isen}}{\eta_{HPP}} \tag{7.217}$$

To determine the specific work for the isentropic HPP we note that the O.W.F.H. mixes a metered amount of steam from the HPT outlet to bring the HPP inlet state to a saturated liquid (see $T-s$ diagram): State 3 is a saturated liquid at pressure P_3. Hence, we have via the expression for reversible adiabatic (i.e. isentropic) pump-work

$$w_{HPP}^{isen} = h_{4_s} - h_3 \approx v_{liq}^{sat}\Big|_{P_3=7bar}(\overset{P_4}{\cancel{P_{4_s}}} - P_3) = 0.0011\frac{m^3}{kg}(100\cancel{bar} - 7\cancel{bar})\frac{100kPa}{1\cancel{bar}} = 10.2\frac{kJ}{kg} \tag{7.218}$$

Returning to the definition of isentropic efficiency (7.217) we have for the HPP

$$w_{HPP} = \frac{w_{HPP}^{isen}}{\eta_{HPP}} = \frac{10.2\frac{kJ}{kg}}{0.8} = 12.8\frac{kJ}{kg} \tag{7.219}$$

Performing a similar analysis on the HPT and LPT we find $w_{HPT} = 509.6\frac{kJ}{kg}$ and $w_{LPT} = 570.6\frac{kJ}{kg}$ where $\dot{W}_{LPT} = (1-\gamma)w_{LPT}\dot{m}_{\substack{steam-\\gen.}}$ and $\dot{W}_{HPT} = w_{HPT}\dot{m}_{\substack{steam-\\gen.}}$. Returning to our expression for efficiency (7.211) utilizing previous results gives

$$\eta = \frac{\dot{W}_{HPT} + \dot{W}_{LPT} - (\dot{W}_{HPP} + \dot{W}_{LPP})}{\dot{Q}_{\substack{steam-\\gen.}}} =$$

$$\frac{w_{HPT}\cancel{\dot{m}_{\substack{steam-\\gen.}}} + w_{LPT}\cancel{\dot{m}_{\substack{steam-\\gen.}}}(1-\gamma) - (w_{HPP}\cancel{\dot{m}_{\substack{steam-\\gen.}}} + w_{LPP}(1-\gamma)\cancel{\dot{m}_{\substack{steam-\\gen.}}})}{(\cancel{\dot{m}_{\substack{steam-\\gen.}}})q_{\substack{steam-\\gen.}}}$$

$$= \frac{w_{HPT} + (1-\gamma)w_{LPT} - (w_{HPP} + (1-\gamma)w_{LPP})}{q_{\substack{steam-\\gen.}}} \tag{7.220}$$

or

$$\eta = \frac{w_{HPT} + (1-\gamma)w_{LPT} - (w_{HPP} + (1-\gamma)w_{LPP})}{q_{\substack{steam-\\gen.}}} \tag{7.221}$$

Hence, we must still determine boiler/steam-generator heat addition per-unit mass and the bleed-fraction $\gamma = \frac{\dot{m}_{bleed}}{\dot{m}_{steam-gen.}}$. Applying the energy balance to the feed heater (O.W.F.H) (an adiabatic and work-less device)

$$\overset{0}{\cancel{\dot{Q}}} - \overset{0}{\cancel{\dot{W}}} = \overset{0,\,S.S.}{\cancel{\frac{dE_{CV}}{dt}}} + \overset{(\dot{m}h)_3}{\cancel{\sum(\dot{m}(h + ke + pe))_{outlets}}} - \overset{(\dot{m}h)_9 + (\dot{m}h)_2}{\cancel{\sum(\dot{m}(h + ke + pe))_{inlets}}}$$

$$\Rightarrow 0 = (\dot{m}h)_3 - (\dot{m}h)_9 - (\dot{m}h)_2 \tag{7.222}$$

Applying the mass-balance to the O.W.F.H gives

$$\overset{0,\,S.S.}{\cancel{\frac{dm_{CV}}{dt}}} = \overset{\dot{m}_9 + \dot{m}_2}{\sum(\cancel{\dot{m}})_{inlets}} - \overset{\dot{m}_3}{\sum(\cancel{\dot{m}})_{outlets}} \Rightarrow \dot{m}_2 = \dot{m}_3 - \dot{m}_9 \tag{7.223}$$

Combining the mass (7.223) and energy-balance (7.222) for the O.W.F.H. gives

$$0 = (\dot{m}h)_3 - (\dot{m}h)_9 - (\dot{m}h)_2 = \dot{m}_3 h_3 - \dot{m}_9 h_9 - (\dot{m}_3 - \dot{m}_9)h_2 = \dot{m}_3(h_3 - h_2) + \dot{m}_9(-h_9 + h_2) \tag{7.224}$$

or

$$0 = \dot{m}_3(h_3 - h_2) + \dot{m}_9(-h_9 + h_2) \tag{7.225}$$

Noting from the device diagram that $\dot{m}_3 = \dot{m}_{\substack{steam- \\ gen.}}$, $\dot{m}_2 = (1-\gamma)\dot{m}_{\substack{steam- \\ gen.}}$ and $\dot{m}_9 = \gamma\dot{m}_{\substack{steam- \\ gen.}}$ we get an expression for the bleed fraction γ via (7.225)

$$0 = (h_3 - h_2)\overset{\dot{m}_{\substack{steam- \\ gen.}}}{\cancel{\dot{m}_3}} + (-h_9 + h_2)\overset{\gamma\dot{m}_{\substack{steam- \\ gen.}}}{\cancel{\dot{m}_9}} = (h_3 - h_2)\dot{m}_{\substack{steam- \\ gen.}} + (-h_9 + h_2)\gamma\dot{m}_{\substack{steam- \\ gen.}} \Rightarrow \gamma = \frac{h_3 - h_2}{h_9 - h_2} \tag{7.226}$$

Application of the energy and mass-balance to the LPP and HPT gives $h_2 = 152.3\frac{kJ}{kg}$ and $h_9 = 2812\frac{kJ}{kg}$. Noting that state 3 is a saturated liquid we have $h_3 = h_{liq}^{sat}\big|_{P=7bar} = 697.2\frac{kJ}{kg}$ we get via (7.226)

$$\gamma = \frac{h_3 - h_2}{h_9 - h_2} = \frac{697.4\frac{kJ}{kg} - 152.3\frac{kJ}{kg}}{2812 - 152.3\frac{kJ}{kg}} = 0.205 \xLeftarrow{\gamma} \tag{7.227}$$

The boiler/steam-generator heat addition per-unit mass is found via the energy-balance noting that $h_5 = h|_{\substack{T_5=753.2^\circ K, \\ P_5=100bar}} = 3321\frac{kJ}{kg}$

$$q_{\substack{steam- \\ gen.}} = h_5 - h_4 = 3321\frac{kJ}{kg} - 710\frac{kJ}{kg} = 2611\frac{kJ}{kg} \tag{7.228}$$

where h_4 is found via the energy-balance applied to the HPP

$$h_4 = w_{HPP} + h_3 = 12.8\frac{kJ}{kg} + 697.2\frac{kJ}{kg} = 710\frac{kJ}{kg} \tag{7.229}$$

Returning to the cycle efficiency (7.221)

$$\eta = \frac{w_{HPT} + (1-\gamma)w_{LPT} - (w_{HPP} + (1-\gamma)w_{LPP})}{q_{\substack{steam- \\ gen.}}} =$$

$$\frac{509.6\frac{kJ}{kg} + (1-0.205)570.6\frac{kJ}{kg} - (12.8\frac{kJ}{kg} + (1-0.205)0.87\frac{kJ}{kg})}{2611\frac{kJ}{kg}} = 36.4\% \xLeftarrow{\eta} \tag{7.230}$$

$\boxed{\gamma = 0.205, \eta = 36.4\%}$

Comment: Recalculation of cycle efficiency without feed heating ($\gamma = 0$) gives an efficiency of 33.8%. Hence, the use of regeneration via the O.W.F.H. yields an increase in efficiency of almost 3%.

□

CHAPTER 8

REFRIGERATION DEVICES

8.1 A POWER CYCLE REVERSED

Thus far we have been concerned with various work producing devices (both closed and open). Now we shift out attention to work consuming devices such as refrigeration cycles, the purpose of which is transfer heat from a cold region to a hotter region.

We begin with the $T - s$ diagram for the ideal Rankine cycle with super-heat as shown in Figure 8.1 a). The ideal cycle is *internally reversible*, hence, the cycle can be reversed yielding the $T - s$ cycle diagram as shown in Figure 8.1 b) where the pump⇒turbine, turbine⇒compressor, boiler⇒condenser and condenser⇒evaporator as shown in Figure 8.1 d). Now so far, for the Rankine power cycle we have typically found the pump work to be insignificant compared to the turbine work: $|w_{pump}| \ll w_{turb}$. This implies that for the reversed cycle, where the pump⇒turbine and the turbine⇒compressor, the turbine work will be insignificant compared to the compressor work: $w_{turbine} \ll |w_{compressor}|$ for the reversed cycle! Since the turbine produces so little work, let's replace it with a no-work, adiabatic device known as a throttle (or expansion valve): A pipe containing an internal restriction used to restrict the flow resulting in a large pressure drop.

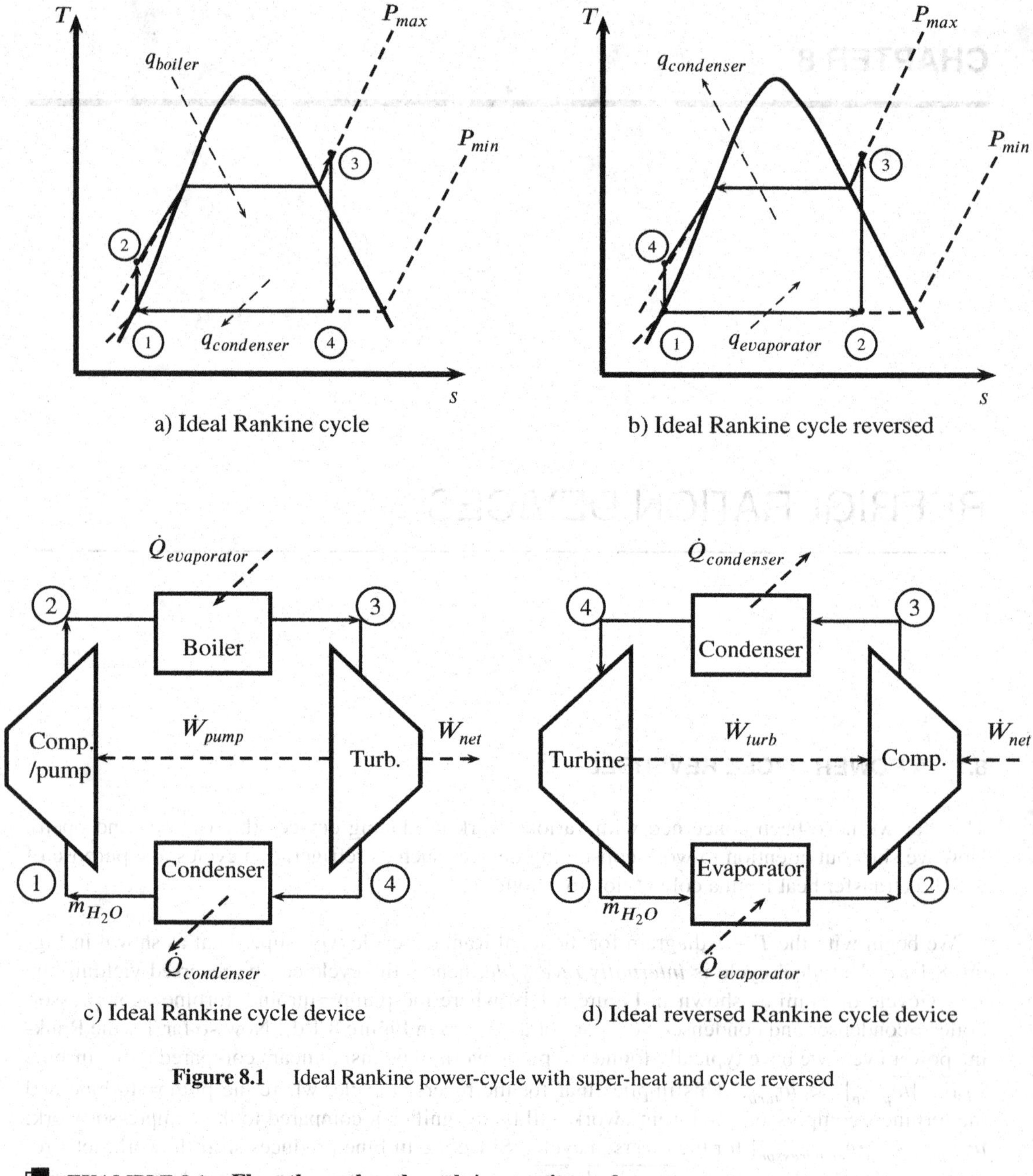

Figure 8.1 Ideal Rankine power-cycle with super-heat and cycle reversed

EXAMPLE 8.1 Flow through a throttle/expansion valve

Problem Statement:

A saturated liquid refrigerant R-134a flows into an adiabatic throttling at a pressure of $0.8 MPa$, the outlet pressure is $0.14 MPa$. Determine the refrigerant quality and temperature at the valve outlet as well as the rate of entropy generation per-unit mass-flow within the throttle. Assume no change in flow specific kinetic or potential energy across the valve.

Given:
Throttle (expansion valve),
refrigerant R-134a,
adiabatic process,
$P_1 = 0.8MPa$,
$x_1 = 0$,
$P_2 = 0.14MPa$,
$\Delta pe_{1\to2} = \Delta ke_{1\to2} = 0$

Find:
x_2,
T_2,
$s_{gen} = \frac{\dot{S}_{gen}}{\dot{m}}$

Assumptions:
Steady-state (S.S.),
steady-flow,
$\dot{W} = 0$

Schematic:

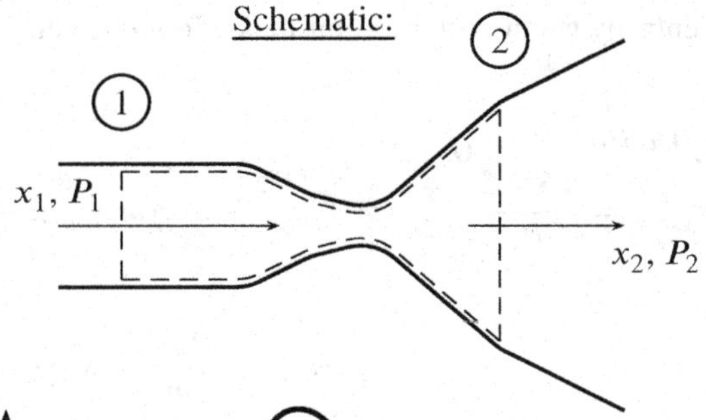

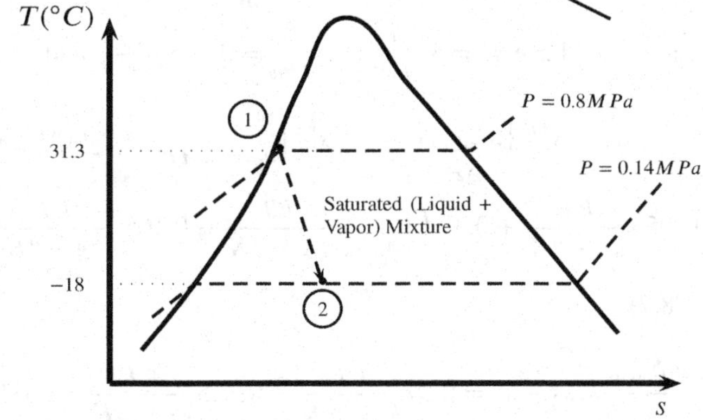

Analysis:
Applying the open system energy-balance (4.85) to the valve (a steady-state, steady-flow, single inlet/single outlet, adiabatic device)

$$\cancelto{0}{\dot{Q}} - \cancelto{0}{\dot{W}} = \cancelto{0,\,S.S.}{\frac{dE_{CV}}{dt}} + \sum(\dot{m}(h + \cancelto{0}{ke + pe}))_{outlets} - \sum(\dot{m}(h + \cancelto{0}{ke + pe}))_{inlets} \tag{8.1}$$

or

$$0 = (\dot{m}h)_2 - (\dot{m}h)_1 \tag{8.2}$$

Now, applying the open system mass-balance (4.71) gives $\dot{m}_1 = \dot{m}_2 = \dot{m}$. Thus, from the energy (8.2) and mass-balance we have an *isenthalpic* process

$$h_2 = h_1 \tag{8.3}$$

Via saturation Tables for R-134a B.6 and (8.3) we have $h_2 = h_1 = h^{sat}_{liq.}\Big|_{P_1=8bar} = 93.42\frac{kJ}{kg}$. Since $h^{sat}_{liq.}\Big|_{P_2=1.4bar} = 25.7\frac{kJ}{kg}$ and $h^{sat}_{vap.}\Big|_{P_2=1.4bar} = 236\frac{kJ}{kg}$ then $h^{sat}_{liq.}\Big|_{P_2=1.4bar} < h_2 < h^{sat}_{vap.}\Big|_{P_2=1.4bar}$ and state 2 is a saturated mix. Next, from the expression for quality

$$x_2 = x|_{P_2=0.14MPa,\ h_2=93.42\frac{kJ}{kg}} = \frac{h_2 - h^{sat}_{liq.}\Big|_{P_2}}{h^{sat}_{vap.}\Big|_{P_2} - h^{sat}_{liq.}\Big|_{P_2}} = \frac{93.42\frac{kJ}{kg} - 25.7\frac{kJ}{kg}}{236\frac{kJ}{kg} - 25.7\frac{kJ}{kg}} = 0.32 \overset{x_2}{\Longleftarrow} \tag{8.4}$$

Since state 2 is a saturated mix, via saturation Tables B.6

$$T_2 = T^{sat}|_{P_2=1.4bar} = -18.8°C \overset{T_2}{\Longleftarrow} \tag{8.5}$$

Finally, the entropy generation per-unit mass-flow is given by the open system entropy-balance (6.76)

$$\sum \left(\frac{\dot{Q}}{T_{int}}\right)_{system}^{0,\ adia.} = \frac{dS_{CV}}{dt}^{0,\ S.S.} + \sum (\dot{m}s)_{outlets}^{\dot{m}s_2} - \sum (\dot{m}s)_{inlets}^{\dot{m}s_1} - \dot{S}_{gen} \Rightarrow 0 = \dot{m}s_2 - \dot{m}s_1 - \dot{S}_{gen} \quad (8.6)$$

or

$$s_{gen} = \frac{\dot{S}_{gen}}{\dot{m}} = s_2 - s_1 \quad (8.7)$$

For state 1 and 2 we have $s_1 = s_{liq}^{sat}\Big|_{P_1=0.8MPa} = 0.345\frac{kJ}{kg-°K}$ and

$$s_2 = s_{mix}^{sat}\Big|_{\substack{P_2=0.14MPa,\\ x_2=0.32}} = s_{liq}^{sat}\Big|_{P_2=0.14MPa} + x_2(s_{vapor}^{sat} - s_{liq}^{sat})\Big|_{P_2=0.14MPa}$$

$$= 0.1055\frac{kJ}{kg-°K} + 0.32\left(0.9322\frac{kJ}{kg-°K} - 0.1055\frac{kJ}{kg-°K}\right) = 0.37\frac{kJ}{kg-°K} \quad (8.8)$$

giving via (8.7)

$$s_{gen} = \frac{\dot{S}_{gen}}{\dot{m}} = s_2 - s_1 = 0.37\frac{kJ}{kg-°K} - 0.345\frac{kJ}{kg-°K} = 0.025\frac{kJ}{kg-°K} \xLeftarrow{s_{gen}} \quad (8.9)$$

$$\boxed{x_2 = 0.32,\ T_2 = -18.8°C,\ s_{gen} = 0.025\frac{kJ}{kg-°K}}$$

Comment: Entropy generation is positive within the throttle, hence the expansion process (un-restrained expansion) of the refrigerant is irreversible!

□

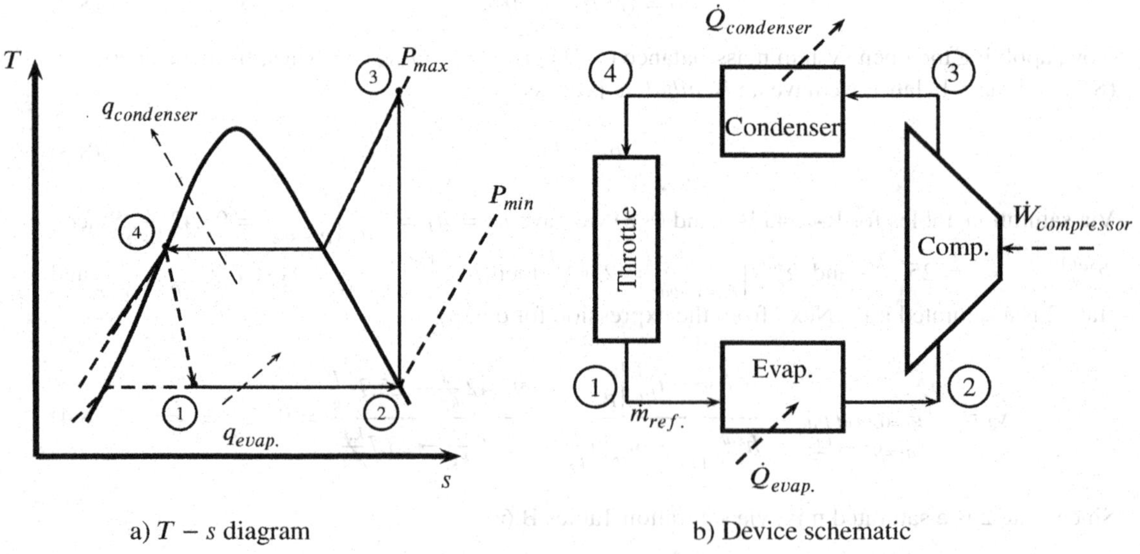

a) $T - s$ diagram b) Device schematic

Figure 8.2 Ideal vapor-compression refrigeration cycle

8.2 THE IDEAL VAPOR-COMPRESSION REFRIGERATION CYCLE

Replacing the turbine in Figure 8.1 d) with an adiabatic throttle gives the ideal vapor compression refrigeration cycle shown in Figure 8.2 b) along with the $T-s$ diagram 8.2 a). Common refrigerants include Ammonia as well as various Chloro-fluorocarbons (CFC's) with trade-names such as R-134a and R-12.

EXAMPLE 8.2 Ideal refrigeration cycle

Problem Statement:
An ideal vapor-compression refrigeration cycle boils refrigerant R-134a within the evaporator at a temperature of $0°C$. If the condenser outlet temperature is $26°K$ and the mass-flow rate of the refrigerant is $0.08kg/s$, determine the compressor power, the refrigeration capacity (rate of heat addition to the evaporator), the coefficient of performance and the coefficient of performance for the equivalent Carnot refrigerator.

Schematic:

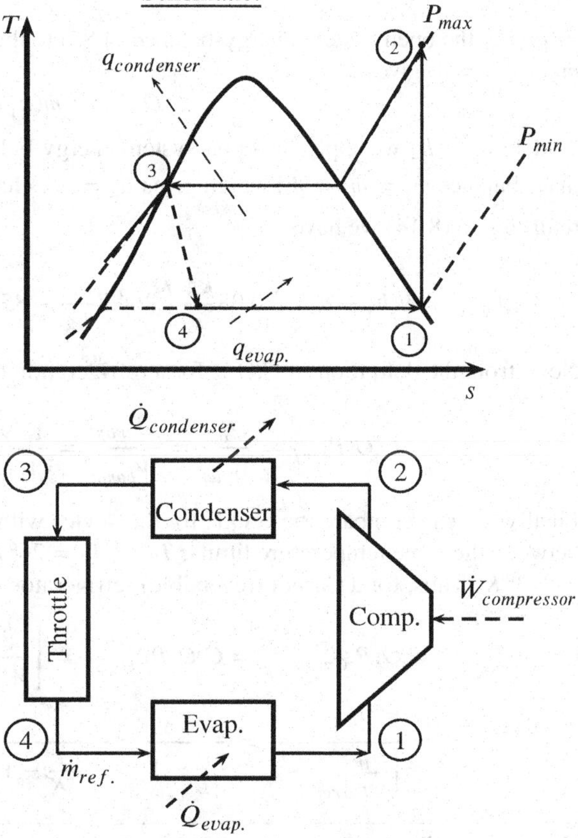

Given:
Ideal vapor-compression refrigeration cycle,
refrigerant R-134a,
$T_1 = 0°C$,
$T_3 = 26°K$,
$\dot{m} = 0.08kg/s$

Find:
$\dot{W}_{comp.}$,
$\dot{Q}_{evap.}$,
$C.O.P._{ref.}$,
$C.O.P._{Carnot_{ref.}}$

Assumptions:
Steady-state S.S.),
steady-flow,
$\dot{W} = 0$

Analysis:
Applying the open system energy-balance (4.85) to the compressor (noting from the mass-balance $\dot{m}_1 = \dot{m}_2 = \dot{m}$)

$$\dot{W}_{comp.} = \dot{m}(h_2 - h_1) \tag{8.10}$$

Furthermore, application of the entropy-balance (6.76) gives

$$\sum \left(\frac{\dot{Q}}{T_{int}}\right)_{system} \overset{0,\,adia.}{} = \frac{dS_{CV}}{dt}\overset{0,\,S.S.}{} + \sum (\dot{m}s)_{outlets}^{\dot{m}s_2} - \sum (\dot{m}s)_{inlets}^{\dot{m}s_1} - \dot{S}_{gen}^{0,\,rev.} \tag{8.11}$$

or

$$0 = \dot{m}(s_2 - s_1) \Rightarrow s_2 = s_1 \tag{8.12}$$

The compressor operates isentropically. From R-134a saturation Table B.5 $h_1 = h_{vapor}^{sat}\Big|_{T_1=0°C} = 247.23\frac{kJ}{kg}$ and $s_1 = s_{vapor}^{sat}\Big|_{T_1=0°C} = 0.919\frac{kJ}{kg-°K}$. Now, the heat rejection process 3 → 4 within the condenser is isobaric. Hence, from saturation Tables $P_2 = P_3 = P^{sat}\big|_{T_3=26°C} = 6.85bar$ and interpolating from super-heated vapor Tables B.7, $h_2 = h\big|_{P_2=6.85bar,\; s_2=s_1=0.919\frac{kJ}{kg-°K}} = 265.7\frac{kJ}{kg}$. Returning to the energy-balance for the compressor:

$$\dot{W}_{comp.} = \dot{m}(h_2 - h_1) = 0.08\frac{kg}{s}\left(265.7\frac{kJ}{kg} - 247.23\frac{kJ}{kg}\right) = 1.47kW \xLeftarrow{\dot{W}_{comp.}} \tag{8.13}$$

Applying the open system energy-balance (4.85) to the evaporator (noting from the mass-balance $\dot{m}_4 = \dot{m}_3 = \dot{m}$) gives

$$\dot{Q}_{evap.} = \dot{m}(h_1 - h_4) \tag{8.14}$$

To determine h_4 we apply the open-system energy-balance (4.85) to the throttle (noting from the mass-balance $\dot{m}_1 = \dot{m}_4 = \dot{m}$) which gives $h_4 = h_3$. Hence, $h_4 = h_3 = h_{liq}^{sat}\Big|_{T_3=26°C} = 85.75\frac{kJ}{kg}$ and returning to (8.14) we have

$$\dot{Q}_{evap.} = \dot{m}(h_1 - h_4) = 0.08\frac{kg}{s}\left(264.7\frac{kJ}{kg} - 85.75\frac{kJ}{kg}\right) = 12.92kW \xLeftarrow{\dot{Q}_{evap.}} \tag{8.15}$$

Next, from the definition of C.O.P. for a refrigeration device (4.141)

$$C.O.P._{ref} = \frac{\dot{Q}_{out}}{\dot{W}_{net}} = \frac{\dot{Q}_{evap.}}{\dot{W}_{comp.}} = \frac{12.92kW}{1.47kW} = 8.79 \xLeftarrow{C.O.P._{ref}} \tag{8.16}$$

Finally, we can compare the C.O.P. for the device with that of a Carnot refrigerator operating between the same temperature limits: $T_H = T_3 = 26°K = 299°K$ and $T_L = T_4 = T_1 = 0°C = 273°K$. Thus, for a Carnot (reversible) refrigerator

$$C.O.P._{Carnot_{ref.}} = C.O.P._{rev._{ref.}} = \left[\frac{\dot{Q}_L}{\dot{W}_{net}}\right]_{rev} = \left[\frac{\dot{Q}_L}{\dot{Q}_H - \dot{Q}_L}\right]_{rev}$$

$$= \frac{1}{\left[\frac{\dot{Q}_H}{\dot{Q}_L}\right]_{rev} - 1} = \frac{1}{\frac{T_H}{T_L} - 1} = \frac{1}{\frac{299°K}{273°K} - 1} = 10.5 \xLeftarrow{C.O.P._{Carnot_{ref.}}} \tag{8.17}$$

$$\boxed{\dot{W}_{comp.} = 1.47kW,\ \dot{Q}_{evap.} = 12.92kW,\ C.O.P._{ref} = 8.79,\ C.O.P._{Carnot_{ref.}} = 10.5}$$

Comment: Note that $C.O.P._{Carnot} = 10.5 > 8.79 = C.O.P.$ demonstrating that the (internally and externally) reversible device, the Carnot refrigerator, is the most efficient.

□

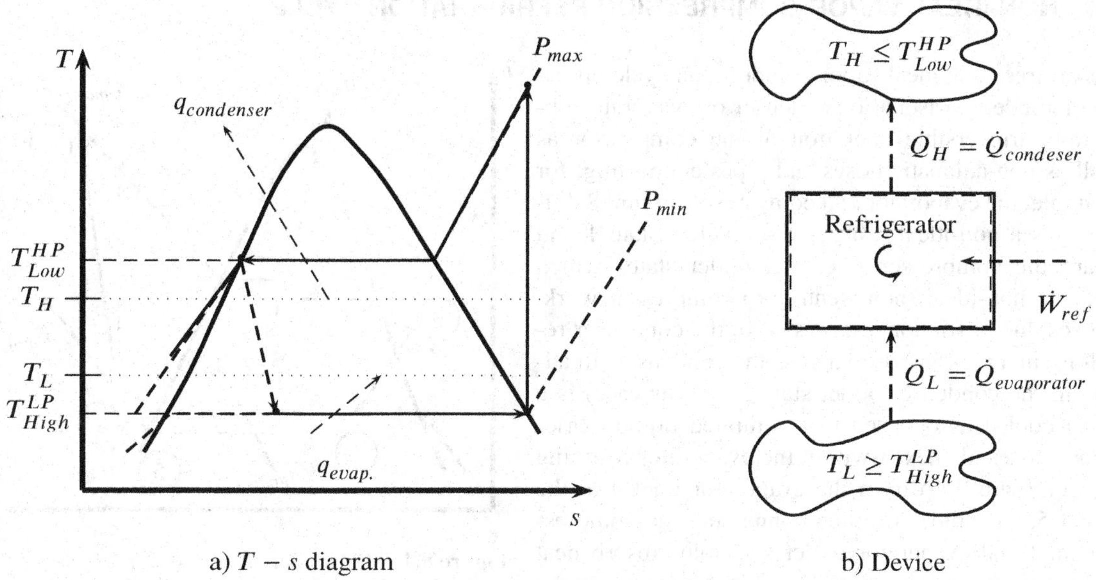

Figure 8.3 Ideal vapor-compression refrigeration cycle and relationship to reservoirs with temperatures T_H and T_L.

8.3 CHOICE OF REFRIGERANTS

Table 8.1 T^{sat} at $3bar$ and $15bar$ for some common refrigerants

Refrigerant	$T^{sat}\|_{3bar}(°C)$	$T^{sat}\|_{15bar}(°C)$
Ammonia	–9.23	38.7
R12	–0.8	59.2
R23	–60.8	–17.77
R134a	0.65	55.2

At this point in our analysis it should be pointed out that the purpose of a refrigeration device is to absorb heat from a cold region and then rejecting this heat to higher temperature region. In the previous example the condenser exit temperature T_3 was the coolest location on the heat-rejection (high-pressure) side of the cycle. Conversely, the evaporator exit T_1 was the warmest on the heat-absorption (low-pressure) side of the cycle. Given that heat flows from hot to cold spontaneously, this implies that the cool region at temperature T_L from which heat is being absorbed by the evaporator must be above the evaporator maximum temperature T_1. Similarly, the condenser exit temperature must be above the high temperature region temperature T_H in order to affect a heat-transfer from the condenser. Figure 8.3 b) shows the refrigeration device inter-acting with a cold and hot region at temperatures T_L and T_H respectively. Figure 8.3 a) shows the ideal cycle with plausible regional/reservoir temperatures included on the temperature axis: Clearly T_H must be below the minimum temperature on the high-pressure side of the cycle, while T_L must be above the maximum temperature on the low-pressure side. For example, for (common) high and low-pressure side pressures of $3bar$ and $15bar$ we have the following list of saturation temperatures given in Table 8.1 for several common refrigerants. If, for example, the ideal refrigeration device operates at these pressures and it is desired to cool a space at $T_L = -2°C$ while rejecting heat to a location at $T_H = 27°C$, the only suitable refrigerant listed in Table 8.1 would be ammonia.

8.4 NON-IDEAL VAPOR-COMPRESSION REFRIGERATION CYCLE

Departures from ideal vapor-compression cycle operation include non-isobaric condenser or evaporator operation, irreversible operation of the compressor as well as non-adiabatic hoses and pipes connecting, for example, the evaporator and compressor. Figure 8.4 illustrates a non-ideal refrigeration cycle: State 1 and 2 are the compressor inlet and outlet states which indicate non-ideal (non-isentropic) compressor work. Losses due to friction are present in the condenser resulting in a pressure drop across the condenser. In addition, the condenser outlet state 3 (in this case) is a super-cooled as opposed to a saturated liquid. Frictional losses also exist within the evaporator resulting in a pressure drop from the evaporator inlet 4 to the outlet 5. A (cold) hose/line connecting the compressor inlet and evaporator outlet also may absorb heat q_{line} resulting in heating of the refrigerant between the evaporator exit state 5 and compressor inlet 1.

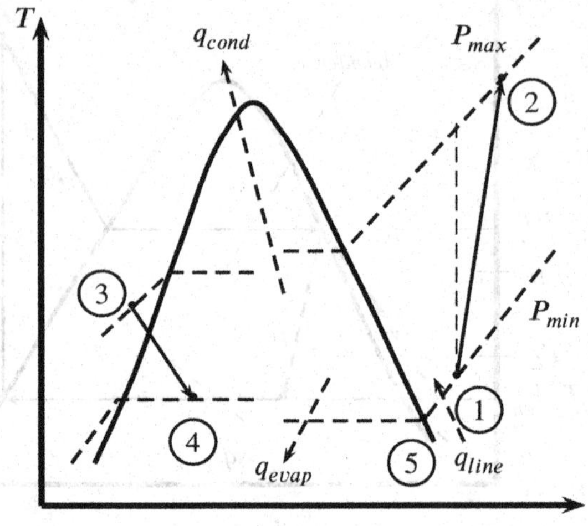

Figure 8.4 Illustration of non-ideal refrigeration cycle.

EXAMPLE 8.3 Availability/Exergy analysis: Real vapor-compression refrigeration cycle

Problem Statement:
A vapor compression cycle using $R134a$ flowing at a rate of $1kg/s$ is used to cool a space at a temperature of $280°K$. The compressor inlet temperature and pressure are $275°K$ and $200kPa$. The compressor is adiabatic and pressurizes the refrigerant to $360°K$ and $1269kPa$. The condenser exhibits friction resulting in a pressure decrease to $1260kPa$ at the outlet with an outlet temperature of $300°K$. An adiabatic throttle expands the coolant to a pressure of $205kPa$. Friction within the evaporator results in a further pressure reduction to $200kPa$. Finally, the environmental temperature is $298°K$. You will perform a general exergy analysis to determine the lost work for each component as well as the overall refrigeration device. Specifically, determine the least power required to achieve the cooling, the rate of availability (or power) destruction and the rate of availability destruction for each component

---

Given:
Real vapor-compression refrigeration cycle,
$R134a$,
$P_1 = 200kPa, T_1 = 275°K$,
$P_2 = 1269kPa, T_2 = 360°K$,
$P_3 = 1260kPa, T_3 = 300°K$,
$P_4 = 205kPa$,
$T_{ref} = 280°K$,
$T_{env} = 298°K$,
$\dot{m} = 1kg/s$

Schematic:

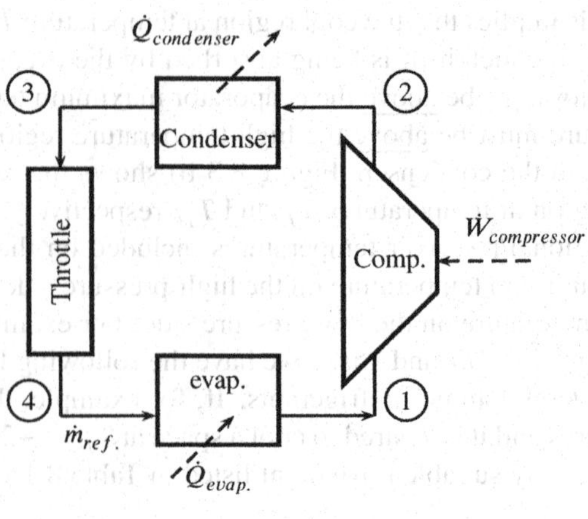

Find:
a) $\dot{W}_{min}$
b) $\dot{W}_{\substack{lost\\overall}}$
c) $\dot{W}_{\substack{lost\\comp.}}$, $\dot{W}_{\substack{lost\\cond.}}$, $\dot{W}_{\substack{lost\\throttle}}$, $\dot{W}_{\substack{lost\\evap.}}$

Assumptions:
Steady-state (S.S.),
Potential and kinetic energy terms are negligible.
Condenser (heat) inter-action with environment only.
Evaporator heat inter-action temperature is T_{ref}

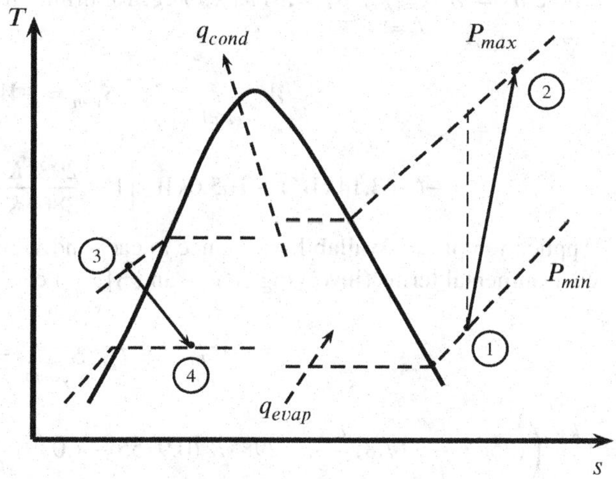

Analysis:
Application of the Availability-balance (6.122) to the overall device, setting $\dot{S}_{gen} = 0$ while noting that the condenser heat-rejection is at the environmental temperature, gives the reversible or minimum power (consumed)

$$\dot{W} = \sum\left[\dot{Q}(1-\frac{T_{env}}{T_{int}})\right] \Rightarrow \dot{W}^{min}_{comp.} = \dot{Q}_{evap}\left[1-\frac{T_{env}}{T_{ref}}\right] - \dot{Q}_{cond.}\left[1-\frac{T_{env}}{T_{env}}\right]^{0} = \dot{Q}_{evap}\left[1-\frac{T_{env}}{T_{ref}}\right] \quad (8.18)$$

Application of the energy-balance to the evaporator ($h_1 = h|_{\substack{P_1=200kPa,\\T_1=275°K}} = 254.6\frac{kJ}{kg}$ and $h_4 = h_3 = h|_{\substack{P_3=1260kPa,\\T_3=300°K}} = 89.05\frac{kJ}{kg}$ where the energy-balance has been applied to the throttle) gives

$$\dot{Q}_{evap.} = \dot{m}(h_1 - h_4) = 1\frac{kg}{s}\left(254.6\frac{kJ}{kg} - 89.05\frac{kJ}{kg}\right) = 165.6kW \quad (8.19)$$

Returning to the Availability-balance (8.18)

$$\dot{W}^{min}_{comp.} = \dot{Q}_{evap}\left[1-\frac{T_{env}}{T_{ref}}\right] = -165.6kW\left[1-\frac{298°K}{280°K}\right] = -10.64kW \xLeftarrow{\dot{W}^{min}_{comp.}} \quad (8.20)$$

Application of the Availability-balance to the overall device and solving for $\dot{W}_{\substack{lost\\overall}} = T_{env}\dot{S}_{gen}$ gives

$$\dot{W} = \sum\left[\dot{Q}(1-\frac{T_{env}}{T_{int}})\right] - T_{env}\dot{S}_{gen} \Rightarrow \dot{W}_{\substack{lost\\overall}} = T_{env}\dot{S}_{gen} = -\dot{W}_{comp.} + \dot{Q}_{evap.}\left[1-\frac{T_{env}}{T_{ref}}\right] \quad (8.21)$$

The compressor power is found via the energy-balance

$$\dot{W}_{comp.} = -\dot{m}(h_2 - h_1) = -1\frac{kg}{s}\left(317.8\frac{kJ}{kg} - 254.6\frac{kJ}{kg}\right) = -63.14kW \quad (8.22)$$

where $h_2 = h|_{P_2=1269kPa,\ T_1=360^\circ K} = 317.8kJ/kg$. Returning to (8.21)

$$\dot{W}_{\substack{lost\\overall}} = T_{env}\dot{S}_{gen} = -\dot{W}_{comp.} + \dot{Q}_{evap.}\left[1 - \frac{T_{env}}{T_{ref}}\right]$$

$$= -(-63.14kW) + 165.6kW\left[1 - \frac{298^\circ K}{280^\circ K}\right] = 52.49kW \xleftarrow{\dot{W}_{\substack{lost\\overall}}} \tag{8.23}$$

Application of the Availability-balance to each individual component, one at a time, noting that the environmental terms (involving $h|_{T_{env},\,P_{env}}$ and $s|_{T_{env},\,P_{env}}$) cancel out, gives

$$\dot{W}_{\substack{lost\\comp.}} = T_{env}\dot{S}_{\substack{gen.\\comp.}} = \dot{m}((h_1 - h_2) - T_{env}(s_1 - s_2)) - \dot{W}_{comp.} =$$

$$1\frac{kg}{s}\left((254.6 - 317.8)\frac{kJ}{kg} - 298^\circ K\,(0.9755 - 1.039)\frac{kJ}{kg-^\circ K}\right) - (-63.14) = 18.84kW \xleftarrow{\dot{W}_{\substack{lost\\comp.}}} \tag{8.24}$$

$$\dot{W}_{\substack{lost\\cond.}} = T_{env}\dot{S}_{\substack{gen.\\cond.}} = \dot{m}((h_2 - h_3) - T_{env}(s_2 - s_3)) + \dot{Q}_{cond.}\left[1 - \frac{T_{env}}{T_{env}}\right]^{\nearrow 0} =$$

$$1\frac{kg}{s}\left((317.8 - 89.05)\frac{kJ}{kg} - 298^\circ K\,(1.039 - 0.3315)\frac{kJ}{kg-^\circ K}\right) = 17.95kW \xleftarrow{\dot{W}_{\substack{lost\\cond.}}} \tag{8.25}$$

$$\dot{W}_{\substack{lost\\throttle}} = T_{env}\dot{S}_{\substack{gen.\\throttle}} = \dot{m}((\cancel{h_3} - \cancel{h_4}^{\,h_3}) - T_{env}(s_3 - s_4)) = \dot{m}(-T_{env}(s_3 - s_4))$$

$$= 1\frac{kg}{s}\left(-298^\circ K\,(0.3315 - 0.3465)\frac{kJ}{kg-^\circ K}\right) = 4.473kW \xleftarrow{\dot{W}_{\substack{lost\\throttle}}} \tag{8.26}$$

$$\dot{W}_{\substack{lost\\evap.}} = T_{env}\dot{S}_{\substack{gen.\\evap.}} = \dot{m}((h_4 - h_1) - T_{env}(s_4 - s_1)) + \dot{Q}_{evap.}\left[1 - \frac{T_{env}}{T_{ref}}\right]$$

$$= 1\frac{kg}{s}\left((89.05 - 254.6)\frac{kJ}{kg} - 298^\circ K\,(0.3465 - 0.9755)\frac{kJ}{kg-^\circ K}\right)$$

$$+165.6kW\left[1 - \frac{298^\circ K}{280^\circ K}\right] = 11.22kW \xleftarrow{\dot{W}_{\substack{lost\\evap.}}} \tag{8.27}$$

$$\boxed{\begin{array}{l}\dot{W}^{min}_{comp.} = -10.64kW,\ \dot{W}_{\substack{lost\\overall}} = 52.49kW,\ \dot{W}_{\substack{lost\\comp.}} = 18.84kW,\ \dot{W}_{\substack{lost\\cond.}} = 17.95kW,\\ \dot{W}_{\substack{lost\\throttle}} = 4.473kW,\ \dot{W}_{\substack{lost\\evap.}} = 11.22kW\end{array}}$$

Comment: In terms of lost work, the compressor, condenser, throttle and evaporator are responsible for 35.9%, 34.2%, 8.5% and 21.4% of the total losses respectively. To increase overall efficiency, effort should be expended primarily on improving the compressor and condenser performance.

□

8.5 REGENERATIVE REFRIGERATION CYCLES: MULTI-STEP COMPRESSION WITH FLASH COOLING

As in the case of work producing devices, a form of regeneration can be implemented in refrigeration cycles using the work consuming equivalent of multi-step expansion with inter-cooling and regeneration: multi-step compression. Example 8.4 gives a modified vapor-compression cycle with compression and expansion performed in two steps via a high and low pressure compressor (HPC and LPC). The refrigeration cycle equivalent of inter-cooling and/or regeneration is affected by expanding the refrigerant through a high-pressure throttle and into a "flash-chamber" where the saturated vapor component is extracted and diverted to the HPC. The remaining saturated liquid is then expanded through the low-pressure throttle and passes into the evaporator. Improvement in the C.O.P. is due to the diversion of high enthalpy fluid (saturated vapor) from the LP expansion valve inlet resulting in a higher heat absorbing capability of the remaining refrigerant routed to the evaporator: Increasing the heat transfer within the evaporator per-unit mass-flow. In addition, the mass-flow rate and, hence power required of the LPC is reduced, thus decreasing cycle power.

EXAMPLE 8.4 Non-ideal multi-stage refrigeration cycle with flash-cooling

Problem Statement:

A vapor-compression refrigeration cycle with R-134a as the working fluid condenses the refrigerant at a pressure of $1200kPa$. The evaporator outlet temperature is $-30°C$ and the refrigeration capacity (heat transfer from the low-temperature source) is $10ton$ ($1ton = 211kJ/min$). The cycle exhibits two stages of compression and expansion with all the saturated vapor extracted at the HP expansion valve (throttle) exit (at $400kPa$) via a flash-chamber and mixed with the HPC inlet flow in a direct-contact heat exchanger (mixer). The HPC and LPC isentropic efficiencies are 0.8 and 0.9 respectively and the LPC inlet state is a saturated vapor. The condenser exit state is also a saturated liquid. Determine the cycle power, C.O.P., the mass-flow bled via the flash-chamber as a fraction of the condenser mass-flow $y = \dot{m}_{bleed}/\dot{m}_{condenser}$ and the condenser mass-flow rate.

Given:

Multi-stage vapor-compression refrigeration cycle, w/flash-cooling, direct contact heat-exchanger, refrigerant R-134a,
$P_4 = 1200kPa$,
$T_1 = -30°C$,
$P_2 = 400kPa$,
$x_1 = 1$, $x_5 = 0$,
$x_7 = 0$, $x_9 = 1$,
$\dot{Q}_{evap.} = 10ton$,
$\eta_{LPC} = 0.9$, $\eta_{HPC} = 0.8$,

Schematic:

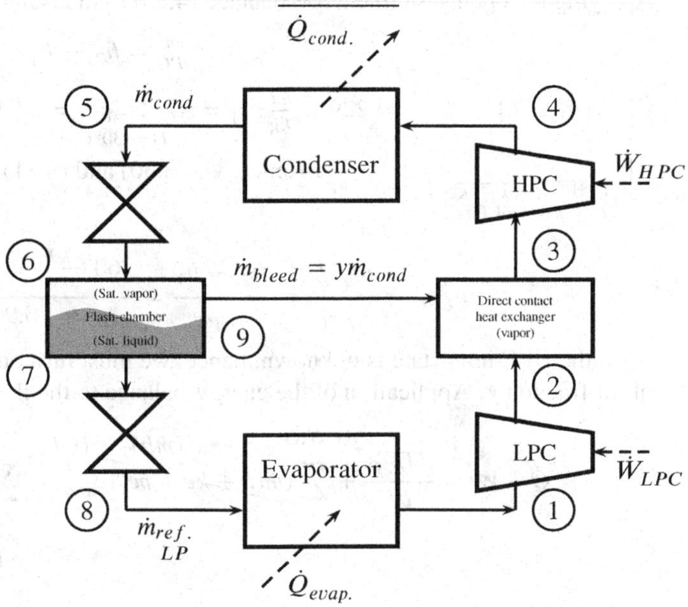

Find:
$\dot{W}_{cycle}$,
C.O.P.,
$\gamma = \frac{\dot{m}_{bleed}}{\dot{m}_{condenser}}$,
$\dot{m}_{condenser}$

Assumptions:
Steady-state,steady-flow, isobaric condenser, isobaric evaporator, isobaric connecting-lines, isobaric flash-chamber & heat-exchanger ($P_7 = P_6 = P_9 = P_3 = P_2$), adiabatic expansions valves, negligible pe and ke

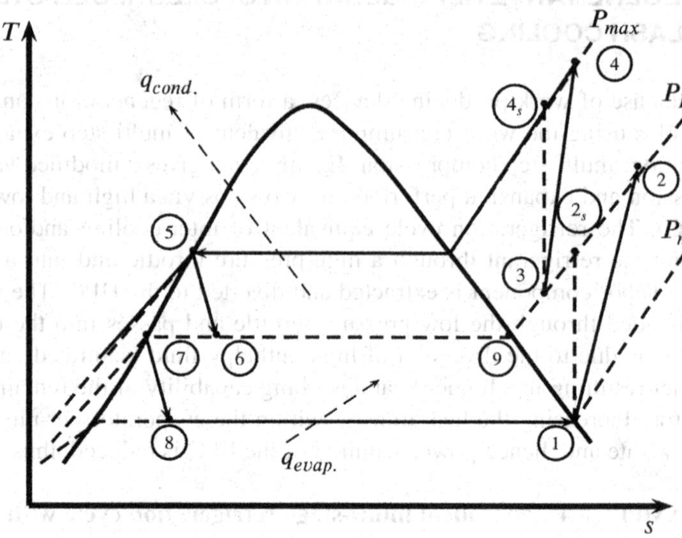

Analysis:
Inspection of the device diagram shows that the cycle power is given by

$$\dot{W}_{cycle} = \dot{W}_{LPC} + \dot{W}_{HPC} = \dot{m}_{LPC}w_{LPC} + \dot{m}_{HPC}w_{HPC} \tag{8.28}$$

Noting that $\dot{m}_{HPC} = \dot{m}_{cond.}$ and $\dot{m}_{LPC} = \dot{m}_{evap.} = (1-\gamma)\dot{m}_{cond.}$ we get

$$\dot{W}_{cycle} = \dot{W}_{LPC} + \dot{W}_{HPC} = \dot{m}_{cond.}[(1-\gamma)w_{LPC} + w_{HPC}] \tag{8.29}$$

From the definition of isentropic efficiency (6.83), we have, for the LPC

$$\eta_{LPC} = \frac{\dot{W}^{isen}_{LPC}}{\dot{W}_{LPC}} = \frac{w^{isen}_{LPC}}{w_{LPC}} \Rightarrow w_{LPC} = \frac{w^{isen}_{LPC}}{\eta_{LPC}} \tag{8.30}$$

Applying the open system energy-balance (4.85) to the isentropic LPC gives

$$w^{isen}_{LPC} = h_{2_s} - h_1 \tag{8.31}$$

where $h_1 = h|_{\substack{x_1=1, \\ T_1=-30°C}} = 229.14\frac{kJ}{kg}$, $s_1 = s|_{\substack{x_1=1, \\ T_1=-30°C}} = 0.943\frac{kJ}{kg-°K}$ and $h_{2_s} = h|_{\substack{s_{2_s}=s_1=0.943\frac{kJ}{kg-°K}, \\ P_{2_s}=P_2=400kPa}} = 260.6\frac{kJ}{kg}$. Hence, via (8.30) and (8.31)

$$w_{LPC} = \frac{w^{isen}_{LPC}}{\eta_{LPC}} = \frac{h_{2_s} - h_1}{\eta_{LPC}} = \frac{260.6\frac{kJ}{kg} - 229.14\frac{kJ}{kg}}{0.9} = 34.96\frac{kJ}{kg} \tag{8.32}$$

Now, the HPC inlet state is unknown, hence, we must turn our attention to the flash-chamber bleed-fraction γ. Application of the energy-balance to the flash-chamber gives

$$\overset{0}{\dot{Q}} - \overset{0}{\dot{W}} = \overset{0,\,S.S.}{\frac{dE_{CV}}{dt}} + \sum \overset{(\dot{m}h)_7 + (\dot{m}h)_9}{(\dot{m}(h + ke + pe))_{outlets}} - \sum \overset{(\dot{m}h)_6}{(\dot{m}(h + ke + pe))_{inlets}}$$

$$\Rightarrow 0 = (\dot{m}h)_7 + (\dot{m}h)_9 - (\dot{m}h)_6 \tag{8.33}$$

Noting that $\dot{m}_6 = \dot{m}_{cond.}$, $\dot{m}_9 = \gamma\dot{m}_{cond.}$ and $\dot{m}_7 = (1-\gamma)\dot{m}_{cond.}$. Combining these results with flash-chamber energy-balance (8.33) gives

$$0 = (1-\gamma)\cancel{\dot{m}_{cond.}}h_7 + \gamma\cancel{\dot{m}_{cond.}}h_9 - \cancel{\dot{m}_{cond.}}h_6 = (1-\gamma)h_7 + \gamma h_9 - h_6 \tag{8.34}$$

or

$$\gamma = \frac{h_6 - h_7}{h_9 - h_7} \tag{8.35}$$

Application of the energy-balance to the HP expansion-valve gives $h_6 = h_5$. Hence, via the $T-s$ diagram and Tables $h_6 = h_5 = h|_{\substack{x_5=0,\\ P_5=1200kPa}} = 115.8\frac{kJ}{kg}$, $h_7 = h|_{\substack{x_7=0,\\ P_7=400kPa}} = 62.0\frac{kJ}{kg}$ and $h_9 = h|_{\substack{x_9=1,\\ P_9=400kPa}} = 252.3\frac{kJ}{kg}$. Hence (8.35) yields

$$\gamma = \frac{h_6 - h_7}{h_9 - h_7} = \frac{115.8\cancel{\frac{kJ}{kg}} - 62.0\cancel{\frac{kJ}{kg}}}{252.3\cancel{\frac{kJ}{kg}} - 62.0\cancel{\frac{kJ}{kg}}} = 0.28 \overset{\gamma}{\Longleftarrow\!=\!=} \tag{8.36}$$

Turning our attention to the HPC we have, from the definition of isentropic efficiency (6.83)

$$\eta_{HPC} = \frac{\dot{W}^{isen}_{HPC}}{\dot{W}_{HPC}} = \frac{w^{isen}_{HPC}}{w_{HPC}} \Rightarrow w_{HPC} = \frac{w^{isen}_{HPC}}{\eta_{HPC}} \tag{8.37}$$

Applying the open system energy-balance (4.85) to the (isentropic) HPC gives

$$w^{isen}_{HPC} = h_{4_s} - h_3 \tag{8.38}$$

To determine h_3 we apply the energy-balance to the direct-contact heat-exchanger (now that the bleed-fraction γ is known) giving

$$\overset{0}{\cancel{\dot{Q}}} - \overset{0}{\cancel{\dot{W}}} = \overset{0,\,S.S.}{\cancel{\frac{dE_{CV}}{dt}}} + \overset{(\dot{m}h)_3}{\cancel{\sum(\dot{m}(h + ke + pe))_{outlets}}} - \overset{(\dot{m}h)_2 + (\dot{m}h)_9}{\cancel{\sum(\dot{m}(h + ke + pe))_{inlets}}}$$

$$\Rightarrow 0 = (\dot{m}h)_3 - (\dot{m}h)_9 - (\dot{m}h)_2 \tag{8.39}$$

Noting that $\dot{m}_3 = \dot{m}_{cond.}$, $\dot{m}_9 = \gamma\dot{m}_{cond.}$ and $\dot{m}_2 = (1-\gamma)\dot{m}_{cond.}$. Combining these results with heat-exchanger energy-balance (8.39) gives

$$0 = (\dot{m}h)_3 - (\dot{m}h)_9 - (\dot{m}h)_2 = \cancel{\dot{m}_{cond.}}h_3 - \gamma\cancel{\dot{m}_{cond.}}h_9 - (1-\gamma)\cancel{\dot{m}_{cond.}}h_2 = h_3 - \gamma h_9 - (1-\gamma)h_2 \tag{8.40}$$

or

$$h_3 = \gamma h_9 + (1-\gamma)h_2 \tag{8.41}$$

Noting that via the energy-balance applied to the LPC

$$h_2 = w_{LPC} + h_1 = 34.96\frac{kJ}{kg} + 229.14\frac{kJ}{kg} = 264.1\frac{kJ}{kg} \tag{8.42}$$

we have via (8.41)

$$h_3 = \gamma h_9 + (1-\gamma)h_2 = (0.280)252.3\frac{kJ}{kg} + (1-0.28)264.1\frac{kJ}{kg} = 260.8\frac{kJ}{kg} \tag{8.43}$$

Next, $h_{4_s} = h|_{\substack{P_4, \\ s_{4_s}=s_3}}$ where $s_{4_s} = s_3 = s|_{\substack{h_3=260.8\frac{kJ}{kg}, \\ P_3=400kPa}} = 0.944\frac{kJ}{kg-^\circ K}$. Thus, via interpolation from Tables, $h_{4_s} = h|_{\substack{P_4=400kPa, \\ s_{4_s}=0.944\frac{kJ}{kg-^\circ K}}} = 284.58\frac{kJ}{kg}$. Returning to (8.37) and using (8.38) we get

$$w_{HPC} = \frac{w^{isen}_{HPC}}{\eta_{HPC}} = \frac{h_{4_s} - h_3}{\eta_{HPC}} = \frac{288.1\frac{kJ}{kg} - 264.1\frac{kJ}{kg}}{0.8} = 29.98\frac{kJ}{kg} \tag{8.44}$$

Finally, application of the energy-balance to the evaporator gives

$$\dot{Q}_{evap.} = (\dot{m}h)_1 - (\dot{m}h)_8 \tag{8.45}$$

Note from the device schematic and mass-balance $\dot{m}_8 = \dot{m}_1 = (1-\gamma)\dot{m}_{cond.}$. Also, application of the energy-balance to the adiabatic LP expansion valve gives $h_8 = h_7$. Hence, from (8.45)

$$\dot{Q}_{evap.} = (\dot{m}h)_1 - (\dot{m}h)_8 = (1-\gamma)\dot{m}_{cond.}h_1 - (1-\gamma)\dot{m}_{cond.}h_8 \tag{8.46}$$

or

$$\dot{m}_{cond.} = \frac{\dot{Q}_{evap.}}{(1-\gamma)(h_1 - h_8)} = \frac{10\cancel{ton}\frac{211\frac{\cancel{kJ}}{min}}{1\cancel{ton}}}{(1-0.28)\left(229.14\frac{\cancel{kJ}}{kg} - 62.0\frac{\cancel{kJ}}{kg}\right)} = 17.53\frac{kg}{min} \xleftarrow{\dot{m}_{cond.}} \tag{8.47}$$

Returning to the cycle power (8.29)

$$\dot{W}_{cycle} = \dot{m}_{cond.}[(1-\gamma)w_{LPC} + w_{HPC}] =$$

$$17.53\frac{kg}{\cancel{min}}\left[(1-0.28)34.96\frac{kJ}{\cancel{kg}} + 29.75\frac{kJ}{\cancel{kg}}\right]\frac{1\cancel{min}}{60s} = 16.2kW \xleftarrow{\dot{W}_{cycle}} \tag{8.48}$$

Finally, the C.O.P. is found via the definition

$$C.O.P._{ref} = \frac{\dot{Q}_{evap.}}{\dot{W}_{cycle}} = \frac{10\cancel{ton}\frac{211\frac{kJ}{min}}{1\cancel{ton}}}{16.2kW} = 130.24\frac{\frac{\cancel{kJ}}{\cancel{min}}}{\frac{\cancel{kJ}}{\cancel{s}}}\frac{1\cancel{min}}{60\cancel{s}} = 2.17 \xleftarrow{C.O.P._{ref}} \tag{8.49}$$

$\dot{W}_{cycle} = 16.2kW, \gamma = 0.28, \dot{m}_{Cond.} = 17.53\frac{kg}{min}, C.O.P._{ref} = 2.17$

Comment: Note, removal of the flash-chamber ($\gamma = 0$) reduces coefficient of performance to 1.735 while increasing the cycle power to 20.27kW. Hence, installation of the flash-chamber increases the coefficient of performance by approximately 20% while decreasing cycle power by 25%.

□